Readings in the History of Evolutionary Theory

Readings in the History of Evolutionary Theory

Selections from Primary Sources

RONALD K. WETHERINGTON
Southern Methodist University

New York Oxford
OXFORD UNIVERSITY PRESS

Oxford University Press, Inc., publishes works that further Oxford University's
objective of excellence in research, scholarship, and education.

Oxford New York
Auckland Cape Town Dar es Salaam Hong Kong Karachi
Kuala Lumpur Madrid Melbourne Mexico City Nairobi
New Delhi Shanghai Taipei Toronto

With offices in
Argentina Austria Brazil Chile Czech Republic France Greece
Guatemala Hungary Italy Japan Poland Portugal Singapore
South Korea Switzerland Thailand Turkey Ukraine Vietnam

For titles covered by Section 112 of the US Higher Education Opportunity Act,
please visit www.oup.com/us/he for the latest information about
pricing and alternate formats.

Published by Oxford University Press, Inc.
198 Madison Avenue, New York, New York 10016
http://www.oup.com

ISBN: 978-0-19-538687-5

Printing number: 9 8 7 6 5 4 3 2 1

Printed in the United States of America
on acid-free paper

For Judith

CONTENTS

Unit 8
The Age of Darwin, II: After the Voyage • 89

Unit 9
The Age of Darwin, III: The Origin • 105

Unit 15

Current Challenges to the Synthetic Theory • 183

PREFACE

The purpose of the book is to provide, through original sources, documents that represent the intellectual history and background of evolutionary concepts. My focus is exclusively on the history of the Western intellectual tradition. Despite its length, and in order to make that length manageable, I have been selective in choosing authors and have heavily abridged many of those I have chosen. I may be justly criticized for shortcomings or omissions in both. Indeed, the authors themselves deserve more biographical detail than I have given them, but such details are commonly available.

The units through Darwin are chronological, and, even though the theoretical road to his concept of evolution is hardly a linear one, changes in scientific ideas were more or less sequential and cumulative. Many books on the history of evolution either concentrate on or culminate with Darwin—after all, what most people mean by "evolution" is *Darwinian* evolution.

But in a very important way, both philosophical and pedagogical, it is what happened with evolutionary ideas *after* Darwin that tells us most about ourselves and our on-again, off-again love affair with science. These are largely chronological as well, but not exclusively.

Darwin lived in the romantic period—the period of Victorian sensibilities fueled with an awe for science and scientific discovery but harboring a cautious skepticism about its intersection with an imbedded devotion to Christian aspirations. By century's end, most in British society had come to terms with the ambiguities, and science and its industrial partnerships moved forward with the new social realities of the emergent middle class.

We arbitrarily subdivide the past into periods designated by predominant ideas or events or other categories embedded in history, such as classifying individuals into species and genera, or technology into the Paleolithic, Neolithic, and Iron ages. Our brains are wired for typological rather than statistical thinking. We can therefore identify revolutions in thought—or at least major changes in thinking brought about through discovery—that carry our narrative on evolution beyond Darwin and down to today, and that is basically the organization of this book.

If we recognize that a major innovation in thinking about nature began with Copernicus, when he removed the earth from the center of the universe and supported a mechanistic cosmos, we can suggest that the next revolution followed with Darwin, who removed humankind from the center of the living world and thus eliminated purpose as the criterion for measuring nature. The genetic revolution followed, then, in introducing both the source of variation and the mechanism of inheritance, and then the biochemical revolution, which reduced both to the molecular level. Finally, the new science of evolutionary developmental biology ("evo-devo") is revolutionizing our understanding of evolution as the re-sorting and timing of activation in developmental genes, rather than the structural changes in the genes themselves. Interestingly, if we examine these "revolutions," we find that—after the first one—they seem to occur in similar intervals: from Darwin (1859) to genetics (1900) to DNA (1953) and finally to gene regulation (2000) is roughly a time line in half-century intervals.

The social and religious collateralities of Darwinism that have been so central in public debate in this century are those that also found some voice in the nineteenth century: creationism, of course, but social Darwinism, eugenics, and even sociobiology, as well. These all deserve treatment, but they do not sequence themselves as readily in any exclusive chronological pattern. Furthermore, in addition to evo-devo, there are new approaches to the evolutionary interfaces of biology and culture and biology and environment. All of these scientific inquiries, many of them non-Darwinian and some on the verge of a different paradigm, became part of the evolutionary conversation in the last thirty years. I have attempted to provide representative exposure to some of them while straining to keep the book to manageable size.

The final units, then, treat this newer body of evidence and debate in a less sequential manner than do the others. In the middle are four units devoted to Darwin's work and the correspondence surrounding it: the voyage of the *Beagle*, the period of thought and writing immediately after the voyage, selections from the *Origin*, and the critical period of its effects in the aftermath.

Rationale for the Book

I wrote this book after having realized that the filter of secondary sources is too coarse to give full appreciation to original ideas. How can we expect—if, indeed, we do—our students to grasp the deep legacy bestowed on science by the thinkers of our past, if we do not ask them to taste the original words? Seminal ideas roll around in the human mind for long periods before they finally find public expression, and this expression reflects both a fear and a conviction (think Copernicus; think Darwin). We ought not to deprive our students of a vicarious walk in those shoes.

After teaching human evolution for over forty years without adequately giving my students this experience, I offered an advanced course in "Concepts of Evolution" that attempted to fulfill this important pedagogic task. The third time I taught it was at Oxford University in the summer of 2009. It was an edifying experience. It led me to substantial revisions in the course and the developing manuscript, based on the experience of teaching in the very place where much of the action in the book had happened. There is no substitute for firsthand experience of the sense of place in understanding how things happen, and to visit the room where the Huxley-Wilberforce debates took place, or to lecture in the space above the original laboratories of Boyle and Hooke, is a heady experience.

This book, I knew, could not confer that sense, but it might be a leg up on summaries of original work not in the original tongue, translated or not. So this was the rationale for the book: "In their own words."

How, then, might the book be used? Who should assign it?

Use of the Book

Since I wrote this for students of evolutionary science, both biological and anthropological, and for students of the history of science, I aim this book toward those faculty who wish to supplement their principal texts—or their lectures—with original texts to provide the more personal and historical context. It will also be useful as a supplemental text in any course on the history of science. Finally, for any course on the history of evolution, this book should be of primary interest.

Introducing each unit is a commentary that attempts to set the sociocultural and historical context of the included ideas and their authors. I introduce each author's selections with a brief biographical sketch to give the reader a better sense of the person. Finally, italicized commentary preceding many of the source materials draws attention to particular relevancies to evolution and connections to ideas in both preceding and following sections.

I have in almost all cases provided imbedded pagination from the cited original document. This is in the form of bracketed page numbers. In cases where original passages had footnotes and/or bibliographic references, I have included these. Where illustrations in the original are omitted, I note this.

Following each unit, I provide a few discussion topics linking the readings. These "Ideas to Think About" are also suitable for reflection paper topics, to stimulate student thinking. I have also attached a brief selection of books "For Further Reading" to each unit. I have intentionally avoided those with more pedantic styles or instructional goals, instead focusing on those I have found invigorating and thought provoking. This list could, of course, be expanded.

Acknowledgments

Book writing is an edifyingly lonely experience, and most of us seek that selfish solitude. We also are willful in our dependence on the help of others. Indeed, we insist upon it as if it were a right! In both cases, we owe a debt. Yet we almost always express gratitude for the assistance we receive and never for those dear souls who leave us alone! The latter frequently deserve our deepest devotion, and so my Judith is atop my list. She called me to dinner but never insisted that I dine on her time; she tolerated chores undone, yet she provided some of the most welcome critiques of style and content. And so I dedicate this to her.

Oxford University Press has a wealth of patient and encouraging people who provided immense help with the organization as well as the copyright permissions. My editor, Janet Beatty, originally encouraged me to turn a course syllabus into a book and provided solicitude and excellent suggestions throughout its production. She has my gratitude.

Students who took the course for which this book in manuscript form was a reader have helped in its evolution. The eager students at Oxford University during the summer of 2009 were especially helpful in identifying threads of ideas and questions as we progressed. Two colleagues, Melissa Dowling and Kathleen Wellman, in Southern Methodist University's history department provided excellent commentary on four units, and I have profited by their suggestions and corrections. Audrey Eads indexed the manuscript, and I thank her for this.

I finally wish to thank the anonymous reviewers—both of the original book proposal and of the penultimate manuscript. They each provided detailed critiques and worthy suggestions. Without their professional insights and expertise, this would have been a poorer effort.

Ronald K. Wetherington
Dallas, Texas
June 15, 2010

INTRODUCTION

> There are stranger things in heaven and earth, Horatio,
> Than are dreamt of in your philosophy.
>
> —Shakespeare

Long before science existed, as we commonly understand science, philosophers wondered about the strange things in heaven and earth. They particularly wondered about humankind's place and purpose in all of this, and how we might go about finding out. These early attempts to discover the meaning of knowledge in the knowledge of meaning were handed down in fragments of writing from classical civilizations of the Middle East and Mediterranean, but there can be little doubt that such questions arose long before writing, as far back as consciousness of self and of mortality. We find this curiosity in the sensual fertility figures and in the somber ochre burials of the Pleistocene. We find it in cave drawings and on antler carvings and in the later megalithic dolmens and stone circles of the elusive Neolithic.

Inevitably, curiosity and casual observation raised questions about ourselves and the rest of living forms—how similar we are in many ways yet how profoundly different—and how they, we, and indeed the stars and planets came to be. But this was a supernatural world, where unicorn and tiger together ranged the forests, where dying stars and living constellations populated the heavens, and where life arose not alone from seed but also spontaneously from decaying matter.

It was in this mythological thought—centered in logic but wary of the reliability of visual evidence—that not only science but also scientific concepts of evolution have their beginning. It is here, therefore, that we, too, must begin, for my purpose is to explore the intellectual growth of evolutionary concepts as these took shape over the millennia before and the centuries after the Darwinian theory of evolution became coherent. The line from that ancient time to Darwin and beyond is not a straight one, and some of what I include here focuses not on evolutionary thought at all, but on the representative organized thinking that formed part of the intellectual heritage of evolution. Scientific thinking began, after all, in the heavens and in physical phenomena such as rainbows and eclipses and in the geometric relations and mathematical proportions by which these could be made comprehensible. Metaphysics has its origin here, for the very nature of reality was a question originating in celestial spheres, and the ancient methodologies for investigating these questions came through this realm, as well.

Science is also, and fundamentally, a social enterprise. Scientific ideas—such as political, economic, or religious ideas—are deeply imbedded in the cultural soil from which they emerge. It is no wonder, then, that new ideas arrive in clusters, from several minds, at roughly the same time. The famous Darwin-Wallace coincidence of discovery is only one example. "Nothing else in the world...," wrote Victor Hugo, "not all the armies... is so powerful as an idea whose time has come." Several of these examples will come to light in the readings included here.

It should be axiomatic, following this coincidence, that the convergence of ideas from different domains of social thought is likewise more or less concurrent. One example is the convergence of the Enlightenment philosophy of Rousseau, the political economy of Adam Smith and Thomas Malthus, and the liberal evolutionary views of Erasmus (and eventually grandson Charles) Darwin. Some of these convergence examples, too, will appear here.

Furthermore, it is not the importance of ideas alone that will help to inform us of the arrival of evolutionary explanations; it is also the importance of place. Universities as centers of learning gave legitimacy to intellectual inquiry in the Middle Ages; scientific societies provided opportunities to share ideas and demonstrate experiments in the seventeenth century; and public museums in the eighteenth century provided taxonomic illustrations of life's commonalities and a visual sense of change through time. The way people viewed nature and the way it was studied were always influenced by, and in turn influenced, the sense of place, philosophically ("Man's place in Nature") and topographically (a fossil's place in the museum display).

All of this is another way of saying that if we are to understand the growth of an idea such as evolution, we must also understand the unfolding of history and the social fabric in which it is interwoven, in addition to the march of personalities (some famous, others not) across history's stage. I earnestly hope that understanding will be served by the original readings included here as well as by the interpretive and biographical commentary I provide.

But there is yet another—less uplifting—lesson to be learned in reading these pioneers on the farther boundaries of the intellect, that place which Shel Silverstein describes as "where the sidewalk ends." These men (alas, few women were permitted here) stood at a lonely place. New ideas are often as dangerous as they are lofty. Those who generate them, discover them, and are absorbed into them, as often suffer ridicule as praise, condemnation as approval. This book is populated with them. Socrates sacrificed himself for his radical teachings. Copernicus had his heretical ideas published after his death. Voltaire escaped the guillotine by his exile to England.

Loren Eiseley (1966, 7) wrote, "The man who learns how difficult it is to step outside the intellectual climate of his or any age has taken the first step on the road to emancipation...," and concluded that this lesson should bring "not only greater tolerance for the ideas of others but a clearer realization that even the scientific atmosphere evolves and changes with the society of which it is a part." Our experience will challenge Eiseley's optimism. Darwin's idea has not been well tolerated, and we will see that the more recent treatment of sociobiology is hardly an exercise in academic restraint.

There is indeed a lesson to be learned from reading the history of scientific ideas "in their own words" and in the contexts of their times. It is a mixed lesson.

UNIT 1

Roots of Science

This initial unit is designed to set the stage for examining the rise of the Western scientific tradition and, within it, concepts of evolution. This stage setting requires a brief study—with examples—of philosophical and intellectual approaches during three periods: classical Greek culture, the European Middle Ages when much of Greek philosophy was rediscovered, and the separation of philosophical and theological traditions in the sixteenth century

I: Classical Greece

> All men by nature desire to know. An indication of this is the delight we take in our senses; for even apart from their usefulness they are loved for themselves.
>
> —ARISTOTLE, *Metaphysics, Bk. 1*

All men by nature desire to know. But what is knowable, and what is not? This was not a trivial question in Greek society of the sixth and fifth centuries BC, but it was one that had not been asked before with such passion that was to consume intellectual discourse for two centuries. Can the knowable be reliably apprehended by the senses, as Aristotle claimed? Or instead must we rely on reflection alone, as Plato insisted? Can the knowable be explained by reference to natural processes and laws alone, as the Ionian philosophers Thales, Anaximander, Empedocles, and Anaxagoras claimed? Or is reality fundamentally a mystical order, apprehensible only through mathematical ratios and their musical resonations, as the followers of Pythagoras claimed?

In this turbulent period when philosophy emerged, the legions of different answers to these questions shared fundamental presuppositions that would, over the following millennia, form the intellectual core of Western science and, ultimately, of concepts of evolution. The natural world was orderly, not chaotic; it regularity displayed an organization, or pattern, subject to investigation; events in the cosmos followed systematic rules, and these rules can be discovered; and what lay behind these rules—whether mystical or not—was rational nature and not the arbitrary acts of gods.

These presuppositions defined (and limited) the kinds of questions that were asked, including those just posed, and many of these questions implied or permitted evolutionary answers. If one asks whether the cosmos is permanent and fixed or impermanent and subject to change, evolutionary concepts hang in the balance of the answers. If one asks whether elements of nature—stars and planets, plants and animals—are each independent or are part of a common order, the idea of evolution lies nascent in the response.

Though we commonly find this classical philosophical body of work dominated by Plato and Aristotle, this is largely because not only did more of their writings survive, but Latin translations of them were more readily available to later scholars than the works of others. It remains, however, that earlier writings are in many respects just as important, and many novel ideas of these pre-Socratic philosophers are acknowledged by Plato and Aristotle. Indeed, most of what we know of some of these early thinkers comes to us *only* through these two prolific writers.

1. PLATO

Born in Athens sometime in the last quarter of the fifth century BC, Plato remains a somewhat elusive historical figure. It is certain that he was schooled in the academic subjects of his day, including grammar, music, and philosophy. His most revered teacher was Socrates, who is prominently featured in many of Plato's *Dialogues*. Plato is perhaps most famous for his theory of forms (ideas) as the only true reality, believing that specific incarnations of the ideal form of something are corrupted and made inaccurate by the senses that apprehend them. To Plato, pure thought is the only route to knowledge

Dialogues: Phaedo (Vol. I)

Numbers in brackets represent numbered paragraphs from the translation by Jowett. The following passage is one of several in which Plato—following the pre-Socratic philosopher Parmenides—rejects the senses as an accurate portrayal of reality.

(Socrates in dialogue with Simmias)

[65] What again shall we say of the actual acquirement of knowledge?—is the body, if invited to share in the enquiry, a hinderer or a helper? I mean to say, have sight and hearing any truth in them? Are they not, as the poets are always telling us, inaccurate witnesses? And yet, if even they are inaccurate and indistinct, what is to be said of the other senses?—for you will allow that they are the best of them?

Certainly, he replied.

Then when does the soul attain truth?—for in attempting to consider anything in company with the body she is obviously deceived.

True.

Then must not true existence be revealed to her in thought, if at all?

Yes.

And thought is best when the mind is gathered into herself and none of these things trouble her—neither sounds nor sights nor pain nor any pleasure,—when she takes leave of the body, and has as little as possible to do with it, when she has no bodily sense or desire, but is aspiring after true being?

Certainly.

Dialogues: Timaeus (Vol. III)

[28] Is the world created or uncreated?—that is the first question. Created, I reply, being visible and tangible and having a body, and therefore sensible; and all sensible things are apprehended by opinion and sense and are in a process of creation and created.

Laws, Bk. VI (Vol. IV)

The following passage is intended to reflect Plato's early ideas about inheritance of character traits. Compare the included passage from Hippocrates on cranial deformation and from Aristotle on his questioning of direct inheritance in On the Generation of Animals, *bk. I, part 18.*

[775] Wherefore, also, the drunken man is bad and unsteady in sowing the seed of increase, and is likely to beget offspring who will be unstable and untrustworthy, and cannot be expected to walk straight either in body or mind. Hence during the whole year and all his life long, and especially while he is begetting children, ought to take care and not intentionally do what is injurious to health, or what involves insolence and wrong; for he cannot help leaving the impression of himself on the souls and bodies of his offspring, and he begets children in every way inferior.

Republic, Bk. V (Vol. I)

The Socratic philosophers such as Plato and Aristotle were concerned not only with metaphysical and epistemological questions, but also with questions of duty and ethics. In this passage, we see the early concept of selection in animal breeding being applied to human breeding and, ultimately, to the state.

(Socrates in dialogue with Glaucon)

[458] You, I said, who are their legislator, having selected the men, will now select the women and give them to them;—they must be as far as possible of like natures with them; and they must live in common houses and meet at common meals. None of them will have anything specially his or her own; they will be together, and will be brought up together, and will associate at gymnastic exercises. And so they will be drawn by a necessity of their natures to have intercourse with each other—necessity is not too strong a word, I think?

Yes, he said;—necessity, not geometrical, but another sort of necessity which lovers know, and which is far more convincing and constraining to the mass of mankind. True, I said; and this, Glaucon, like all the rest, must proceed after an orderly fashion; in a city of the blessed, licentiousness is an unholy thing which the rulers will forbid.

Yes, he said, and it ought not to be permitted.

Then clearly the next thing will be to make matrimony sacred in the highest degree, and what is most beneficial will be deemed sacred?

[459] Exactly.

And how can marriages be made most beneficial?—that is a question which I put to you, because I see in your house dogs for hunting, and of the nobler sort of birds not a few. Now, I beseech you, do tell me, have you ever attended to their pairing and breeding?

In what particulars?

Why, in the first place, although they are all of a good sort, are not some better than others?

True.

And do you breed from them all indifferently, or do you take care to breed from the best only?

From the best.

And do you take the oldest or the youngest, or only those of ripe age?

I choose only those of ripe age.

And if care was not taken in the breeding, your dogs and birds would greatly deteriorate?

Certainly.

And the same of horses and animals in general?

Undoubtedly.

Good heavens! my dear friend, I said, what consummate skill will our rulers need if the same principle holds of the human species!

Certainly, the same principle holds; but why does this involve any particular skill?

Because, I said, our rulers will often have to practise upon the body corporate with medicines. Now you know that when patients do not require medicines, but have only to be put under a regimen, the inferior sort of practitioner is deemed to be good enough; but when medicine has to be given, then the doctor should be more of a man.

That is quite true, he said; but to what are you alluding?

I mean, I replied, that our rulers will find a considerable dose of falsehood and deceit necessary for the good of their subjects: we were saying that the use of all these things regarded as medicines might be of advantage.

And we were very right.

And this lawful use of them seems likely to be often needed in the regulations of marriages and births.

How so?

Why, I said, the principle has been already laid down that the best of either sex should be united with the best as often, and the inferior with the inferior, as seldom as possible; and that they should rear the offspring of the one sort of union, but not of the other, if the flock is to be maintained in first-rate condition. Now these goings on must be a secret which the rulers only know, or there will be a further danger of our herd, as the guardians may be termed, breaking out into rebellion.

2. Hippocrates

Hippocrates, born about 460 BC on the Greek island of Kos, just off the Turkish coast, became a famous practitioner and teacher of medicine. He rejected mythology and superstition as a cause of disease, emphasizing the role of diet and environment. Medicine, he asserted, must be kept separate from religion and was more appropriately aligned with philosophy. Most specifically, he wrote, the gods do not punish men with disease, as his treatise on epilepsy—the "sacred disease"—testifies.

On the other hand, almost nothing about the human body and its organs was known to his age, as Greek culture forbade human dissection. Consequently, his knowledge was riddled with mistaken beliefs. This is particularly true with his understanding of sexual reproduction, where false notions of male and female semen and their origins and characteristics were transmitted to the West in the Middle Ages and continued to be held well into the scientific revolution. His successor, Galen, the Roman physician of the second century AD, also a Greek, had the advantage of better scientific knowledge and had dissection experience as well.

Airs, Waters, Places, Para. 14

The following passage comes from The Genuine Writings of Hippocrates, *translation by Francis Adams. The "long headed" recipients of artificial cranial deformation to which Hippocrates refers were peoples from Anatolia. His belief that cultural custom led to inheritance is, of course, wrong. Inheritance of acquired characteristics was reincarnated in the eighteenth century by Lamarck.*

I will pass over the smaller differences among the nations, but will now treat of such as are great either from nature, or custom; and, first, concerning the Macrocephali. There is no other race of men which have heads in the least resembling theirs. At first, usage was the principal cause of the length of their head, but now nature cooperates with usage. They think those the most noble who have the longest heads. It is thus with regard to the usage: immediately after the child is born, and while its head is still tender, they fashion it with their hands, and constrain it to assume a lengthened shape by applying bandages and other suitable contrivances whereby the spherical form of the head is destroyed, and it is made to increase in length. Thus, at first, usage operated, so that this constitution was the result of force: but, in the course of time, it was formed naturally; so that usage had nothing to do with it; for the semen comes from all parts of the body, sound from the sound parts, and unhealthy from the unhealthy parts. If, then, children with bald heads are born to parents with bald heads; and children with blue eyes to parents who have blue eyes; and if the children of parents having distorted eyes squint also for the most part; and if the same may be said of other forms of the body, what is to prevent it from happening that a child with a long head should be produced by a parent having a long head? But now these things do not happen as they did formerly, for the custom no longer prevails owing to their intercourse with other men. Thus it appears to me to be with regard to them.

3. Aristotle

The great aristocratic philosopher, Aristotle, was Plato's most esteemed student. He was born in Stageira (near today's Thessaloniki), in 384 BC, but moved to Athens as a teen to join Plato at his academy in the north of Athens, were he stayed until after Plato's death. His subsequent travels allowed him to study the broad diversity of nature and fed his strong appetite for learning. At age forty, he became tutor to Alexander the Great, and to Alexander's eventual successor, Ptolemy.

At fifty, Aristotle established his own school in Athens's east side, the Lyceum, where he taught for over a decade. He was a true polymath, studying and writing about virtually every intellectual topic available to Greek knowledge. Although both he and his mentor, Plato, revered the teaching of Socrates and sought an understanding of universals in their two schools, their approaches to knowledge were quite distinct. For Aristotle, universals were most profitably sought in the study of particulars, and only through such particulars could the knowledge of universal "essences" be acquired. He thus championed "natural philosophy," the study of the natural world, and placed little emphasis on the search for mathematical relationships that characterized much of Plato's interest.

Aristotle's two major contributions to later science—including evolutionary ideas—were to establish the foundations of biological inquiry and to establish the formality of logic as an analytic approach to such inquiry.

On the Heavens, Bk. 1, Ch. 3 (270^b [14–20])

The bold-faced numbers in the following selections refer to page numbers of the standard Berlin Greek text, whereas the superscript a–d refer to the columns therein; the bracketed numbers are the original line numbers. All of these follow the translations published by Ross and Smith.

The translation that immediately follows is by J. L. Stocks. In this passage, Aristotle affirms that the cosmos had no beginning, and that the universal element "ether" subsumes the functional or compositional four elements. It is also referred to as the "fifth element."

For in the whole range of time past, so far as our inherited records reach, no change appears to have taken place either in the whole scheme of the outermost heaven or in any of its proper parts. The common name, too, which has been handed down from our distant ancestors even to our own day, seems to show that they conceived of it in the fashion which we have been expressing. The same ideas, one must believe, recur in men's minds not once or twice but again and again. And so, implying that the primary body is something else beyond earth, fire, air, and water, they gave the highest place a name of its own, *aither*, derived from the fact that it 'runs always' for an eternity of time.

On the Soul, Bk. 1, Ch. 1 (403^a [10, 15])

This and the following quote were translated by J. A. Smith, in Ross and Smith. In these passages, Aristotle logically infers that the soul cannot have an existence separate from the body.

[10] If there is any way of acting or being acted upon proper to soul, soul will be capable of separate existence; if there is none, its separate existence is impossible.

[15] . . . it cannot be so divorced at all, since it is always found in a body. It therefore seems that all the affections of soul involve a body—passion, gentleness, fear, pity, courage, joy, loving, and hating; in all these there is a concurrent affection of the body.

On the Soul, Bk. II, Ch. 2 [25–30]

We have no evidence as yet about mind or the power to think; it seems to be a widely different kind of soul, differing as what is eternal from what is perishable; it alone is capable of existence in isolation from all other psychic powers. All the other parts of the soul, it is evident from what we have said, are, in spite of certain statements to the contrary, incapable of separate existence though, of course, distinguishable by definition.

On the Parts of Animals, Bk. 1, Ch. 5 (645^b [20–28])

The following seven selections detail Aristotle's arguments on principles of biological classification. This first one describes levels of comparison from analogy (at the most inclusive level), to generic (equivalent to our class or order), to the specific (our species, when undivided into subspecies).

We have, then, first to describe the common functions, common, that is, to the whole animal kingdom, or to certain large groups, or to members of a species. In other words, we have to describe the attributes common to all animals, or to assemblages, like the class of Birds, or closely allied groups differentiated by gradation, or to groups like Man not differentiated into subordinate groups. In the first case the common attributes may be called analogies, in the second generic, and in the third specific.

Metaphysics, Bk. V, Part 28 (Translated by W. D. Ross, in Ross and Smith)

The term 'race' or 'genus' is used (1) if generation of things which have the same form is continuous, e.g. 'while the race of men lasts' means 'while the generation of them goes on continuously'. (2) It is used with reference to that which first brought things into existence; for it is thus that some are called Hellenes by race and others Ionians, because the former proceed from Hellen and the latter from Ion as their first begetter. And the word is used in reference to the begetter more than to the matter, though people also get a race-name from the female, e.g. 'the descendants of Pyrrha'. (3) There is genus in the sense in which 'plane' is the genus of plane figures and 'solid' of solids; for each of the figures is in the one case a plane of such and such a kind, and in the other a solid of such and such a kind; and this is what underlies the differentiae. Again (4) in definitions the first constituent element, which is included in the 'what', is the genus, whose differentiae the qualities are said to be. 'Genus' then is used in all these ways, (1) in reference to continuous generation of the same kind, (2) in reference to the first mover which is of the same kind as the things it moves, (3) as matter; for that to which the differentia or quality belongs is the substratum, which we call matter.

Those things are said to be 'other in genus' whose proximate substratum is different, and which are not analysed the one into the other nor both into the same thing (e.g. form and matter are different in genus); and things which belong to different categories of being (for some of the things that are said to 'be' signify essence, others a quality, others the other categories we have before distinguished); these also are not analysed either into one another or into some one thing.

The History of Animals, Bk. VIII, Part 1 (588^b [5–15])

Aristotle recognized three broad classes of life forms: plants, animals, and humans. Each was distinguished in possessing unique souls: plants, a vegetative soul; animals, a vegetative as well as sensitive (sensory) soul; and humans, a vegetative and sensitive as well as a rational soul. While these three—and their various subdivisions (except for man) constituted a hierarchy of life from simple to complex, the categories were themselves fixed, and hence their relationship in analogies was not one of common ascent. We see this in the following selection.

Nature proceeds little by little from things lifeless to animal life in such a way that it is impossible to determine the exact line of demarcation, nor on which side thereof an intermediate form should lie. Thus, next after lifeless things in the upward scale comes the plant, and of plants one will differ from another as to its amount of apparent vitality; and, in a word, the whole genus of plants, whilst it is devoid of life as compared with an animal, is endowed with life as compared with other corporeal entities. Indeed, as we just remarked, there is observed in plants a continuous scale of ascent towards the animal. So, in the sea, there are certain objects concerning which one would be at a loss to determine whether they be animal or vegetable. For instance, certain of these objects are fairly rooted, and in several cases perish if detached; thus the pinna is rooted to a particular spot, and the solen (or razor-shell) cannot survive withdrawal from its burrow. Indeed, broadly speaking, the entire genus of testaceans have a resemblance to vegetables, if they be contrasted with such animals as are capable of progression.

On the Generation of Animals, Bk. 1, Ch. 17–18

In this long passage, Aristotle seeks to understand how new life is generated through sexual intercourse. The question of the role of semen, and whether from both male and female, and whether from all parts of the body or not, occupied discourse for millennia after Hippocrates, Aristotle, and other Greek philosophers contemplated the question. We see here Aristotle's keen sense of logic, and his argument not only from evidence but also from analogy. His "proofs," in the last paragraph of chapter 17 and the first of chapter 18, constitute a typical form of argument. We see it much later, for example in Copernicus's Revolutions of the Heavenly Bodies*—particularly the paragraphs on proof that the universe and the earth are spherical. It is also significant that little empirical progress was made on human reproductive biology and physiology for well over a millennium after Aristotle wrote this. William Harvey, in 1651, followed Aristotle's arguments almost precisely in his* On Animal Generation.

[17] Some animals manifestly emit semen, as all the sanguinea, but whether the insects and cephalopoda do so is uncertain. Therefore this is a question to be considered, whether all males do so, or not all; and if not all, why some do and some not; and whether the female also contributes any semen or not; and, if not semen, whether she does not contribute anything else either, or whether she contributes something else which is not semen. We must also inquire what those animals which emit semen contribute by means of it to generation, and generally what is the nature of semen, and of the so-called catamenia in all animals which discharge this liquid.

Now it is thought that all animals are generated out of semen, and that the semen comes from the parents. Wherefore it is part of the same inquiry to ask whether both male and female produce it or only one of them, and to ask whether it comes from the whole of the body or not from the whole; for if the latter is true it is reasonable to suppose that it does not come from both parents either. Accordingly, since some say that it comes from the whole of the body, we must investigate this question first.

The proofs from which it can be argued that the semen comes from each and every part of the body may be reduced to four. First, the intensity of the pleasure of coition; for the same state of feeling is more pleasant if multiplied, and that which affects all the parts is multiplied as compared with that which affects only one or a few. Secondly, the alleged fact that mutilations are inherited, for they argue that since the parent is deficient in this part the semen does not come from thence, and the result is that the corresponding part is not formed in the offspring. Thirdly, the resemblances to the parents, for the young are born like them part for part as well as in the whole body; if then the coming of the semen from the whole body is cause of the resemblance of the whole, so the parts would be like because it comes from each of the parts. Fourthly, it would seem to be reasonable to say that as there is some first thing from which the whole arises, so it is also with each of the parts, and therefore if semen or seed is cause of the whole so each of the parts would have a seed peculiar to itself. And these opinions are plausibly supported by such evidence as that children are born with a likeness to their parents, not in congenital but also in acquired characteristics; for before now, when the parents have had scars, the children have been born with a mark in the form of the scar in the same place, and there was a case at Chalcedon where the father had a brand on his arm and the letter was marked on the child, only confused and not clearly articulated. That is pretty much the evidence on which some believe that the semen comes from all the body.

[18] On examining the question, however, the opposite appears more likely, for it is not hard to refute the above arguments and the view involves impossibilities. First, then, the resemblance of children to parents is no proof that the semen comes from the whole body, because the resemblance is found also in voice, nails, hair, and way of moving, from which nothing comes. And men generate before they yet have certain characters, such as a beard or grey hair. Further, children are like their more remote ancestors from whom nothing has come, for the resemblances recur at an interval of many generations, as in the case of the woman in Elis who had intercourse with the Aethiop; her daughter was not an Aethiop but the son of that daughter was. The same thing applies also to plants, for it is clear that if this theory were true the seed would come from all parts of plants also; but often a plant does not possess one part, and another part may be removed, and a third grows afterwards. Besides, the seed does not come from the pericarp, and yet this also comes into being with the same form as in the parent plant.

We may also ask whether the semen comes from each of the homogeneous parts only, such as flesh and bone and sinew, or also from the heterogeneous, such as face and hands. For if from the former only, we object that resemblance exists rather in the heterogeneous parts, such as face and hands and feet; if then it is not because of the semen coming from all parts that children resemble their parents in *these*, what is there to stop the homogeneous parts also from being like for some other reason than this? If the semen comes from the heterogeneous alone, then it does not come from all parts; but it is more fitting that it should come from the homogeneous parts, for they are prior to the heterogeneous which are composed of them; and as children are born like their parents in face and hands, so they are, necessarily, in flesh and nails. If the semen comes from both, what would be the manner of generation? For the heterogeneous parts are composed of the homogeneous, so that to come from the former would be to come from the latter and from their composition. To make this clearer by an illustration, take a written name; if anything came from the whole of it, it would be from each of the syllables, and if from these, from the letters and their composition. So that if really flesh and bones are composed of fire and the like elements, the semen would come rather from the elements than anything else, for how can it come from their composition? Yet without this composition there would be no resemblance. If again something creates this composition later, it would be this that would be the cause of the resemblance, not the coming of the semen from every part of the body....

The History of Animals, Bk. VIII, Ch. 28 (607[a])

The concept of hybrid species was of general interest among Greek philosophers, as this seemed to violate the neat taxonomic and phenomenological categories that held their cosmos in order. Aristotle, Herodotus, and others remarked on the quasi-sterility of the mule, for example. In the following passage are some good observations and some fairly wild speculations—in this case all from hearsay. His Libyan references are correct, and it is common that the overlapping ranges of the Olive Baboon (Papio cynocephalus) *and Hamadryas Baboon* (P. hamadryas) *lead to not infrequent hybridization. Lion/tiger hybrids are known, but no credibility may be given to the fox/dog or tiger/Indian dog story.*

It would appear that in that country [Libya] animals of diverse species meet, on account of the rainless climate, at the watering-places, and there pair together; and that such pairs will often breed if they be nearly of the same size and have periods of gestation of the same length. For it is said that they are tamed down in their behaviour towards each other by extremity of thirst.... Elsewhere also bastard-animals are born to heterogeneous pairs; thus in Cyrene the wolf and the bitch will couple and breed; and the Laconian hound is a cross between the fox and the dog. They say that the Indian dog is a cross between the tiger and the bitch, not the first cross, but a cross in the third generation; for they say that the first cross is a savage creature. They take the bitch to a lonely spot and tie her up: if the tiger be in an amourous mood he will pair with her; if not he will eat her up and this casualty is of frequent occurrence.

The History of Animals, Bk. V, Ch. 1

(539[a]) Now there is one property that animals are found to have in common with plants. For some plants are generated from the seed of plants, whilst other plants are self-generated through the formation of some elemental principle similar to a seed; and of these latter plants some derive their nutriment from the ground, whilst others grow inside other plants, as is mentioned, by the way, in my treatise on Botany. So with animals, some spring from parent animals according to their kind, whilst others grow spontaneously and not from kindred stock; and of these instances of spontaneous generation some come from putrefying earth or vegetable matter, as is the case with a number of insects, while others are spontaneously generated in the inside of animals out of the secretions of their several organs.

(539[b]) But whensoever creatures are spontaneously generated, either in other animals, in the soil, or on plants, or in the parts of these, and when such are generated male and female, then from the copulation of such spontaneously generated males and females there is generated a something—a something never identical in shape with the parents, but a something imperfect. For instance, the issue of copulation in lice is nits; in flies, grubs; in fleas, grubs egg-like in shape; and from these issues the parent-species is never reproduced, nor is any animal produced at all, but the like nondescripts only.

Physics, Bk. II, Ch. 8

(198[b]) We must explain then (1) that Nature belongs to the class of causes which act for the sake of something; (2) about the necessary and its place in physical problems, for all writers ascribe things to this cause, arguing that since the hot and the cold, &c., are of such and such a kind, therefore certain things necessarily are and come to be—and if they mention any other cause (one his 'friendship and strife', another his 'mind'), it is only to touch on it, and then good-bye to it. A difficulty presents itself: why should not nature work, not for the sake of something, nor because it is better so, but just as the sky rains, not in order to make the corn grow, but of necessity? What is drawn up must cool, and what has been cooled must become water and descend, the result of this being that the corn grows. Similarly if a man's crop is spoiled on the threshing-floor, the rain did not fall for the sake of this—in order that the crop might be spoiled—but that result just followed. Why then should it not be the same with the parts in nature, e.g. that our teeth should come up of necessity—the front teeth sharp, fitted for tearing, the molars broad and useful for grinding down the food—since they did not arise for this end, but it was merely a coincident result; and so with all other parts in which we suppose that there is purpose? Wherever then all the parts came about just what they would have been if they had come to be for an end, such things survived, being organized spontaneously in a fitting way; whereas those which grew otherwise perished and continue to perish, as Empedocles says his 'man-faced ox-progeny' did.

Such are the arguments (and others of the kind) which may cause difficulty on this point. Yet it is impossible that this should be the true view. For teeth and all other natural things either invariably or normally come about in a given way; but of not one of the results of chance or spontaneity is this true. We do not ascribe to chance or mere coinci- (199[a]) dence the frequency of rain in winter, but frequent rain in summer we do; nor heat in the dog-days, but only if we have it in winter. If then, it is agreed that things are either the result of coincidence or for an end, and these cannot be the result of coincidence or spontaneity, it follows that they must be for an end; and that such things are all due to nature even the champions of the theory which is before us would agree. Therefore action for an end is present in things which come to be and are by nature....

(199[b]) It is absurd to suppose that purpose is not present because we do not observe the agent deliberating. Art does not deliberate. If the ship-building art were in the wood, it would produce the same results by nature. If, therefore, purpose is present in art, it is present also in nature. The best illustration is a doctor doctoring himself: nature is like that.

It is plain then that nature is a cause, a cause that operates for a purpose.

"Therefore action for an end is present in things which come to be and are by nature." This is teleological reasoning if by "for an end" is meant "for a designed purpose"; on the other hand, if by this phrase is meant "for its function," it comes close to being a simple statement of form = function, in which the end is in this case "that for which the action is meant to serve." When Aristotle says, "It is absurd to think that purpose is not present because we do not observe the agent deliberating," could we not read this as "because we do not know the function which it serves"? How might this translational alternative be judged?

Ideas to Think About

1. There is a curious mixture here of direct observation and pure speculation. Why? How does this relate to the notion of "reason"?
2. Where do we find animal comparisons based on analogy (similarity in function), but not on homology (similarity from descent)? Why did the idea of homology not enter the mind-set?
3. Give examples that suggest a duality (soul–body, essence–form, fixity–change).

For Further Reading

Presocratics: Natural Philosophers before Socrates, James Warren. Berkeley: University of California Press (2007). Engaging reading! This book is an excellent treatment of the roots of later thought in the works of Heraclitus, Parmenides, Anaxagoras, Empedocles, Leucippus, Democritus, and more.

Sailing the Wine-Dark Sea: Why the Greeks Matter, Thomas Cahill. New York: Knopf Doubleday (2004). Always a compelling writer, Cahill peppers generous selections of original writings with accounts of the life, people, and ideas of classical Greece (and Rome, as well). Whimsical and intimate accounts make ancient Greece accessible and desirous of study.

II: Medieval Society and Science

Reason is the greatest enemy that faith has; it never comes to the aid of spiritual things, but—more frequently than not—struggles against the divine Word, treating with contempt all that emanates from God.

—MARTIN LUTHER

Reason was not the first to disappear, but all of those things that bring despair—loss of security, predictability, comfort—began disappearing throughout Europe as the final vestiges of Roman civilization, and the protection it provided, slipped away in the fifth century AD and introduced Western history to the dark medieval world. As Camus said, "To those who despair of everything, reason cannot provide a faith." Reason thus slipped away as well, replaced by a faith that sought rejection of worldly things with the promise of salvation in a much better life beyond.

That long period commonly known as the Early Middle Ages, from the fifth to the eleventh centuries, witnessed a decline of almost every aspect of culture inherited from Greek and Roman society, including intellectual life and literacy itself. Manchester (1992, 3) reminds us that even Charlemagne was illiterate and that literacy was distrusted long afterward, from 1000–1500.

Society during this earlier period was introverted, the small villages closed in upon themselves, circumscribed with suspicion. They seldom had names, these hamlets, and the people within them had no surnames—frequently only descriptive nicknames: In daily face-to-face contact, terms of kinship reference are of less use than terms of personal address.

This long insularity made a larger social order almost unthinkable, as settlements only a few miles apart came to speak mutually unintelligible dialects, commerce was virtually nonexistent, prior technologies were lost, and only the institution of the Roman Catholic Church survived to integrate the continent. Even this integration was laborious, as the population of Europe was now a curious mix of dozens of immigrants from the marauding tribes to the east, and those they had displaced.

In this cultural mix, the Catholic Church found in Platonic metaphysics a compatible link to ancient teachings. In its less secular form—Neo-Platonism—pure ideas, rather than the flawed copies available to the senses, became a legitimate focus of theology, although not without problems. Rejection of material reality in favor of spiritual reality pretty much left science out of the mix, and with a diverse and remotely scattered populace, the common desire for education that characterized the Roman Empire lost both its rationale and its logistic foundation. Instead, a watered-down education for churchmen and illiterate emperors took its place, fostered by Charlemagne in the eighth century through the growth of cathedral and monastic schools. Here, education focused largely on those liberal arts defined in the *trivium*: grammar, rhetoric, and logic, and to a lesser extent those of the *quadrivium*: arithmetic, geometry, astronomy, and music.

All of this had begun to change by the start of the eleventh century, as Europe struggled to emerge from the oppressive darkness of the spirit. It was well underway by the twelfth, as education became detached from the church and as classical Greek literature became broadly available. Several convergent threads led to this change and the rebirth of intellectual curiosity that accompanied it. These threads include Muslim scholarship, urbanization, and paper.

Islamic culture, and Hebraic culture as well, had early on "acquired Aristotle's philosophy and natural science, they had also absorbed Euclid's mathematics, Ptolemy's astronomy and optics, Archimedes's engineering principles, the medical sciences of Hippocrates and Galen, and other classical treasures" (Rubenstein 20 03, 6–7). As they expanded their empire into the western Mediterranean, they brought this learning with them: to Toledo, Cordoba, and Seville, in Spain, and to Sicily and Italy. Their translations from Arabic into Latin began to trickle and then flood into Europe as early as the ninth century.

Throughout the Middle Ages, the population of Europe grew dramatically. Between 1000 and 1200 "the population of Europe may have doubled, tripled, or even quadrupled, while the city-dwelling portion of this population increased even more rapidly" (Lindberg 2007, 204). This urbanization created new centers apart from the ancient cathedral cities and increased the demand for schools apart from them, as well.

The Arabs brought with them a manual papermaking technology unheard of in parchment-dependent Europe. Originating in China during the Han Dynasty of the second century AD and spreading to Baghdad by the eighth, Arabic scholars brought the technology with them to Europe. The process was more efficient than parchment production and vastly increased the output of scribes.

Water mills, used to produce flour early in the medieval period throughout Europe, acquired more commercial application as population increased, and because their use depends upon reliable waterpower, the mills helped dictate where urban centers would arise. The rise of paper mills, in consequence, was a critical automation in providing an impetus for more manuscripts. The first known European paper mill comes from Fabriano, Italy, in 1276, and at about the same time, the technology appears in Spain. It reached France by the 1300s, Germany in 1390, and Eastern Europe and Britain by 1500 (Reynolds 1983, 84–85).

The rapid rise of schools—soon to become universities—and the rapid reproduction of translated classical writings from Greek antiquity (as well as greater availability of Roman literature) were mutually reinforcing. With this came, by the eleventh century, a new interest in the application of reason to the solution of a wide variety of problems, from the practical to the intellectual: accounting and law found aid in ancient mathematics and Roman civil procedure just as Plato's *Timaeus* influenced cosmology and Aristotle's *Treatises* influenced natural philosophy.

But even more critically, we find a sense of urgency in applying ancient methods of philosophy to theological questions. Scholars such as Anselm of Canterbury and Peter Abelard of Paris were willing to "explore what unaided reason could achieve in the theological realm, to ask whether certain fundamental theological doctrines were true as judged by rational or philosophical criteria" (Lindberg 2007, 207).

Thus, we find in the High Middle Ages a new premium placed on education and on the value of intellectual inquiry. We also find both the material and emotional investment in what will later become a true scientific revolution, centered in the very institutions from which evolutionary science would later arise: the University of Paris, established in 1200, the University of Oxford in 1220, Cambridge University in 1225, and Galileo's University of Padua in 1222.

Passages below from Augustine reflect the earlier, Lower Middle Ages. The selection from Roger Bacon represents the later intellectual currents.

---

4. St. Augustine of Hippo

Augustine is the preeminent church scholar in the development of Western Christianity. He also stands on the cusp of the transition from the classical to medieval ages and was thus an important framer of philosophical thought at a time when learning almost ceased to exist. Born in Algeria in 354, his seventy-six years of life spanned the collapse of Rome, and the chaos that followed strongly influenced his philosophical and theological direction.

Augustine had early schooling in rhetoric and taught grammar and rhetoric in Rome. At age thirty, he was appointed professor of rhetoric to the imperial court in Milan. Discouraged by the Gnostic sect of Manichaeism, which he had devoutly followed, and at the urging of the Bishop of Milan and Augustine's own mother, he converted to Catholicism in 386, abandoned his teaching career, and returned to north Africa intent to follow a monastic life of study and self-denial. He was ordained a priest in Hippo Regius (Algeria) in 391 and remained there until he died in 430.

His reading of Greek and Latin scholars had given him exposure to Stoicism, Platonism, and particularly the Neo-Platonism espoused by Plotinus. In this latter philosophy, he found compatibility with much of Christian teaching, and in his prolific writing, he helped transform an essentially mystical Christianity into one of faith combined with reason. His thoughtful and logical treatises on original sin, free will, the nature of time, and the relation of empirical experience to biblical exegesis remain important contributions to philosophy and theology today.

The selected passages come from *Of the City of God* (*De civitate Dei*), written between 423 and 426. In it, he developed the concept of the church as the spiritual city of God, as opposed to the material city of humankind, and urged faithful and skeptic alike to seek solace (and salvation) from worldly ills in the promise of Christianity. Thus, the *City of God* was an attempt to address the crisis of the fall of Rome in asserting that the worldly city should be of no concern to the faithful, who should look only to the heavenly city to come.

It is parenthetically interesting that Plato's *Republic* was analogously written to address the crisis of Athenian moral decay, but instead of escaping the material world Plato argued for a replacement of its political fabric.

Within the twenty-two books of the *City of God*, we find many of the arguments that were to occupy scientists and philosophers alike in the coming millennia. All of the extracts that follow are from *City of God*, translated by Marcus Dods. The *City of God* was probably begun in 413 and was completed in 426.

City of God, Bk. XI, Ch. 4

In this first selection, Augustine illustrates the struggle to overturn the Greek notion (not shared by Plato) that the universe is eternal and thus was not created. The Greek "soul," whether likewise eternal, as Aristotle held, or created by an unconcerned deity, as Plato believed, thus needed to be reinterpreted in Christian theology.

CHAPTER 4. THAT THE WORLD IS NEITHER WITHOUT BEGINNING, NOR YET CREATED BY A NEW DECREE OF GOD, BY WHICH HE AFTERWARDS WILLED WHAT HE HAD NOT BEFORE WILLED.

But why did God choose then to create the heavens and earth which up to that time He had not made? If they who put this question wish to make out that the world is eternal and without beginning, and that consequently it has not been made by God, they are strangely deceived, and rave in the incurable madness of impiety....

As for those who own, indeed, that it was made by God, and yet ascribe to it not a temporal but only a creational beginning, so that in some scarcely intelligible way the world should always have existed a created world they make an assertion which seems to them to defend God from the charge of arbitrary hastiness, or of suddenly conceiving the idea of creating the world as a quite new idea, or of casually changing His will, though He be unchangeable. But I do not see how this supposition of theirs can stand in other respects, and chiefly in respect of the soul; for if they contend that it is co-eternal with God, they will be quite at a loss to explain whence there has accrued to it new misery, which through a previous eternity had not existed.

City of God, Bk. XII, Ch. 10

The following passage is aimed at Platonic cosmology, which envisioned a cyclical history. In Timaeus, *Critias recounts to Socrates a story of the founding of Athens: [22] "There have been, and will be again, many destructions of mankind arising out of many causes; the greatest have been brought about by the agencies of fire and water, and other lesser ones by innumerable other causes." [23] "In the first place, you remember a single deluge only, but there were many previous ones; in the next place, you do not know that there formerly dwelt in your land the fairest and noblest race of men which ever lived, and that you and your whole city are descended from a small seed or remnant of them which survived.... She [the goddess Athene] founded your city a thousand years before ours, receiving from the Earth and Hephaestus the seed of your race, and afterwards she founded ours. Of which the constitution is recorded in our sacred registers to be eight thousand years ago."*

CHAPTER 10. OF THE FALSENESS OF THE HISTORY WHICH ALLOTS MANY THOUSAND YEARS TO THE WORLD'S PAST.

Let us, then, omit the conjectures of men who know not what they say, when they speak of the nature and origin of the human race. For some hold the same opinion regarding men that they hold regarding the world itself, that they have always been. Thus Apuleius says when he is describing our race, "Individually they are mortal, but collectively, and as a race, they are immortal." And when they are asked, how, if the human race has always been, they vindicate the truth of their history, which narrates who were the inventors, and what they invented, and who first instituted the liberal studies and the other arts, and who first inhabited this or that region, and this or that island? they reply, that most, if not all lands, were so desolated at intervals by fire and flood, that men were greatly reduced in numbers, and from these, again, the population was restored to its former numbers, and that thus there was at intervals a new beginning made, and though those things which had been interrupted and checked by the severe devastations were only renewed, yet they seemed to be originated then; but that man could not exist at all save as produced by man. But they say what they think, not what they know.

They are deceived, too, by those highly mendacious documents which profess to give the history of many thousand years, though, reckoning by the sacred writings, we find that not 6000 years have yet passed.

5. Roger Bacon

Roger Bacon stands alongside Thomas Aquinas and Albertus Magnus as the three most important contributors to learning during the High Middle Ages. Among them, he provided the most cutting criticism of the formal schools, both for their ignorance of true scientific inquiry and for the grave errors in teaching on the part of the professors.

Virtually everything we know about this mysterious scholar comes from his own works. Bacon was born in either 1214 or 1220 in Ilchester, Somerset, England. His family was apparently wealthy, but lost their property under the reign of Henry III. At the age of thirteen, Roger attended Oxford and became a master there. Around 1233, he studied at the University of Paris, where he read Aristotle and the other classical Greek scholars and became thoroughly acquainted with Islamic writers. It was here that he came to realize the handicap in learning due to inadequate preparation of the faculty and appallingly bad Latin translations of Greek. Natural science was particularly inadequately known and taught, with teachers failing to follow the Aristotelian way of experience and logic. His vehement critique of the profound inadequacies of scholarship and teaching were published in his *Compendium Studii Philosophiae*, published in 1271.

By 1250, he had returned to Oxford, but his disenchantment with scholastic practices led him to withdraw from teaching and join the Franciscan order—at that time in strong competition with the Dominicans. His writing had attracted wide attention both in England and on the continent, and, though deeply respected, it had garnered some suspicion as being unorthodox. Some of his writing supported astrology and alchemy, although he condemned magic as being contrary to the investigation of natural science. Nevertheless, his popularity was enhanced by close friendships with powerful people, including another strong contributor to medieval science, Robert Grosseteste, bishop of Lincoln.

In 1257, the Franciscan order forbade any friar from publishing without prior approval, and Bacon specifically fell into disfavor. His writing was salvaged, however, by Pope Clement IV who, as a papal legate in England before his election, was attracted to Bacon's ideas. When he urged Bacon to write in secret and send him summaries of his arguments on the role of philosophy in theology, Bacon completed three important works: the *Opus Majus, Opus Minus,* and *Opus Tertium*.

Bacon's critiques and his philosophical position on experimental science presaged the later work of his namesake, Francis Bacon, and his legacy remained strong through the eighteenth century. Roger Bacon died at the Franciscan House at Oxford late in the thirteenth century (1284–1294).

On Experimental Science

In the following selection, Bacon contrasts Plato's emphasis on "reason" with Aristotle's on "experience" (which is what Bacon means by "experiment") and emphasizes the need for both.

In the sixth paragraph, Bacon's reference to Algazel *refers to Muhammad al-Ghazālī, renowned Persian philosopher and theologian of the eleventh century. Bacon here exhibits his eclecticism, for al-Ghazālī rejected the Greek philosophers as unbelievers, whose philosophical methods corrupted Islam.*

Finally, Bacon's extensive comments on the rainbow follow a popular pattern of detailed comparison and generalization in analysis that was introduced by the extensive translations of Aristotle (particularly his books on natural philosophy), as well as those of Euclid and Ptolemy. These works helped transform the University of Paris and the University of Oxford into first-class sources of learning supported by an unexpected freedom of inquiry (see Grant 1984).

Having laid down the main points of the wisdom of the Latins as regards language, mathematics and optics, I wish now to review the principles of wisdom from the point of view of experimental science, because without experiment it is impossible to know anything thoroughly.

There are two ways of acquiring knowledge, one through reason, the other by experiment. Argument reaches a conclusion and compels us to admit it, but it neither makes us certain nor so annihilates doubt that the mind rests calm in the intuition of truth, unless it finds this certitude by way of experience. Thus many have arguments toward attainable facts, but because they have not experienced them, they overlook them and neither avoid a harmful nor follow a beneficial course. Even if a man that has never seen fire, proves by good reasoning that fire burns, and devours and destroys things, nevertheless the mind of one hearing his arguments would never be convinced, nor would he avoid fire until he puts his hand or some combustible thing into it in order to prove by experiment what the argument taught. But after the fact of combustion is experienced, the mind is satisfied and lies calm in the certainty of truth. Hence argument is not enough, but experience is.

This is evident even in mathematics, where demonstration is the surest. The mind of a man that receives that clearest of demonstrations concerning the equilateral triangle without experiment will never stick to the conclusion nor act upon it till confirmed by experiment by means of the intersection of two circles from either section of which two lines are drawn to the ends of a given line. Then one receives the conclusion without doubt. What Aristotle says of the demonstration by the syllogism being able to give knowledge, can be understood if it is accompanied by experience, but not of the bare demonstration. What he says in the first book of the Metaphysics, that those knowing the reason and cause are wiser than the experienced, he speaks concerning the experienced who know the bare fact only without the cause. But I speak here of the experienced that know the reason and cause through their experience. And such are perfect in their knowledge, as Aristotle wishes to be in the sixth book of the Ethics, whose simple statements are to be believed as if they carried demonstration, as he says in that very place.

Whoever wishes without proof to revel in the truths of things need only know how to neglect experience. This is evident from examples. Authors write many things and the people cling to them through arguments which they make without experiment, that are utterly false. It is commonly believed among all classes that one can break adamant only with the blood of a goat, and philosophers and theologians strengthen this myth. But it is not yet proved by adamant being broken by blood of this kind, as much as it is argued to this conclusion. And yet, even without the blood it can be broken with ease. I have seen this with my eyes; and this must needs be because gems cannot be cut out save by the breaking of the stone. Similarly it is commonly believed that the secretions of the beaver that the doctors use are the testicles of the male, but this is not so, as the beaver has this secretion beneath its breast and even the male as well as the female produces a secretion of this kind. In addition also to this secretion the male has its testicles in the natural place and thus again it is a horrible lie that, since hunters chase the beaver for this secretion, the beaver knowing what they are after, tears out his testicles with his teeth and throws them away. Again it is popularly said that cold water in a vase freezes more quickly than hot; and the argument for this is that contrary is excited by the contrary, like enemies running together. They even impute this to Aristotle in the second book of Meteorology, but he certainly did not say this, but says something like it by which they have been deceived, that if both cold and hot water are poured into a cold place as on ice, the cold freezes quicker (which is true), but if they are placed in two vases, the hot will freeze quicker. It is necessary, then, to prove everything by experience.

Experience is of two kinds. One is through the external senses: such are the experiments that are made upon the heaven through instruments in regard to facts there, and the facts on earth that we prove in various ways to be certain in our own sight. And facts that are not true in places where we are, we know through other wise men that have experienced them. Thus Aristotle with the authority of Alexander, sent 2,000 men throughout various parts of the earth in order to learn at first hand everything on the surface of the world, as Pliny says in his Natural History. And this experience is human and philosophical just as far as a man is able to make use of the beneficent grace given to him, but such experience is not enough for man, because it does not give full certainty as regards corporeal things because of their complexity and touches the spiritual not at all. Hence man's intellect must be aided in another way, and thus the patriarchs and prophets who first gave science to the world secured inner light and did not rest entirely on the senses. So also many of the faithful since Christ. For grace makes many things clear to the faithful, and there is divine inspiration not alone concerning spiritual but even about corporeal things. In accordance with which Ptolemy says in the Centilogium that there is a double way of coming to the knowledge of things, one through the experiments of science, the other through divine inspiration, which latter is far the better as he says.

Of this inner experience there are seven degrees, one through spiritual illumination in regard to scientific things. The second grade consists of virtue, for evil is ignorance as Aristotle says in the second book of the Ethics. And Algazel says in the logic that the mind is disturbed by faults, just as a rusty mirror in which the images of things cannot be clearly seen, but the mind is prepared by virtue like a well polished mirror in which the images of things show clearly. On account of this, true philosophers have accomplished more in ethics in proportion to the soundness of their virtue, denying to one another that they can discover the cause of things unless they have minds free from faults. Augustine relates this fact concerning Socrates in Book VIII, chapter III, of the City of God: to the same purpose Scripture says, to an evil mind, etc., for it is impossible that the mind should lie calm in the sunlight of truth while it is spotted with evil, but like a parrot or magpie it will repeat words foreign to it which it has learned through long practice. And this is our experience, because a known truth draws men into its light for love of it, but the proof of this love is the sight of the result. And indeed he that is busy against truth must necessarily ignore this, that it is permitted him to know how to fashion many high sounding words and to write sentences not his own, just as the brute that imitates the human voice or an ape that attempts to carry out the works of men, although he does not understand their purpose. Virtue, then, clears the mind so that one can better understand not only ethical, but even scientific things. I have carefully proved this in the case of many pure youths who, on account of their innocent minds, have gone further in knowledge than I dare to say, because they have had correct teaching in religious doctrine, to which class the bearer of this treatise belongs, to whose knowledge of principles but few of the Latins rise. Since he is so young (about twenty years old) and poor besides, not able to have masters nor the length of any one year to learn all the great things he knows, and since he neither has great genius or a wonderful memory, there can be no other cause, save the grace of God, which, on account of the clearness of his mind, has granted to him these things which it has refused to almost all students, for a pure man, he has received pure things from me. Nor have I been able to find in him any kind of a mortal fault, although I have searched diligently, and he has a mind so clear and far seeing that he receives less from instruction than can be supposed. And I have tried to lend my aid to the purpose that these two youths may be useful implements for the Church of God, inasmuch as they have with the Grace of God examined the whole learning of the Latins....

And indeed, since all speculative thought proceeds through arguments which either proceed through a proposition by authority or through other propositions of argument, in accordance with this which I am now investigating, there is a science that is necessary to us, which is called experimental. I wish to explain this, not only as useful to philosophy, but to the knowledge of God and the understanding of the whole world: as in a former book I followed language and science to their end, which is the Divine wisdom by which all things are ordered.

And because this experimental science is a study entirely unknown by the common people, I cannot convince them of its utility, unless its virtue and characteristics are shown. This alone enables us to find out surely what can be done through nature, what through the application of art, what through fraud, what is the purport and what is mere dream in chance, conjuration, invocations, imprecations, magical sacrifices and what there is in them; so that all falsity may be lifted and the truths we alone of the art retained. This alone teaches us to examine all the insane ideas of the magicians in order not to confirm but to avoid them, just as logic criticizes the art of sophistry. This science has three great purposes in regard to the other sciences: the first is that one may criticize by experiment the noble conclusions of all the other sciences, for the other sciences know that their principles come from experiment, but the conclusions through arguments drawn from the principles discovered, if they care to have the result of their conclusions precise and complete. It is necessary that they have this through the aid of this noble science. It is true that mathematics reaches conclusions in accordance with universal experience about figures and numbers, which indeed apply to all sciences and to this experience, because no science can be known without mathematics. If we would attain to experiments precise, complete and made certain in accordance with the proper method, it is necessary to undertake an examination of the science itself, which is called experimental on our authority. I find an example in the rainbow and in like phenomena, of which nature are the circles about the sun and stars, also the halo beginning from the side of the sun or of a star which seems to be visible in straight lines and is called by Aristotle in the third book of the Meteorology a perpendicular, but by Seneca a halo, and is also called a circular corona, which have many of the colors of the rainbow. Now the natural philosopher discusses these things, and in regard to perspective has many facts to add which are concerned with the operation of seeing which is pertinent in this place. But neither Aristotle or Avicenna have given us knowledge of these things in their books upon Nature, nor Seneca, who wrote a special book concerning them. But experimental science analyzes such things.

The experimenter considers whether among visible things, he can find colors formed and arranged as given in the rainbow. He finds that there are hexagonal crystals from Ireland or India which are called rainbow-hued in Solinus Concerning the Wonders of the World and he holds these in a ray of sunlight falling through the window, and finds all the colors of the rainbow, arranged as in it in the shaded part next the ray. Moreover, the same experimenter places himself in a somewhat shady place and puts the stone up to his eye when it is almost closed, and beholds the colors of the rainbow clearly arranged, as in the bow. And because many persons making use of these stones think that it is on account of some special property of the stones and because of their hexagonal shape the investigator proceeds further and finds this in a crystal, properly shaped, and in other transparent stones. And not only are these Irish crystals in white, but also black, so that the phenomenon occurs in smoky crystal and also in all stones of similar transparency. Moreover, in stones not shaped hexagonally, provided the surfaces are rough, the same as those of the Irish crystals, not entirely smooth and yet not rougher than those—the surfaces have the same quality as nature has given the Irish crystals, for the difference of roughness makes the difference of color. He watches, also, rowers and in the drops falling from the raised oars he finds the same colors, whenever the rays of the sun penetrate the drops.

The case is the same with water falling from the paddles of a water-wheel. And when the investigator looks in a summer morning at the drops of dew clinging to the grass in the field or plane, he sees the same colors. And, likewise, when it rains, if he stands in a shady place and the sun's rays beyond him shine through the falling drops, then in some rather dark place the same colors appear, and they can often be seen at night about a candle. In the summer time, as soon as he rises from sleep while his eyes are not yet fully opened, if he suddenly looks at a window through which the light of the sun is streaming, he will see the colors. Again, sitting outside of the sunlight, if he holds his head covering beyond his eyes, or, likewise, if he closes his eyes, the same thing happens in the shade at the edges, and it also takes place through a glass vase filled with water, sitting in the sunlight. Similarly, if any one holding water in his mouth suddenly sprinkles the water in jets and stands at the side of them; or if through a lamp of oil hanging in the air the rays shine in the proper way, or the light shines upon the surface of the oil, the colors again appear. Thus, in an infinite number of ways, natural as well as artificial, colors of this kind are to be seen, if only the diligent investigator knows how to find them.

Ideas to Think About

1. Describe what you believe Augustine is attempting to do in the readings. What is his source of authority and his view of nature?
2. Do you see any Greek influence in his writing?
3. Roger Bacon represents the period of Scholasticism and the rise of the universities. What intellectual and spiritual goals does his writing reveal?
4. Scholasticism represented both a threat and a promise to the Christian church. Can you suggest why?

For Further Reading

Aristotle's Children: How Christians, Muslims, and Jews Rediscovered Ancient Wisdom and Illuminated the Dark Ages, Richard Rubenstein. New York: Harcourt Brace (2003). This is a critical and impeccable inquiry into the burst of intellectual fervor when translations introduced the clash between Platonic spirituality and Aristotelian materialism.

The Closing of the Western Mind: the Rise of Faith and the Fall of Reason, Charles Freeman. London: Vintage (2005). This is an excellent account of the pivotal sea change in Western history when the Roman Empire became Christian by decree of Emperor Constantine.

III: The Mechanical World of the Sixteenth Century

To teach mankind about nature is not the purpose of Holy Scripture, which speaks to people about these matters in a human way in order to be understood by them and uses popular concepts.... Why is it surprising then, that Scripture also talks the language of human senses in situations where the reality of things differs from the perception?

—JOHANNES KEPLER

It was in 1609 that the great mathematician and astronomer penned this statement. It was heretical, but Kepler, a devout Protestant, had no fear of persecution from Rome. He openly accepted the Copernican view. Here in these men and in others during this century, we find a subtle but fundamental disunion of science from classical philosophy (and its theological reincarnation). Thus, in these final breaths of Scholasticism, we see a withdrawal of science as handmaiden to scripture.

This was not a complete turnaround, though, and we find natural theology in various forms continuing into the current century, but it did constitute a profound change in assumptions about nature and the way nature should be investigated. Therefore, the sixteenth century may be viewed as the beginning of a revolution in science.

The change was not so much in scientific methodology (e.g., the rise of empirically driven experimentation, which would awaken interest in the next century) as in the metaphysical foundations of science: The old teleology (everything exists for a purpose) was replaced by a mechanistic view of the world. Aristotle—however much the coming centuries would continue to pay homage to his logic and only reluctantly dismiss his explanations—slowly became part of an obsolete organic world to be critiqued and challenged by a colder world of physical cause and effect, as Lindberg (2007, 364–67) eloquently discusses.

Experiment became freed from the constraint of searching for some mystical inner purpose. Causality was reduced to the efficient cause—the immediate act responsible for anything—with Aristotle's final and formal causes eliminated and material cause no longer a "cause," but a quality or component.

The ancient philosophy of Epicureanism, revived from its Greek roots in Democritus's "atomism" and its later expression in Lucretius's *De Rerum Natura*, played an important role in this. Lifeless atoms, bouncing in random collisions, well suited this metaphysics. In the seventeenth century, its application by Galileo, Descartes, Boyle, and Newton pushed still further the transition to early modern science from the world of Scholasticism.

For our purposes, the new ways of looking at nature provided by Copernicus and Da Vinci are relevant. These show a willingness to break with the past and, in Da Vinci, early insights into geology using both observation and measurement. Universities also needed to break from the past: the more limited ecclesiastical purposes of medieval institutions slowly began to respond to the growing pains of pure intellectual curiosity.

Martin Luther, at the cusp of this response, recognized this. "The universities need a sound and thorough reformation," he claimed. "I must say so no matter who takes offence. Everything that the papacy has instituted or ordered is directed solely towards the multiplication of sin and error. Unless they are completely altered from what they have been hitherto, the universities will fit exactly what is said in the Book of Maccabees: 'Places for the exercise of youth, and for the Greekish fashion....' Nothing could be more wicked, or serve the devil better, than unreformed universities" (quoted in Spitz 1984, 42).

6. NICHOLAS COPERNICUS

If it had not been for the medieval universities, many revolutionary scholars would not have left us their legacies, and Copernicus is a good example. Born in 1473, in Torun, Poland, he studied mathematics, philosophy, and astronomy at the University of Krakow (established in 1364 and now known as Jagiellonian University); pursued more liberal arts at the University of Bologna (the earliest of them all, established in 1088); studied medicine at the University of Padua (established in 1222); and, in 1503, received a doctorate in canon law from the University of Ferrara (in northern Italy, established in 1391).

The education served him well, for he served as canon of the cathedral in Frauenberg, Poland, practiced medicine, and wrote on monetary reform, before turning his attention to astronomy. Here, his calculations from Ptolemy's planetary orbits suggested that an alternative model would be simpler and more harmonious.

His preliminary treatise on this, circulated privately in 1514 because it was so radical, led to encouragement by friends and colleagues to publish it. Copernicus hesitated. Despite the facts that his theory conservatively followed Ptolemy's planetary motions via epicycles and did not dispute the Aristotelian cosmology asserting solid celestial spheres, nor his theory of motion, and left unchanged the Aristotelian and Ptolemaic assumption of perfectly circular planetary orbits, he recognized that having the sun replace the earth as the center necessitated an earth that orbited the sun—a direct contradiction of Ptolemy and, more seriously, the scriptures' claim that the earth remains motionless.

Copernicus was prudent to delay publication. The Spanish Inquisition was already well underway, and the Roman Inquisition began the year before his death. At his death in 1543, his treatise came into print and led to vehement ecclesiastical objection lasting well into Galileo's time. Yet it led to future developments in the following century, as Kepler revealed the elliptical nature of planetary orbits, Galileo developed a new theory of motion that came very close to inertia, and Newton explained gravity.

Copernicus, however, used no significant mathematics or geometry in his theory: it was, he argued, a simpler way to explain the angular positions of all planets, including earth, by holding the sun motionless. It was an aesthetically pleasing solution. It was not new science. But it dared to suggest that reality need not agree with scripture, and that the search for empirical truth should be independent of "biblical truth."

Copernicus knew well that he was taking a huge risk in submitting his manuscript, albeit it was not submitted for general publication. One can see this clearly in his dedication, in which he justifies his transgression against received wisdom by asserting the logic of his method and the simplicity (over Ptolemy) of his explanation.

In the second selection, one should note the careful use of analogy in which he generalizes his conclusion to other affects of nature.

On the Revolutions of the Heavenly Bodies, Dedication

To Pope Paul III

I can readily imagine, Holy Father, that as soon as some people hear that in this volume, which I have written about the revolutions of the spheres of the universe, I ascribe certain motions to the terrestrial globe, they will shout that I must be immediately repudiated together with this belief. For I am not so enamored of my own opinions that I disregard what others may think of them. I am aware that a philosopher's ideas are not subject to the judgment of ordinary person's, because it is his endeavor to seek the truth in all things, to the extent permitted to human reason by God. Yet I hold that completely erroneous views should be shunned. Those who know that the consensus of many centuries has sanctioned the conception that the earth remains at rest in the middle of the heaven as its center would, I reflected, regard it as an insane pronouncement if I made the opposite assertion that the earth moves. Therefore I debated with myself for a long time whether to publish the volume which I wrote to prove the earth's motion or rather to follow the example of the Pythagoreans and certain others, who used to transmit philosophy's secrets only to kinsmen and friends, not in writing but by word of mouth....

Having thus assumed the motions which I ascribe to the earth later on in the volume, by long and intense study I finally found that if the motions of the other planets are correlated with the orbiting of the earth, and are computed for the revolution of each planet, not only do their phenomena follow therefrom but also the order and size of all the planets and spheres, and heaven itself is so linked together that in no portion of it can anything be shifted without disrupting the remaining parts and the universe as a whole.

On the Revolutions of the Heavenly Bodies, Selection

That the universe is spherical. FIRST WE must remark that the universe is spherical in form, partly because this form being a perfect whole requiring no joints, is the most complete of all, partly because it makes the most capacious form, which is best suited to contain and preserve everything; or again because all the constituent parts of the universe, that is the sun, moon and the planets appear in this form; or because everything strives to attain this form, as appears in the case of drops of water and other fluid bodies if they attempt to define themselves. So no one will doubt that this form belongs to the heavenly bodies.

That the earth is also spherical. That the earth is also spherical is therefore beyond question, because it presses from all sides upon its center. Although by reason of the elevations of the mountains and the depressions of the valleys a perfect circle cannot be understood, yet this does not affect the general spherical nature of the earth. This appears in the following manner. To those who journey towards the North the North pole of the daily revolution of the heavenly sphere seems gradually to rise, while the opposite seems to sink. Most of the stars in the region of the Bear seem not to set, while some of the Southern stars seem not to rise at all. So Italy does not see Canopus which is visible to the Egyptians. And Italy sees the outermost star of the Stream, which our region of a colder zone does not know. On the other hand to those who go towards the South the others seem to rise and those to sink which are high in our region. Moreover, the inclination of the poles to the diameter of the earth bears always the same relation, which could happen only in the case of a sphere. So it is evident that the earth is included between the two poles, and is therefore spherical in form. Let us add that the inhabitants of the East do not observe the eclipse of the sun or of the moon which occurs in the evening, and the inhabitants of the West those which occur in the morning, while those who dwell between see those later and these earlier. That the water also has the same form can be observed from the ships, in that the land which cannot be seen from the deck, is visible from the mast-tree. And conversely if a light be placed at the masthead it seems to those who remain on the shores gradually to sink and at last still sinking to disappear. It is clear that the water also according to its nature continually presses like the earth downward, and does not rise above its banks higher than its convexity permits. So the land extends above the ocean as much as the land happens to be higher.

Whether the earth has a circular motion, and concerning the location of the earth. As it has been already shown that the earth has the form of a sphere, we must consider whether a movement also coincides with this form, and what place the earth holds in the universe. Without this there will be no secure results to be obtained in regard to the heavenly phenomena. The great majority of authors of course agree that the earth stands still in the center of the universe, and consider it inconceivable and ridiculous to suppose the opposite. But if the matter is carefully weighed it will be seen that the question is not yet settled and therefore by no means to be regarded lightly. Every change of place which is observed is due, namely, to a movement of the observed object or of the observer, or to movements of both, naturally in different directions, for if the observed object and the observer move in the same manner and in the same direction no movement will be seen. Now it is from the earth that the revolution of the heavens is observed and it is produced for our eyes. Therefore if the earth undergoes no movement this movement must take place in everything outside of the earth, but in the opposite direction than if everything on the earth moved, and of this kind is the daily revolution. So this appears to affect the whole universe, that is, everything outside the earth with the single exception of the earth itself. If, however, one should admit that this movement was not peculiar to the heavens, but that the earth revolved from west to east, and if this was carefully considered in regard to the apparent rising and setting of the sun, the moon and the stars, it would be discovered that this was the real situation. Since the sky, which contains and shelters all things, is the common seat of all things, it is not easy to understand why motion should not be ascribed rather to the thing contained than to the containing, to the located rather than to the location. From this supposition follows another question of no less importance, concerning the place of the earth, although it has been accepted and believed by almost all, that the earth occupies the middle of the universe. But if one should suppose that the earth is not at the center of the universe, that, however, the distance between the two is not great enough to be measured on the orbits of the fixed stars, but would be noticeable and perceptible on the orbit of the sun or of the planets: and if one was further of the opinion that the movements of the planets appeared to be irregular as if they were governed by a center other than the earth, then such an one could perhaps have given the true reasons for the apparently irregular movement. For since the planets appear now nearer and now farther from the earth, this shows necessarily that the center of their revolutions is not the center of the earth: although it does not settle whether the earth increases and decreases the distance from them or they their distance from the earth.

Refutation of the arguments of the ancients that the earth remains still in the middle of the universe, as if it were its center. From this and similar reasons it is supposed that the earth rests at the center of the universe and that there is no doubt of the fact. But if one believed that the earth revolved, he would certainly be of the opinion that this movement was natural and not arbitrary. For whatever is in accord with nature produces results which are the opposite of those produced by force. Things upon which force or an outside power has acted, must be injured and cannot long endure: what happens by nature, however, preserves itself well and exists in the best condition. So Ptolemy feared without good reason that the earth and all earthly objects subject to the revolution would be destroyed by the act of nature, since this latter is opposed to artificial acts, or to what is produced by the human spirit. But why did he not fear the same, and in a much higher degree, of the universe, whose motion must be as much more rapid as the heavens are greater than the earth? Or has the heaven become so immense because it has been driven outward from the center by the inconceivable power of the revolution; while if it stood still, on the contrary, it would collapse and fall together? But surely if this is the case the extent of the heavens would increase infinitely. For the more it is driven higher by the outward force of the movement, so much the more rapid will the movement become, because of the ever increasing circle which must be traversed in twenty-four hours; and conversely if the movement grows the immensity of the heavens grows. So the velocity would increase the size and the size would increase the velocity unendingly. According to the physical law that the endless cannot wear away nor in any way move, the heavens must necessarily stand still. But it is said that beyond the sky no body, no place, no vacant space, in fact nothing at all exists; then it is strange that some thing should be enclosed by nothing. But if the heaven is endless and is bounded only by the inner hollow, perhaps this establishes all the more clearly the fact that there is nothing outside the heavens, because everything is within it, but the heaven must then remain unmoved. The highest proof on which one supports the finite character of the universe is its movement. But whether the universe is endless or limited we will leave to the physiologues; this remains sure for us that the earth enclosed between the poles is bounded by a spherical surface. Why therefore should we not take the position of ascribing to a movement conformable to its nature and corresponding to its form, rather than suppose that the whole universe whose limits are not and cannot be known moves? And why will we not recognize that the appearance of a daily revolution belongs to the heavens, but the actuality to the earth; and that the relation is similar to that of which one says: "We run out of the harbor, the lands and cities retreat from us." Because if a ship sails along quietly, everything outside of it appears to those on board as if it moved with the motion of the boat, and the boatman thinks that the boat with all on board is standing still, this same thing may hold without doubt of the motion of the earth, and it may seem as if the whole universe revolved.

7. Leonardo Da Vinci

He has no last name, this most revered polymath in Western history, nor was his birth legitimate. He was simply known as Leonardo of Vinci, the small Tuscan town west of Florence where he was born. His father, Piero of Vinci, was a notary and his mother, Caterina, a peasant woman. His multiple talents, however, were never as humble as his birth. His only schooling was in the studio of the Florentine artist, Verrocchio, and his fame as an artist spread as he worked in Milan, Rome, Venice, and elsewhere.

His interests and talents spread themselves over numerous fields: engineering, architecture, botany, geology and paleontology, music, anatomy, and more. Had he focused, he would doubtless have preempted much later discoveries in any of these fields. As it was, his seemingly limitless intelligence was matched only by his very limited patience, and much of his work and ideas were left in incomplete passages and notes and sketches. These filled over 4,000 surviving manuscript pages in his notebooks, and it is from these pages that we find his precocious scientific observations.

In his observations of mountain formations and riverine deposits, Leonardo grasped the principles of sedimentary rock formation in water, the differential erosive action of water on different size gravels, and—in observing hillside strata—came upon the realization that older sediments are at the bottom whereas younger deposits are near the top. This law of superposition was only later articulated by Steno in the seventeenth century.

Leonardo's knowledge of fossils occupied much of his enlightened scientific speculation. Over the centuries, observations of shells and fossil teeth and bones on mountaintops had led to various interpretations. Many claimed that apparent fossils in rocks could not have once been living forms, because they could not have found their way inside stone. Shell deposits at high elevations were widely seen as having resulted from the rising waters of the biblical flood.

"Of the Deluge and of Marine Shells"

In the following selection, we have Leonardo's astute scientific reasoning on these. His conclusions are based on acute insight and empirical measurement, representative of scientific methodology and conceptualizations of the natural world that originated during this final medieval period. The observations probably date to the late fifteenth century, based on his 1473 drawing of geological formations in the Arno valley.

If you were to say that the shells which are to be seen within the confines of Italy now, in our days, far from the sea and at such heights, had been brought there by the deluge which left them there, I should answer that if you believe that this deluge rose 7 cubits above the highest mountains—as he who measured it has written—these shells, which always live near the sea-shore, should have been left on the mountains; and not such a little way from the foot of the mountains; nor all at one level, nor in layers upon layers. And if you were to say that these shells are desirous of remaining near to the margin of the sea, and that, as it rose in height, the shells quitted their first home, and followed the increase of the waters up to their highest level; to this I answer, that the cockle is an animal of not more rapid movement than the snail is out of water, or even somewhat slower; because it does not swim, on the contrary it makes a furrow in the sand by means of its sides, and in this furrow it will travel each day from 3 to 4 braccia; therefore this creature, with so slow a motion, could not have travelled from the Adriatic sea as far as Monferrato in Lombardy, which is 250 miles distance, in 40 days; which he has said who took account of the time. [*Note: a* braccio *varied in length in different parts of Italy; in Florence it was the equivalent of about 22.9 inches. RKW*] And if you say that the waves carried them there, by their gravity they could not move, excepting at the bottom. And if you will not grant me this, confess at least that they would have to stay at the summits of the highest mountains, in the lakes which are enclosed among the mountains, like the lakes of Lario, or of Como and il Maggiore and of Fiesole, and of Perugia, and others.

And if you should say that the shells were carried by the waves, being empty and dead, I say that where the dead went they were not far removed from the living; for in these mountains living ones are found, which are recognisable by the shells being in pairs; and they are in a layer where there are no dead ones; and a little higher up they are found, where they were thrown by the waves, all the dead ones with their shells separated, near to where the rivers fell into the sea, to a great depth; like the Arno which fell from the Gonfolina near to Monte Lupo, where it left a deposit of gravel which may still be seen, and which has agglomerated; and of stones of various districts, natures, and colours and hardness, making one single conglomerate. And a little beyond the sandstone conglomerate a tufa has been formed, where it turned towards Castel Florentino; farther on, the mud was deposited in which the shells lived, and which rose in layers according to the levels at which the turbid Arno flowed into that sea. And from time to time the bottom of the sea was raised, depositing these shells in layers, as may be seen in the cutting at Colle Gonzoli, laid open by the Arno which is wearing away the base of it; in which cutting the said layers of shells are very plainly to be seen in clay of a bluish colour, and various marine objects are found there.

And if the earth of our hemisphere is indeed raised by so much higher than it used to be, it must have become by so much lighter by the waters which it lost through the rift between Gibraltar and Ceuta; and all the more the higher it rose, because the weight of the waters which were thus lost would be added to the earth in the other hemisphere. And if the shells had been carried by the muddy deluge they would have been mixed up, and separated from each other amidst the mud, and not in regular steps and layers—as we see them now in our time.

Ideas to Think About

1. Why was Copernicus so reluctant to publish his views? Why would he have probably thought differently if he had lived in Leonardo Da Vinci's time?
2. What influences can you see in Copernicus's thoughts from Plato (whose philosophy related well to mathematics and music), and from Aristotle (whose philosophy related well with geometry and physics)?
3. Be prepared to discuss the questions raised in reading Da Vinci. A historian of science has said, "Science has not occurred in cultures uninterested in the fine arts." Do you have a notion why this claim is made?

For Further Reading

Uncentering the Earth: Copernicus and the Revolutions of the Heavenly Spheres, William T. Vollmann. New York: W. W. Norton: Atlas Books [Great Discoveries series] (2006). Fresh and magnetically energetic, this little book is worth its weight in any metal.

The Notebooks of Leonardo da Vinci (Oxford World's Classics), Leonardo Da Vinci. New York: Oxford University Press (1999). Consider this from his comments on science: "what trust can we place in the ancients, who tried to define what the Soul and Life are—which are beyond proof—whereas those things which can at any time be clearly known and proved by experience remained for many centuries unknown or falsely understood."

UNIT 2

Seventeenth Century: Public Science and the New Methodologies

> Men have sought to make a world from their own conception and to draw from their own minds all the material which they employed, but if, instead of doing so, they had consulted experience and observation, they would have the facts and not opinions to reason about, and might have ultimately arrived at the knowledge of the laws which govern the material world.
>
> —FRANCIS BACON

The science protagonists of the Age of Reason, as the seventeenth century is sometimes called, lived in a new social and intellectual world that they created and to which they responded. This was a new landscape of inquiry, and it had a feature that had not existed before: science became public. This was occasioned by the increased voyages of discovery, which introduced exotic nature into European society. This in turn stimulated the rise of public displays, from the initial cabinets of curiosities in private homes to public museums (the first public museum, the Ashmolean at Oxford, opened in 1683).

Public lecture halls and scientific demonstrations followed, including gatherings in the growing scientific societies. The Royal Society was established in 1660; the French Académie des Sciences in 1666; and, earlier still, Rome's Accademia dei Lincei began in 1603. The Accademia del Cimento, in Florence, dates to 1657—established by the de Medicis to foster Galileo's experimental methods. Discoveries, inventions, and experiments brought a heady sense of achievement to scientists, encouraged by public support and patronage.

Finally, there arose the botanical gardens. The first in the English-speaking world was at Oxford in 1621. The earliest in Europe were the gardens at Padua, in 1598. London's Kew Gardens were established in the eighteenth century. All of these open venues further stimulated a public love affair with natural science.

The achievement in ideas included both an expanded mechanistic view of the physical world and methodologies with which to investigate it: the inductive approach of Francis Bacon and the deductive logic of René Descartes. The first results of this new effort—one that will be emphasized in Unit 3—were various theories of the earth. These contributed greatly to the necessary preconditions to evolutionary theory. Studies of and theories about the living world during this time also produced progress.

From a more explicit metaphysical perspective, this period made stronger the distinction between philosophy and religion, reaching the final detachment of the two in the eighteenth century. Looking back from that century, Kant drew an even more decisive philosophical distinction between Descartes and Bacon, identifying them, and the traditions they began, as either rationalist or empiricist, respectively. This exaggerates the tendency of Cartesian epistemology to emphasize the power of reason alone and hence to espouse mathematics as the route to knowledge and the tendency of Baconian epistemology to focus on observation and the physical sciences.

This unit focuses on the progress in the physical sciences during the seventeenth century—but most illustratively on the new methodologies that were developed in the first truly mechanistic conceptualization of nature.

8. Francis Bacon

Francis Bacon was both humanist and scientist. He was ambitiously wedded to political intrigue and a resolute enemy of barren Aristotelian science. He wrote both utopian fiction, *The New Atlantis* (bookended by Sir Thomas More's *Utopia* in the sixteenth century and Jonathan Swift's *Gulliver's Travels* in the eighteenth), and intellectual essays. Indeed, Bacon was one of the most prolific and diverse of all authors of his century.

Politically, he served variously as a member of Parliament, Queen's Counsel to Elizabeth I, and Lord Chancellor under James I. Born in 1561 and dying of pneumonia in 1626, Bacon lived under three monarchs and through substantial political intrigue. As a devout Protestant in the Parliament of 1586, he argued for the execution of Mary, Queen of Scots. He was knighted in 1603, given a baronage in 1618, and became Viscount St. Alban in 1621. His political career ended that year with charges of bribery and corruption. Though he claimed innocence, he was convicted and fined. King James freed him from the Tower of London after a few days. Disgraced, he spent the rest of his life writing.

Bacon's contributions to science were neither empirical nor experimental: He made no significant discoveries. Rather, his influence lay in his strong support of observation and testing through inductive hypotheses. Although others before him—notably Roger Bacon—had called for the same kind of observation and experiment, Francis was not only much more thorough and explicit, but divorced himself from a theological purpose.

Novum Organum, Second Part, Preface

[343] They who have presumed to dogmatize on Nature, as on some well-investigated subject, either from self-conceit or arrogance, and in the professorial style, have inflicted the greatest injury on philosophy and learning. For they have tended to stifle and interrupt inquiry exactly in proportion as they have prevailed in bringing others to their opinion: and their own activity has not counterbalanced the mischief they have occasioned by corrupting and destroying that of others. They again who have entered upon a contrary course, and asserted that nothing whatever can be known, whether they have fallen into this opinion from their hatred of the ancient sophists, or from the hesitation of their minds, or from an exuberance of learning, have certainly adduced reasons for it which are by no means contemptible. They have not, however, derived their opinion from true sources, and, hurried on by their zeal, and some affectation, have certainly exceeded due moderation. But the more ancient Greeks (whose writings have perished) held a more prudent mean, between the arrogance of dogmatism, and the despair of skepticism; and though too frequently intermingling complaints and indignation at the difficulty of inquiry, and the obscurity of things, and champing, as it were, the bit, have still persisted in pressing their point, and pursuing their intercourse with nature: thinking, as it seems, that the better method was not to dispute upon the very point of the possibility of any thing being known, but to put it to the test of experience. Yet they themselves, by only employing the power of the understanding, have not adopted a fixed rule, but have laid their whole stress upon intense meditation, and a continual exercise and perpetual agitation of the mind.

Our method, though difficult in its operation, is easily explained. It consists in determining the degrees of certainty, whilst we, as it were, restore the senses to their former rank, but generally reject that operation of the mind which follows close upon the senses, and open and establish a new and certain course for the mind from the first actual perceptions of the senses themselves. This no doubt was the view taken by those who have assigned so much to logic; showing clearly thereby that they sought some support for the mind, and suspected its natural and spontaneous mode of action. But this is now employed too late as a remedy, when all is clearly lost, and after the mind, by the daily habit and intercourse of life, has become prepossessed with corrupted doctrines, and filled with the vainest idols. The art of logic therefore being (as we have mentioned) too late a precaution, and in no way remedying the matter, has tended more to confirm errors, than to disclose truth. Our only remaining hope and salvation is to begin the whole labour of the mind again; not leaving it to itself, but directing it perpetually from the very first, and attaining our end as it were by mechanical aid. If men, for instance, had attempted mechanical labours with their hands alone, and without the power and aid of instruments, as they have not hesitated to carry on the labours of their understanding with the unaided efforts of their mind, they would have been able to move and overcome but little, though they had exerted their utmost and united powers. And, just to pause a while on this comparison, and look into it as a mirror; let us ask, if any obelisk of a remarkable size were perchance required to be moved, for the purpose of gracing a triumph or any similar pageant, and men were to attempt it with their bare hands, would not any sober spectator avow it to be an act of the greatest madness? And if they should increase the number of workmen, and imagine that they could thus succeed, would he not think so still more? But if they chose to make a selection, and to remove the weak, and only employ the strong and vigorous, thinking by this means, at any rate, to achieve their object, would he not say that they were more fondly deranged? Nay, if, not content with this, they were to determine on consulting the athletic art, and were to give orders for all to appear with their hands, arms, and muscles regularly oiled and prepared, would [344] he not exclaim that they were taking pains to rave by method and design? Yet men are hurried on with the same senseless energy and useless combination in intellectual matters, so long as they expect great results either from the number and agreement, or the excellence and acuteness of their wits; or even strengthen their minds with logic, which may be considered as an athletic preparation, but yet do not desist (if we rightly consider the matter) from applying their own understandings merely with all this zeal and effort. Whilst nothing is more clear, than that in every great work executed by the hand of man without machines or implements, it is impossible for the strength of individuals to be increased, or for that of the multitude to combine.

The following passage (Aphorism 8) is succinct in pointing out Bacon's departure from classic and scholastic methods of understanding. To "begin with physics" is, in other words, to jump-start inquiry with Aristotelian realism in observing the material, natural world; to "end in mathematics" is to reach conclusions on these observations with Platonic rational thought. Scholasticism was essentially the reverse in its sequence. Aphorism 9, in addition to rejecting the teleology of the scholastic appetite, further clarifies the vast divide between philosophy and science, the former a focus on the theoretical and general, the latter on the observable and particular. Through induction, the latter leads to the former. Bacon's use of the term "magic" here is more a reference to nonempirical inquiry than to supernaturalism, although elsewhere it is apparent that he does believe in the occult.

[374] 8. This method will not bring us to atoms, which takes for granted the vacuum, and the immutability of matter (neither of which hypotheses is correct;) but to the real particles, such as we discover them to be. Nor is there any ground for alarm at this refinement, as if it were inexplicable, for, on the contrary, the more inquiry is directed to simple natures, the more will every thing be placed in a plain and perspicuous light; since we transfer our attention from the complicated to the simple, from the incommensurable to the commensurable, from surds to rational quantities, from the indefinite and vague to the definite and certain: as when we arrive at the elements of letters, and the simple tones of concords, The investigation of nature is best conducted when mathematics are applied to physics. Again, let none be alarmed at vast numbers and fractions; for, in calculation, it is as easy to set down or to reflect upon a thousand as a unit, or the thousandth part of an integer as an integer itself.

9. From the two kinds of axioms above specified arise the two divisions of philosophy and the sciences, and we will use the commonly adopted terms, which approach the nearest to our meaning, in our own sense. Let the investigation of forms, which (in reasoning at least, and after their own laws) are eternal and immutable, constitute *metaphysics*, and let the investigation of the efficient cause of matter, latent process, and latent conformation (which all relate merely to the ordinary course of nature, and not to her fundamental and eternal laws) constitute *physics*. Parallel to these let there be two practical divisions; to *physics* that of *mechanics*, and to *metaphysics* that of *magic*, in the purest sense of the term, as applied to its ample means and its command over nature.

In the following passage, Aphorism 27, Bacon's methodological preference for seeking similarities rather than differences within nature, and thus identifying analogies, is presented. This was an important step in the scientific development of comparative studies, and mere paces away from inferring homologies (traits shared by common descent).

[391] ... Lastly, we must particularly recommend and suggest, that man's present industry in the investigation and compilation of natural history be entirely changed, and directed to the reverse of the present system. For, it has hitherto been active and curious in noting the variety of things and explaining the accurate differences of animals, vegetables, and minerals, most of which are the mere sport of nature, rather than of any real utility as concerns the sciences. Pursuits of this nature are certainly agreeable, and sometimes of practical advantage, but contribute little or nothing to the thorough investigation of nature. Our labour must, therefore, be directed towards inquiring into, and observing resemblances and analogies, both in the whole, and in its parts, for, they unite nature, and lay the foundation of the sciences.

Here, however, a severe and rigorous caution must be observed, that we only consider as similar and proportionate instances, those which (as we first observed) point out physical resemblances: that is, real and substantial resemblances, deeply founded in nature, and not casual and superficial, much less superstitious or curious; such as those which are constantly put forward by the writers on natural magic, (the most idle of men, and who are scarcely fit to be named in connection with such serious matters as we now treat of,) who, with much vanity and folly, describe, and sometimes, too, invent unmeaning resemblances and sympathies.

Bacon devotes much of Novum Organum *to prerogative instances, by which he means approaches to understanding that allow experimental comparisons of a "best fit" as alternative hypotheses are matched against observations. His reference in the following Aphorism (29) to "latent process" is to the "generative process," which is roughly the equivalent of our genes (and what John Locke refers to as the "nominal essence"). Note how close the last sentence comes to Darwin's identification of varieties.*

[392] In the eighth rank of prerogative instances, we will place *deviating* instances; such as the errors of nature, or strange and monstrous objects, in which nature deviates and turns from her ordinary course. For the errors of nature differ from singular instances, inasmuch as the latter are the miracles of species, the former of individuals. Their use is much the same, for they rectify the understanding in opposition to habit, and reveal common forms. For, with regard to these, also, we must not desist from inquiry till we discern the cause of the deviation. The cause does not, however, in such cases, rise to a regular form, but only to the latent process towards such a form. For he who is acquainted with the paths of nature will more readily observe her deviations,

and, vice versa, he who has learnt her deviations, will be able more accurately to describe her paths.

They differ again from singular instances, by being much more apt for practice, and the operative branch. For it would be very difficult to generate, but less so to vary known species, and thus produce many rare and unusual results....

Bacon's references to bordering instances combine our notions of hybrid species and transitional species in biology. This is Aphorism 30.

[392] In the ninth rank of prerogative instances we will place *bordering* instances, which we are also wont to term participants. They are such as exhibit those species of bodies which appear to be composed of two species, or to be the rudiments between the one and the other. They may well be classed with the singular or heteroclite [abnormal] instances; for in the whole system of things they are rare and extraordinary. Yet from their dignity they must be treated of and classed separately, for they point out admirably the order and constitution of things, and suggest the causes of the number and quality of the more common species in the universe, leading the understanding from that which is, to that which is possible.

We have examples of them in moss, which is something between putrescence and a plant; in some comets, which hold a place between stars and ignited meteors; in flying fishes, between fishes and birds; and in bats, between birds and quadrupeds.† Again,

Simia quam similis turpissima bestia nobis; [*"How like to us is the degraded ape." From the Latin poet Ennius, as quoted by Cicero.—RKW*]

Of the Proficience and Advancement of Learning, Human and Divine, Vol. I, Bk. II

[194] This science being therefore first placed as a common parent, like unto Berecynthia, which had so much heavenly issue,

Omnes Coelicolas, omnes supera alta tenentes, [*"All dwellers in the heaven and upper sky." From Virgil,* Aeneid, VI. *Berecynthia is Cybele, a goddess of nature. Science, in other words, concerns earthly or natural matters.—RKW*] we may return to the former distribution of the three philosophies, divine, natural, and human.

And as concerning Divine Philosophy or Natural Theology, it is that knowledge or rudiment of knowledge concerning God, which may be obtained by the contemplation of His creatures; which knowledge may be truly termed divine in respect of the object, and natural in respect of the light. The bounds of this knowledge are, that it sufficeth to convince atheism, but not to inform religion: and therefore there was never miracle wrought by God to convert an atheist, because the light of nature might have led him to confess a God: but miracles have been wrought to convert idolators and the superstitious, because no light of nature extendeth to declare the will and true worship of God. For as all works do show forth the power and skill of the workman, and not his [195] image; so it is of the works of God, which do show the omnipotency and wisdom of the Maker, but not His image: and therefore therein the heathen opinion differeth from the sacred truth; for they supposed the world to be the image of God, and man to be an exact or compendious image of the world, but the Scriptures never vouchsafe to attribute to the world that honour, as to be the image of God, but only the work of His hands neither do they speak of any other image of God, but man: wherefore by the contemplation of nature to induce and perforce the acknowledgment of God, and to demonstrate His power, providence, and goodness, is an excellent argument, and hath been excellently handled by divers.

But on the other side, out of the contemplation of nature, or ground of human knowledge, to induce any verity or persuasion concerning the points of faith, is in my judgment not safe: "Da fidei quæ fidei sunt" [*"Give unto faith the things that are faith's"—RKW*]. For the heathen themselves conclude as much, in that excellent and divine fable of the golden chain: "That men and gods were not able to draw Jupiter down to earth; but contrariwise, Jupiter was able to draw them up to heaven."

[208]...Secondly, the induction which the logicians speak of, and which seemeth familiar with Plato (whereby the principles of sciences may be pretended to be invented, and so the middle propositions by derivation from the principles), their form of induction, I say, is utterly vicious and incompetent: wherein their error is the fouler, because it is the duty of art to perfect and exalt nature; but they contrariwise have wronged, abused, and traduced nature....For to conclude upon an enumeration of particulars, without instance contradictory, is no conclusion, but a conjecture; for who can assure, in many subjects, upon those particulars which appear of a side, that there are not other on the contrary side which appear not?...

Novum Organum, Vol. III, Bk. I Aphorisms 9–13, 19–20, 22–23, 40–45, 71

9. The sole cause and root of almost every defect in the sciences is this; that whilst we falsely admire and extol the powers of the human mind, we do not search for its real helps.

10. The subtilty [*sic*] of nature is far beyond that of sense or of the understanding: so that the specious meditations, speculations, and theories of mankind, are but a kind of insanity, only there is no one to stand by and observe it.

11. As the present sciences are useless for the discovery of effects, so the present system of logic is useless for the discovery of the sciences.

12. The present system of logic rather assists in confirming and rendering inveterate the errors founded on vulgar notions, than in searching after truth; and is therefore more hurtful than useful.

†There is, however, no real approximation to birds in either the flying fish or bat, any more than a man approximates to a fish because he can swim. The wings of the flying fish and bat are mere expansions of skin, bearing no resemblance to those of birds.

13. The syllogism is not applied to the principles of the sciences, and is of no avail in intermediate axioms, as being very unequal to the subtilty of nature. It forces assent, therefore, and not things.

[345] 19. There are and can exist but two ways of investigating and discovering truth. The one hurries on rapidly from the senses and particulars to the most general axioms; and from them as principles and their supposed indisputable truth derives and discovers the intermediate axioms. This is the way now in use. The other constructs its axioms from the senses and particulars, by ascending continually and gradually, till it finally arrives at the most general axioms, which is the true but unattempted way.

20. The understanding when left to itself proceeds by the same way as that which it would have adopted under the guidance of logic, namely, the first. For the mind is fond of starting off to generalities, that it may avoid labour, and after dwelling a little on a subject is fatigued by experiment. But these evils are augmented by logic, for the sake of the ostentation of dispute.

22. Each of these two ways begins from the senses and particulars, and ends in the greatest generalities. But they are immeasurably different; for the one merely touches cursorily the limits of experiment, and particulars, whilst the other runs duly and regularly through them; the one from the very outset lays down some abstract and useless generalities, the other gradually rises to those principles which are really the most common in nature.

23. There is no small difference between the *idols* of the human mind, and the *ideas* of the divine mind; that is to say, between certain idle dogmas, and the real stamp and impression of created objects, as they are found in nature.

[347] 39. Four species of idols beset the human mind: to which (for distinction's sake) we have assigned names: calling the first idols of the tribe; the second idols of the den; the third idols of the market; the fourth idols of the theatre.

40. The formation of notions and axioms on the foundation of true induction, is the only fitting remedy, by which we can ward off and expel these idols. It is, however, of great service to point them out. For the doctrine of idols bears the same relation to the interpretation of nature, as that of confutation of sophisms does to common logic.

41. The idols of the tribe are inherent in human nature, and the very tribe or race of man. For man's sense is falsely asserted to be the standard of things. On the contrary, all the perceptions, both of the senses and the mind, bear reference to man, and not to the universe, and the human mind resembles those uneven mirrors, which impart their own properties to different objects, from which rays are emitted, and distort and disfigure them.

42. The idols of the den are those of each individual. For everybody (in addition to the errors common to the race of man) has his own individual den or cavern, which intercepts and corrupts the light of nature; either from his own peculiar and singular disposition, or from his education and intercourse with others, or from his reading, and the authority acquired by those whom he reverences and admires, or from the different impressions produced on the mind, as it happens to be preoccupied and predisposed, or equable and tranquil, and the like: so that the spirit of man (according to its several dispositions) is variable, confused, and as it were actuated by chance; and Heraclitus said well that men search for knowledge in lesser worlds, and not in the greater or common world.

The following aphorism very nearly anticipates the Sapir-Whorf hypothesis that the cultural construction of reality influences and limits cognition as reflected in language.

43. There are also idols formed by the reciprocal intercourse and society of man with man, which we call idols of the market, from the commerce and association of men with each other. For men converse by means of language; but words are formed at the will of the generality; and there arises from a bad and unapt formation of words a wonderful obstruction to the mind. Nor can the definitions and explanations, with which learned men are wont to guard and protect themselves in some instances, afford a complete remedy: words still manifestly force the understanding, throw every thing into confusion, and lead mankind into vain and innumerable controversies and fallacies.

44. Lastly, there are idols which have crept into men's minds from the various dogmas of peculiar systems of philosophy, and also from the perverted rules of demonstration, and these we denominate idols of the theatre. For we regard all the systems of philosophy hitherto received or imagined, as so many plays brought out and performed, creating fictitious and theatrical worlds. Nor do we speak only of the present systems, or of the philosophy and sects of the ancients, since numerous other plays of a similar nature can be still composed and made to agree with each other, the causes of the most opposite errors being generally the same. Nor, again, do we allude merely to general systems, but also to many elements and axioms of sciences, which have become inveterate by tradition, implicit credence, and neglect. We must, however, discuss each species of idols more fully and distinctly, in order to guard the human understanding against them.

45. The human understanding, from its peculiar nature, easily supposes a greater degree of order and equality in things than it really finds; and although many things in nature be sui generis, and most irregular, will yet invent parallels and conjugates, and relatives where no such thing is. Hence the fiction, that all celestial bodies were in perfect circles, thus rejecting entirely spiral and serpentine lines, (except as explanatory terms.) Hence, also, the element of fire is introduced with its peculiar orbit, to keep square with those other three which are objects of our senses. The relative rarity of the elements (as they are called) is arbitrarily made to vary in tenfold progression, with many other dreams of the like nature. Nor is this folly confined to theories, but it is to be met with even in simple notions.

[354] 71. The sciences we possess have been principally derived from the Greeks: for the addition of the Roman,

Arabic, or more modern writers are but few, and of small importance: and, such as they are, are founded on the basis of Greek invention. But the wisdom of the Greeks was professional and disputatious, and thus most adverse to the investigation of truth. The name, therefore, of sophists, which the contemptuous spirit of those who deemed themselves philosophers, rejected and transferred to the rhetoricians, Gorgias, Protagoras, Hippias, Polus, might well suit the whole tribe, such as Plato, Aristotle, Zone, Epicurus, Theophrastus, and their successors, Chrysippus, Carneades, and the rest. There was only this difference between them, the former were mercenary vagabonds, travelling about to different states, making a show of their wisdom and requiring pay; the latter, more dignified and noble, in possession of fixed habitations, opening schools, and teaching philosophy gratuitously. Both, however, (though differing in other respects,) were professorial, and reduced every subject to controversy, establishing and defending certain sects and dogmas of philosophy: so that their doctrines were nearly (what Dionysius not unaptly [*sic*] objected to Plato) "the talk of idle old men to ignorant youths." But the more ancient Greeks, as Empedocles, Anaxagoras, Leucippus, Democritus, Parmenides, Heraclitus, Xenophanes, Philolaus, and the rest, (for I omit Pythagoras, as being superstitious,) did not (that we are aware) open schools; but betook themselves to the investigation of truth with greater silence, and with more severity and simplicity: that is, with less affectation and ostentation. Hence, in our opinion, they acted more advisedly, however their works may have been eclipsed in course of time by those lighter productions which better correspond with and please the apprehensions and passions of the vulgar: for time, like a river, bears down to us that which is light and inflated, and sinks that which is heavy and solid....

9. René Descartes

With his decisive break from Scholasticism and its theological trappings, and from the Aristotelian philosophy on which it was largely based, René Descartes is deservedly called the "father of modern philosophy." His contributions to science derive from his philosophical opposition to the reliance on the senses promoted by the Scholastics and the sterility of Aristotelian "substantial forms" as a teleological explanation of natural objects. It is both the conflict between Cartesian and Baconian scientific methodology and the mechanistic view of the universe, which both methodologies supported, that heralds the fresh wind of science sweeping across England and the continent during the seventeenth century. To Descartes, mathematical physics was the framework for scientific knowledge.

Descartes was born in La Haye, France, in 1596. His early schooling (from 1607 to 1614) was in the traditional trivium and quadrivium. His university training at Poitiers earned him a degree in civil and canon law (1616), with license to practice, but his interest in science and love of the purity and elegance of mathematics led him to seek a new method of discovery in which the latter could be applied to the former. It was in geometry, for example, that theorems deduced from unambiguous axioms provided a clear route to certainty and truth. The axioms, derived intellectually and intuitively, lie in the realm beyond doubt and are not compromised by unreliable sensation. For Descartes, the removal of all doubt is the absolute point of departure for any inquiry, and this epistemological necessity explains his attraction to a mechanistic model of the universe.

An important component of his mechanistic perspective led Descartes to visualize the universe and all of its component parts as consisting of particles in motion, with motion derived from physical collisions among their infinitely divisible parts. This view is not far removed from the atomic nature of things claimed by the Epicureans, although Descartes believed that nothing is indivisible, and hence atoms cannot exist. For Descartes, there is no action at a distance, as seen in the Newtonian force of gravity. Because all particles collide, these collisions must eventually return to their origins, which implies a circular motion at the highest, most inclusive level. This "Cartesian vortex" explained the origin and nature of the universe.

His three laws of nature, which focused on motion, departed from the medieval view and anticipated Newton. The first law states that everything continues in the same state unless in contact with something that changes it: An object in motion continues to move, and an object at rest continues to be unmoved. Galileo had proposed this already. The second law states that all motion is in a straight line: An object moving in one direction continues in that direction until contacting another object, which will change the direction but not change the rectilinear nature of the motion. These two were later incorporated into Newton's first law. Descartes's third law concerns the change of motive force in two colliding bodies in relation to their respective strengths of motion.

Even though he had intended to write about the origin of animals, including humans, he never tackled this difficult subject. His mind-body dualism that distinguished the human soul from the body, for example, was difficult to integrate into his motion principles. Although both body and mind were considered to be substances, they had completely opposite natures—the body being material and the mind immaterial. The difficulty arises when the mind must contact the body (as in willing the hand to move). Princess Elizabeth of Bohemia, a friend and patron, asked how such contact—necessary for motion to occur—is possible, as the mind has no surface with which the body may come into contact, and the brain is that surface (with his posited connection with the pineal gland).

Descartes was a prolific writer. His first major treatise, *Le Monde*, revealed his thoughts on physics, which contradicted Aristotle and the Scholastics and proposed the evolution of the world from chaos. As it neared completion, Galileo was condemned by the Inquisition. Descartes held it from publication in a prudent move to avoid the same fate, and it saw partial publication only after his death in 1650. These included work on optics (*La Dioptrique*), meteorology (*Les Météores*), and geometry (*La Géométrie*). The work on meteorology influenced Boyle and Hooke at Oxford. His *Meditations on First Philosophy* (1641), written principally for theologians and philosophers, included his mind-body treatise and found considerable opposition, particularly from the English materialist, Thomas Hobbes. Descartes's *Discourse on Method* explicated his scientific views and his philosophy—despite their challenge to established authority—in the vernacular for all literate Frenchmen, and Frenchwomen, to read.

Descartes's major work, which summarized his mechanical views of nature, was *Principia Philosophiæ*, published in 1644. Despite problems with his ideas, the French supported them—even a century after Newton demonstrated the fallacies of his vortex theory. His *Rules for the Direction of the Mind*, extracts of which follow, was begun in 1619.

Rules for the Direction of the Mind

RULE II

Only those objects should engage our attention, to the sure and indubitable knowledge of which our mental powers seem to be adequate.

... [O]f all the sciences known as yet, Arithmetic and Geometry alone are free from any taint of falsity or uncertainty. We must further observe that while our inferences are frequently fallacious, deduction, or the pure illation [inference] of one thing from another, though it may be passed over, if it is not seen through, cannot be erroneous when performed by an understanding that is in the least degree rational.

RULE III

In the subjects we propose to investigate, our inquiries should be directed, not to what others have thought, nor to what we ourselves conjecture, but to what we can clearly and perspicuously behold and with certainty deduce; for knowledge is not won in any other way.

... But in case anyone may be put out by this new use of the term intuition and of other terms which in the following pages I am similarly compelled to dissever from their current meaning, I here make the general announcement that I pay no attention to the way in which particular terms have of late been employed in the schools, because it would have been difficult to employ the same terminology while my theory was wholly different. All that I take note of is the meaning of the Latin of each word, when, in cases where an appropriate term is lacking, I wish to transfer to the vocabulary that expresses my own meaning those that I deem most suitable.

Hence now we are in a position to raise the question as to why we have, besides intuition, given this supplementary method of knowing, viz. knowing by *deduction*, by which we understand all necessary inference from other facts that are known with certainty. This, however, we could not avoid, because many things are known with certainty, though not by themselves evident, but only deduced from true and known principles by the continuous and uninterrupted action of a mind that has a clear vision of each step in the process. It is in a similar way that we know that the last link in a long chain is connected with the first, even though we do not take in by means of one and the same act of vision all the intermediate links on which that connection depends, but only remember that we have taken them successively under review and that each single one is united to its neighbour, from the first even to the last. Hence we distinguish this mental intuition from deduction by the fact that into the conception of the latter there enters a certain movement or succession, into that of the former there does not. Further deduction does not require an immediately presented evidence such as intuition possesses; its certitude is rather conferred upon it in some way by memory. The upshot of the matter is that it is possible to say that those propositions indeed which are immediately deduced from first principles are known now by intuition, now by deduction, i.e. in a way that differs according to our point of view. But the first principles themselves are given by intuition alone, while, on the contrary, the remote conclusions are furnished only by deduction....

RULE VII

If we wish our science to be complete, those matters which promote the end we have in view must one and all be scrutinized by a movement of thought which is continuous and nowhere interrupted: they must also be included in an enumeration which is both adequate and methodical.

... This enumeration or induction is thus a review or inventory of all those matters that have a bearing on the problem raised, which is so thorough and accurate that by its means we can clearly and with confidence conclude that we have omitted nothing by mistake. Consequently as often as we have employed it, if the problem defies us, we shall at least be wiser in this respect, viz. that we are quite certain that we know of no way of resolving it. If it chance, as often it does, that we have been able to scan all the routes leading to it which lie open to the human intelligence, we shall be entitled boldly to assert that the solution of the problem lies outside the reach of human knowledge.

Furthermore we must note that by adequate enumeration or induction is only meant that method by which we may attain surer conclusions than by any other type of proof, with the exception of simple intuition. But when the knowledge of some matter cannot be reduced to this, we must cast aside all syllogistic fetters and employ induction, the only method left us, but one in which all confidence should be reposed....

10. Isaac Newton

Amicus Plato amicus Aristoteles magis amica veritas.
[I am a friend of Plato, I am a friend of Aristotle, but I am more a friend of truth]
—Isaac Newton

Isaac Newton was born on Christmas day, 1642. That same year, Galileo died, and one might be forgiven for finding an omen here: Not only was Newton to be strongly influenced by reading Galileo, the two men pioneered the use of the telescope—a refracting model of Galileo's invention and the much improved reflecting model devised by Newton. But one could also find other omens in this year: The English Civil War begins, signaling a growing democratic spirit on both sides of the Atlantic, and the following year Louis XIV becomes king of France, signaling an end to privileges of the old aristocracy and heralding needed tax reforms. Before the end of the century, freedom of speech and the press came to England.

Newton was, from the beginning, a study in contradictions. He was small for his age and frail, yet quick to challenge bigger boys who would torment him. He led a solitary, withdrawn life, yet possessed a strong ego that made him unwilling to assent to any intellect superior to his. He championed the inductive method of Bacon, demanding of himself multiple tests of any hypothesis before accepting experimental results, yet was a closet alchemist. Whereas his arrogance and vindictiveness cost him close friends, his superb mind and its products earned him highest honors during his lifetime, yet despite his intelligence, Newton often performed poorly in school.

His admission to Trinity College, Cambridge, in 1661, put him in contact with like-minded scholars (and a few like-minded students), giving him a sense of belonging added to an already strong sense of direction. Thus began a feverish pace of unrelenting investigation. The lengthy philosophical notebook begun at that time is a mixture of notes from classical scholars and thoughts of his own, often in reflection upon others—including Descartes's writings.

Though he differed from Descartes on a number of issues, including the infinite divisibility of matter, the requirement of physical contact for motion, and the impossibility of a vacuum, there were several points of agreement. Most profoundly, however, Newton disagreed with the central role of intuitive knowledge and the deductive method for deriving natural explanations. Like Descartes, however, Newton distrusted reliance on the senses. He wrote in his notebook, under "Certain Philosophical Questions":

> The nature of things is more securely & naturally deduced from their operacions one upon another than upon our senses. And when by the former Experiments we have found the nature of bodys, by the latter wee may more clearely find the nature of our senses. But so long as wee are ignorant of the nature of both soule & body wee cannot clearely distinguish how far an act of sensation proceeds from the soule & how far from the body &c. [*Isaac Newton, University Library Cambridge, Additional Ms. 3996, p. 29; online at http://www.newtonproject.sussex.ac.uk/view/texts/normalized/THEM00092*]

Ironically, Newton's most prolific period came during 1665 to 1667, when the bubonic plague caused the closing of Cambridge. Spending his time back home in Woolsthorpe, Lincolnshire, he invented calculus, devised the laws of motion and gravity, and in an elegant (and physically painful) series of experiments discovered the optics of colors. In these latter two, Newton upset the models dating back to Aristotle and continued in modified form by Descartes.

Newton progressed rapidly upon returning to Cambridge. He was elected a fellow of Trinity in 1667, completed his master of arts degree the following year, and in October 1669, at age twenty-seven, was appointed Cambridge's second Lucasian Professor of Mathematics. Shortly afterward, he was elected membership in the Royal Society.

Newton's greatest work, perhaps, and surely a sign of his great genius in analysis, is his *Principia Mathematica*, published in three volumes in 1687. It resulted, as did many of his comprehensive analyses, from both his obstinacy and his unwillingness to be bested. Newton had exchanged correspondence in 1679 with Robert Hooke—with whom he had previously had professional disagreements—regarding the nature of planetary orbits and whether they properly described spirals or ellipses. In the ensuing years, Hooke, Christopher Wren, and Edmund Halley (all at Oxford) had worked out the mathematics of circular orbits, but had not the calculus to define the elliptical. In 1684, Halley rode to Cambridge to ask Newton for help, to which Newton replied that he had already calculated the equations but had lost them. Upon being pressed to reconstruct them, Newton became obsessed with expanding the orbital phenomena to incorporate his nascent ideas about gravitation.

It was at Halley's personal expense and his insistence, as secretary of the Royal Society, that a manuscript be completed and sent in for publication, that Newton began work at a feverish pace. By April 1686, the first volume of *Principia* was sent to the society and the remainder a year later. In his preface, Newton wrote:

> In the publication of this work the most acute and universally learned Mr. Edmund Halley not only assisted me with his pains in correcting the press and taking care of the schemes, but it was to his solicitations that its becoming public is owing; for when he had obtained of me my demonstrations of the figure of the celestial orbits, he continually pressed me to communicate the same to the *Royal Society*, who afterwards, by their kind encouragement and entreaties, engaged me to think of publishing them.

Finally, we find in the *Principia* Newton's four rules for scientific reasoning, which have endured through the centuries. They are:

1. Admit no more causes of natural things than are both true and sufficient to explain their appearances,
2. To the same natural effect, assign the same causes,
3. Qualities of bodies, which are found to belong to all bodies within experiments, are to be esteemed universal, and
4. Propositions collected from observation of phenomena should be viewed as accurate or very nearly true until contradicted by other phenomena.

We know the first of these as the principle of parsimony, appearing long before in the writings of William of Occam in the fourteenth century. It would be instructive to compare these rules—and the final paragraph of the following selection—with the arguments of Descartes.

Opticks, Bk. III, Part 1

[375] All these things being considered, it seems probable to me, that God in the beginning formed matter in solid, massy, hard, impenetrable, moveable particles, of such sizes and figures, and with such other properties, and in such proportion [376] to space, as most conduced to the end for which he formed them; and that these primitive particles, being solids, are incomparably harder than any porous bodies compounded of them; even so very hard, as never to wear or break in pieces; no ordinary power being able to divide what God himself made one in the first creation. While the particles continue entire, they may compose bodies of one and the same nature and texture in all ages: But should they wear away, or break in pieces, the nature of things depending on them would be changed. Water and earth, composed of old worn particles and fragments of particles would not be of the same nature and texture now, with water and earth composed of entire particles in the beginning. And therefore, that nature may be lasting, the changes of corporeal things are to be placed only in the various separations and new associations and motions of these permanent particles; compound bodies being apt to break, not in the midst of solid particles, but where those particles are laid together, and only touch in a few points.

It seems to me, further, that those particles have not only a *vis inertiæ*, accompanied with such passive laws of motion as naturally result from that force, but also that they are moved by certain active principles, such as is that of gravity, and that which causes fermentation, and the cohesion of bodies. These principles I consider, not as occult qualities, supposed to result from the specific forms of things, but [377] as general laws of nature, by which the things themselves are formed; their truth appearing to us by phenomena, though their causes be not yet discovered. For these are manifest qualities, and their causes only are occult. And the Aristotelians gave the name of occult qualities, not to manifest qualities, but to such qualities only as they supposed to lie hid in bodies, and to be the unknown causes of manifest effects: Such as would be the causes of gravity, and of magnetic and electric attractions, and of fermentations, if we should suppose that these forces or actions arose from qualities unknown to us, and uncapable of being discovered and made manifest. Such occult qualities put a stop to the improvement of natural philosophy, and therefore of late years have been rejected. To tell us that every species of things is endowed with an occult specific quality by which it acts and produces manifest effects, is to tell us nothing: But to derive two or three general principles of motion from phenomena, and afterwards to tell us how the properties and actions of all corporeal things follow from those manifest principles, would be a very great step in philosophy, though the causes of those principles were not yet discovered: And therefore I scruple not to propose the principles of motion abovementioned, they being of very general extent, and leave their causes to be found out.

Now by the help of these principles, all material things seem to have been composed of [378] the hard and solid particles abovementioned, variously associated in the first creation by the counsel of an intelligent agent. For it became Him who created them to set them in order. And if He did so, it's unphilosophical to seek for any other origin of the world, or to pretend that it might arise out of a chaos by the mere laws of nature; though being once formed, it may continue by those laws for many ages. For while comets move in very eccentric orbs in all manner of positions, blind fate could never make all the planets move one and the same way in orbs concentric, some inconsiderable irregularities excepted, which may have risen from the mutual actions of comets and planets upon one another, and which will be apt to increase, till this System wants a reformation. Such a wonderful uniformity in the planetary system must be allowed the effect of choice. And so much the uniformity in the bodies of animals, they having generally a right and a left side shaped alike, and on either side of their bodies two legs behind, and either two arms, or two legs, or two wings before their shoulders, a neck running down into a backbone, and a head upon it; and in the head two ears, two eyes, a nose, a mouth, and a tongue, alike situated. Also the first contrivance of those very artificial parts of animals, the eyes, ears, brain, muscles, heart, lungs, midriff, glands, larynx, hands, wings, swimming bladders, na- [379] tural spectacles, and other organs of sense and motion; and the instinct of brutes and insects, can be the effect of nothing else than the wisdom and skill of a powerful everliving agent, who being in all places, is more able by his will to move the bodies within his boundless uniform sensorium, and thereby to form and reform the parts of the universe, than we are by our will to move the parts of our bodies. And yet we are not to consider the world as the body of God, or the several parts thereof, as the parts of God. He is a uniform being, void of organs, members or parts, and they are his creatures subordinate to him, and subservient to his will; and he is no more the soul of them, than the soul of man is the soul of the species of things carried through the organs of sense into the place of its sensation, where it perceives them by means of its immediate presence, without the intervention of any third thing. The organs of sense are not for enabling the soul to perceive the species of things in its sensorium, but only for conveying them thither; and God has no need of such organs, he being everywhere present to the things themselves. And since space is divisible *in infinitum* and matter is not necessarily in all places, it may be also allowed that God is able to create particles of matter of several sizes and figures, and in several proportions to space, and perhaps of different densities and forces, and thereby to vary the laws of nature, and make worlds of several sorts in [380] several parts of the universe. At least, I see nothing of contradiction in all this.

As in mathematics, so in natural philosophy, the investigation of difficult things by the method of analysis, ought ever to precede the method of composition. This analysis consists in making experiments and observations, and in drawing general conclusions from them by induction and admitting of no objections against the conclusions, but such as are taken from experiments, or other certain truths. For hypotheses are not to be regarded in experimental philosophy. And although the arguing from experiments and observations by induction be no demonstration of general conclusions; yet it is the best way of arguing which the nature of things admits of, and may be looked upon as so much the stronger, by how much the induction is more general. And if` no exception occur from phenomena, the conclusion may be pronounced generally. But if` at any time afterwards any exception shall occur from experiments, it may then begin to be pronounced with such exceptions as occur. By this way of analysis we may proceed from compounds to ingredients, and from motions to the forces producing them; and in general, from effects to their causes, and from particular causes to more general ones, till the argument ends in the most general. This is the method of analysis: And the synthesis consists in assuming the causes discovered, and established as principles, and by them explaining the phe- [381] nomena proceeding from them, and proving the explanations....

11. VOLTAIRE

Born François-Marie Arouet in 1694, Voltaire became a central figure of the French Enlightenment. The essayist was outspoken in his support for civil liberties, joining with colleague Jean-Jacques Rousseau, John Locke, and other champions of social and religious reform. His wry wit and his cutting satire, which endeared him to the aristocratic elements of French society—to which he belonged—were anathema to the French government and Catholic church, both of whom he attacked.

Had he lived a generation later, during the Reign of Terror in the French Revolution, he doubtless would have been introduced to the guillotine, as were Lavoisier and other intellectuals. As it was, he was merely imprisoned in the Bastille for almost a year (for insulting a French nobleman) and then exiled. For almost three years he resided in England. There, he found the freedom of speech, religion, and press refreshing, and although he found monarchy threatening to liberty, he much preferred the constitutional form in England to the absolute form in France. His ideal (and idealistic) governmental form, though, was enlightened despotism!

When he returned to France, Voltaire published a series of reviews of British attitudes and intellectual thought in the form of letters, *Lettres philosophique sur les Anglais*, including biting criticism of their French contrasts. Controversy stirred once again, copies of the work were burned, and Voltaire was once more forced to leave France. Such condemnation was only fire for his ego and fuel for his pen, as the caricatured evils of government, theology, and philosophy in his novel *Candide* aptly demonstrate. Voltaire lived to a ripe age of eighty-four; he died in Paris in 1778.

In the selection that follows, Voltaire succinctly captures the contrasts between Newton and Descartes.

Letters Concerning the English Nation, Letter XIV: On Descartes and Sir Isaac Newton

A Frenchman who arrives in London, will find philosophy, like everything else, very much changed there. He had left the world a plenum [*matter-filled space.— RKW*], and he now finds it a vacuum. At Paris the universe is seen composed of vortices of subtile matter; but nothing like it is seen in London. In France, it is the pressure of the moon that causes the tides; but in England it is the sea that gravitates towards the moon; so that when you think that the moon should make it flood with us, those gentlemen fancy it should be ebb, which very unluckily cannot be proved. For to be able to do this, it is necessary the moon and the tides should have been inquired into at the very instant of the creation.

You will observe farther, that the sun, which in France is said to have nothing to do in the affair, comes in here for very near a quarter of its assistance. According to your Cartesians, everything is performed by an impulsion, of which we have very little notion; and according to Sir Isaac Newton, it is by an attraction, the cause of which is as much unknown to us. At Paris you imagine that the earth is shaped like a melon, or of an oblique figure; at London it has an oblate one. A Cartesian declares that light exists in the air; but a Newtonian asserts that it comes from the sun in six minutes and a half. The several operations of your chemistry are performed by acids, alkalies and subtile matter; but attraction prevails even in chemistry among the English.

The very essence of things is totally changed. You neither are agreed upon the definition of the soul, nor on that of matter. Descartes, as I observed in my last, maintains that the soul is the same thing with thought, and Mr. Locke has given a pretty good proof of the contrary.

Descartes asserts farther, that extension [*taking up space.*—RKW] alone constitutes matter, but Sir Isaac adds solidity to it.

How furiously contradictory are these opinions!

> "Non nostrum inter vos tantas componere lites."
> Virgil, Eclog. III.
> "'Tis not for us to end such great disputes."

This famous Newton, this destroyer of the Cartesian system, died in March, anno 1727. His countrymen honoured him in his lifetime, and interred him as though he had been a king who had made his people happy.

The English read with the highest satisfaction, and translated into their tongue, the Elogium of Sir Isaac Newton, which M. de Fontenelle spoke in the Academy of Sciences. M. de Fontenelle presides as judge over philosophers; and the English expected his decision, as a solemn declaration of the superiority of the English philosophy over that of the French. But when it was found that this gentleman had compared Descartes to Sir Isaac, the whole Royal Society in London rose up in arms. So far from acquiescing with M. Fontenelle's judgment, they criticised his discourse. And even several (who, however, were not the ablest philosophers in that body) were offended at the comparison, and for no other reason but because Descartes was a Frenchman.

It must be confessed that these two great men differed very much in conduct, in fortune, and in philosophy.

Nature had indulged Descartes with a shining and strong imagination, whence he became a very singular person both in private life and in his manner of reasoning. This imagination could not conceal itself even in his philosophical works, which are everywhere adorned with very shining, ingenious metaphors and figures. Nature had almost made him a poet; and indeed he wrote a piece of poetry for the entertainment of Christina, Queen of Sweden, which however was suppressed in honour to his memory.

He embraced a military life for some time, and afterwards becoming a complete philosopher, he did not think the passion of love derogatory to his character. He had by his mistress a daughter called Froncine, who died young, and was very much regretted by him. Thus he experienced every passion incident to mankind.

He was a long time of opinion that it would be necessary for him to fly from the society of his fellow creatures, and especially from his native country, in order to enjoy the happiness of cultivating his philosophical studies in full liberty.

Descartes was very right, for his contemporaries were not knowing enough to improve and enlighten his understanding, and were capable of little else than of giving him uneasiness.

He left France purely to go in search of truth, which was then persecuted by the wretched philosophy of the schools. However, he found that reason was as much disguised and depraved in the universities of Holland, into which he withdrew, as in his own country. For at the time that the French condemned the only propositions of his philosophy which were true, he was persecuted by the pretended philosophers of Holland, who understood him no better; and who, having a nearer view of his glory, hated his person the more, so that he was obliged to leave Utrecht. Descartes was injuriously accused of being an atheist, the last refuge of religious scandal: and he who had employed all the sagacity and penetration of his genius, in searching for new proofs of the existence of a God, was suspected to believe there was no such Being.

Such a persecution from all sides, must necessarily suppose a most exalted merit as well as a very distinguished reputation, and indeed he possessed both. Reason at that time darted a ray upon the world through the gloom of the schools, and the prejudices of popular superstition. At last his name spread so universally, that the French were desirous of bringing him back into his native country by rewards, and accordingly offered him an annual pension of a thousand crowns. Upon these hopes Descartes returned to France; paid the fees of his patent, which was sold at that time, but no pension was settled upon him. Thus disappointed, he returned to his solitude in North Holland, where he again pursued the study of philosophy, whilst the great Galileo, fourscore years of age, was groaning in the prisons of the Inquisition, only for having demonstrated the earth's motion.

At last Descartes was snatched from the world in the flower of his age at Stockholm. His death was owing to a bad regimen, and he expired in the midst of some literati who were his enemies, and under the hands of a physician to whom he was odious.

The progress of Sir Isaac Newton's life was quite different. He lived happy, and very much honoured in his native country, to the age of fourscore and five years.

It was his peculiar felicity, not only to be born in a country of liberty, but in an age when all scholastic impertinences were banished from the world. Reason alone was cultivated, and mankind could only be his pupil, not his enemy.

One very singular difference in the lives of these two great men is, that Sir Isaac, during the long course of years he enjoyed, was never sensible to any passion, was not subject to the common frailties of mankind, nor ever had any commerce with women—a circumstance which was assured me by the physician and surgeon who attended him in his last moments.

We may admire Sir Isaac Newton on this occasion, but then we must not censure Descartes.

The opinion that generally prevails in England with regard to these new philosophers is, that the latter was a dreamer, and the former a sage.

Very few people in England read Descartes, whose works indeed are now useless. On the other side, but a small number peruse those of Sir Isaac, because to do this the student must be deeply skilled in the mathematics, otherwise those works will be unintelligible to him. But notwithstanding this, these great men are the subject of everyone's discourse. Sir Isaac Newton is allowed every advantage, whilst Descartes is not indulged a single one. According to some, it is to the former that we owe the discovery of a vacuum, that the air is a heavy body, and the invention of telescopes. In a word, Sir Isaac Newton is here as the Hercules of fabulous story, to whom the ignorant ascribed all the feats of ancient heroes.

In a critique that was made in London on M. de Fontenelle's discourse, the writer presumed to assert that Descartes was not a great geometrician. Those who make such a declaration may justly be reproached with flying in their master's face. Descartes extended the limits of geometry as far beyond the place where he found them, as Sir Isaac did after him. The former first taught the method of expressing curves by equations. This geometry which, thanks to him for it, is now grown common, was so abstruse in his time, that not so much as one professor would undertake to explain it; and Schotten in Holland, and Format in France, were the only men who understood it.

He applied this geometrical and inventive genius to dioptrics, which, when treated of by him, became a new art. And if he was mistaken in some things, the reason of that is, a man who discovers a new tract of land cannot at once know all the properties of the soil. Those who come after him, and make these lands fruitful, are at least obliged to him for the discovery. I will not deny but that there are innumerable errors in the rest of Descartes' works.

Geometry was a guide he himself had in some measure fashioned, which would have conducted him safely through the several paths of natural philosophy. Nevertheless, he at last abandoned this guide, and gave entirely into the humour of forming hypotheses; and then philosophy was no more than an ingenious romance, fit only to amuse the ignorant. He was mistaken in the nature of the soul, in the proofs of the existence of a God, in matter, in the laws of motion, and in the nature of light. He admitted innate ideas, he invented new elements, he created a world; he made man according to his own fancy; and it is justly said, that the man of Descartes is, in fact, that of Descartes only, very different from the real one.

He pushed his metaphysical errors so far, as to declare that two and two make four for no other reason but because God would have it so. However, it will not be making him too great a compliment if we affirm that he was valuable even in his mistakes. He deceived himself, but then it was at least in a methodical way. He destroyed all the absurd chimeras with which youth had been infatuated for two thousand years. He taught his contemporaries how to reason, and enabled them to employ his own weapons against himself. If Descartes did not pay in good money, he however did great service in crying down that of a base alloy.

I indeed believe that very few will presume to compare his philosophy in any respect with that of Sir Isaac Newton. The former is an essay, the latter a masterpiece. But then the man who first brought us to the path of truth, was perhaps as great a genius as he who afterwards conducted us through it.

Descartes gave sight to the blind. These saw the errors of antiquity and of the sciences. The path he struck out is since become boundless. Robault's little work was, during some years, a complete system of physics; but now all the Transactions of the several academies in Europe put together do not form so much as the beginning of a system. In fathoming this abyss no bottom has been found. We are now to examine what discoveries Sir Isaac Newton has made in it.

Ideas to Think About

1. Mysticism and empiricism remained occasionally undistinguished during the seventeenth century, as reason often slipped backward from observation to imagination, or reliance on previous claims. How does Newton respond to this?
2. Identify attempts to criticize and go beyond the Greeks and lapses into Greek thought in the readings by Bacon and Descartes.
3. Compare Descartes's distrust of the senses in scientific perception to Bacon's ultimate trust in only the senses. Does Da Vinci's position (Unit 1) reflect the same reasoning as Bacon's?
4. Describe and give quotes to show how Bacon distanced himself from both the Scholastics of the Middle Ages and Aristotle's system of inquiry. How does he employ Aristotelian causality to distinguish science from philosophy?
5. What are the major differences in comparing Newton, Bacon, and Descartes in (a) their view of reality and (b) their methodology to understand it?
6. To what extent can you detect cynicism in Voltaire's letter? How objective is his treatment of Descartes? Is there evidence that his position was influenced by political discord?

For Further Reading

Hostage to Fortune: The Troubled Life of Francis Bacon, Lisa Jardine and Alan Stewart. New York: Hill and Wang (1998). This is a heavy book (525 text pages), but Lisa Jardine's typical storytelling style makes it very readable. I include it for its excellent treatment of the political and historical context it provides, and for the intrigue Bacon became a part of in the royal court.

The Fellowship: Gilbert, Bacon, Harvey, Wren, Newton, and the Story of a Scientific Revolution, John Gribbin. New York: Overlook Press (2007). A master of popular science writing, Gribbin weaves a remarkable tale of the world on the threshold of modern science. This book travels across this and the following unit.

Isaac Newton: The Last Sorcerer, Michael White. Jackson, TN: Perseus (1997). This is a compelling account of this genius who practiced alchemy while making mathematical and scientific discoveries. This inventor of the calculus stood on the divide between magic and science.

Descartes' Bones, Russell Shorto. New York: Doubleday (2008). The unearthing of Descartes's remains and their serpentine travels over the following three centuries provides the fascinating backdrop for an account of the heresy of the new science in facing strong religious opposition. There are scores of books on Descartes: This is one of the clearest and best!

Putting Science in its Place: Geographies of Scientific Knowledge, David N. Livingstone. Chicago: University of Chicago Press (2003). This small jewel of a book (186 pages plus an excellent bibliographical essay) explores the local contexts in which science was practiced and became available: the rise of museums, laboratories, botanical gardens, coffeehouses, and other public places.

Servants of Nature: A History of Scientific Institutions, Enterprises, and Sensibilities, Lewis Pyenson and Susan Sheets Pyenson. New York: Norton (1999). Larger than the previous book (440 pages), but a complement to it, this book focuses on rise of scientific societies and institutions of higher learning from the seventeenth century to today and how these reflect and shape the way science is conducted and conceptualized by the scientists themselves.

UNIT 3

Seventeenth Century: Reconciling Past and Present

> Let it not suffice us to be Book-learned, to read what others have written, and to take upon Trust more Falsehood than Truth: But let us ourselves examine things as we have opportunity, and converse with Nature as well as Books.
>
> —John Ray

As science began to peel back the layers of time in the seventeenth century, revealing fossils and strata in ancient deposits, and as the mysteries of life began to unfold under the microscope and contradict the conventional wisdom of a respected past, a sense of unease in science led to attempts to reconcile these secular truths with biblical truth. This was particularly true among English natural scientists.

Thus, as Nicholas Steno recognized (with Da Vinci) that "tongue stones" were sharks' teeth, and Robert Hooke demonstrated that "formed stones" were early forms of animal life, Robert Plot argued that there could be no early fossils, because God would not allow any forms to become extinct. Whereas Descartes's mechanistic earth arrived through spinning atoms in cooling layers of hot liquid, Thomas Burnet's *Sacred Theory of the Earth* produced a more pious history that aligned with the account in Genesis.

In this unit, then, we focus on writings about the living forms and the fossils that preceded them. In the historical context, technology and exploration contributed to this new thrust. Just as Newton's functional reflecting telescope allowed better views of the heavens and revealed new planetary moons, so Robert Hook's improved microscope allowed him to identify and elaborate on the biological cell and other hitherto uninvestigated specimens.

The growing collections of plants and animals from explorations beyond England and the continent not only raised questions about the origin of life; they also made systematic classification a necessity. Particularly in botany, collections from the New World added to the herbalists' inventory and stimulated their careful organization. Rare imported plants likewise encouraged experimentation in already popular English gardens. Both John Tradescents, father and son, transformed the art of gardening, and their collections of plants helped to establish the Ashmolean Museum in Oxford. Robert Plot became the first keeper of the Ashmolean at its founding in 1683. John Ray found Plot's two natural history books useful in his own efforts in classification. A cataloguer of all things natural, Plot classified fossils found around Oxfordshire as stones formed by nature in accidental imitation of living things. Steno would correct this mistake.

In addition to the new ways of looking at the natural world, given to us by Bacon, Descartes, and Newton, we find in the seventeenth century a strain on the previously strong marriage of science and religion. Though it remained true that each sought to affirm its validity through the other and thus minimize discord, it had been a shotgun marriage under Scholasticism, and the new scientific discoveries of immense change over immense time constituted grounds for divorce. The ongoing attempt to reveal the handiwork of God in the laws of nature, however, did not abate.

12. Nicholas Steno

There are numerous scholars who have been celebrated as "Father of Modern Geology," and Nicholas Steno is one of them. Earth's revealed crustal formations were, in fact, popular subjects of study and speculation during the seventeenth century and earlier, and many individuals contributed to their understanding. Steno was equally interested in both the formations themselves and in the fossils found within them, and he justly deserves recognition for important inferences in stratigraphy and paleontology.

Born in Copenhagen as Niels Stensen in 1638, he was one of the fortunate half of his early school population to survive the plague of 1654. A voracious reader as a youth, he digested Galileo and Kepler, Pascal and Gassendi, the philosopher Descartes and the geographer Varen. He went to the University of Leiden in 1660—the year London's Royal Society was founded—to study medicine. He excelled in anatomy there (Stensen's salivary duct of the parotid gland is named after him), but failed to receive a professorship. He moved first to the University of Paris, then Montpellier, where medicine had become established as a specialized study in the thirteenth century, and finally to Florence in 1665.

His anatomical studies here led to the patronage of the Grand Duke of Tuscany, Ferdinand II de Medici (who, along with his father, Cosimo II, had been the patron of Galileo). Steno was shortly admitted to the Accademia del Cimento in Florence, established in 1657 by Ferdinand II. The duke was impassioned with the study of medicine and had the occasion to have the head of a large shark, caught near Livorno, to be sent to Steno for dissection. Steno immediately recognized the shark's teeth as identical to "tongue stones" (*glossopetre)* found as fossils throughout Europe and Britain over the centuries. The anatomical study was appended to a paper on muscles in 1667.

This dissection stimulated Steno's interest in fossil and stratigraphic context, an interest that led to his famous *Prodromus to a Dissertation concerning Solids naturally Contained within Solids*, published in 1669. In this, he recognized a series of basic principles that guides one to correct interpretations of strata, their content, and their occasional bending and folding. In applying these principles, Steno was the first to realize the chronological implication of older, deeper strata and younger, more superficial layers, and this led to the obvious conclusion (to Steno) that not only was the earth very, very ancient, but that some fossil deposits had occurred before the great flood of Noah.

It is instructive here to recall the importance of the accident of place in forming, or provoking, one's beliefs and attitudes. To a mind already bending to the call of science and pricked by a natural curiosity, Steno's travels from Copenhagen to Paris to Montpellier to Florence took him across a tortured landscape from which Dante, in the fourteenth century, may have found inspiration. Scherz identifies this for us (see the following selection): "Denmark and the surrounding terrain, through northern Germany, was a land of horizontal Tertiary and Cretaceous subsoils, landmarked by later Pleistocene glacial sediments."

As Steno travelled to Paris and then to Montpellier, he would have encountered a few Mesozoic volcanic formations, but from Montpellier to Florence, he would have found the strikingly different formations that enthralled Da Vinci in our earlier selection: the mighty Western Alps and the fossil-bearing limestone sediments of the Mesozoic to the south of these, as well as the Tertiary sediments and eruptive foundations so rich in molluscan species found in the Tuscan region. Steno's collections here were followed a generation later with visits by Charles Lyell. In the Apennine Mountains near Florence, Steno recognized the lower, metamorphic strata were without fossils, whereas the upper, sedimentary strata were abundant in them. He hypothesized that the earlier had formed before life existed, and that the upper, later, formations were postdiluvian.

Steno was on shaky ideological ground here, as he advanced ideas that contested the scriptures. Against popular conviction, he proffered these as his opinions only. Charles Lyell, writing in 1830 (*Principles of Geology*, vol. I, 28–29), gives him faint praise for this:

> His generalizations were for the most part comprehensive and just; but such was his awe of popular prejudice, that he only ventured to throw them out as mere conjectures, and the timid reserve of his expressions must have raised doubts as to his own confidence in his opinions, and deprived them of some of the authority due to them.

Steno thus represents the happy convergence of a bright and curious observer of the natural world and a variable environment beckoning investigation into its patterns of formation. But his personal life—of which we have regrettably little to inform us—intervened to interrupt his scientific career. Steno converted from Lutheran to Roman Catholic in 1667 and was ordained as a priest in 1675. His scientific work and intellectual hunger ceased thereafter, and he returned to northern Europe to minister to the spiritual hunger of the minority Catholic populations there.

The Prodromus

There seem to me two main reasons underlying the fact that in the solution of natural problems not only are many doubts left undecided but also most often the doubts multiply with the number of writers.

The first is that few take it on themselves to examine all those difficulties without whose resolution the solution of the investigation itself is left marred and imperfect. The present investigation is an obvious example of this point. Only one such difficulty troubled the ancients, namely the way in which marine objects had been left in places far from the sea, nor was the question ever asked whether similar bodies had been produced in places other than the sea. In more recent times the difficulty of the ancients received less emphasis since almost everyone was busy inquiring into the origin of the said bodies. Those who ascribed them to the sea did their best to show that bodies of such type could not have been produced in any other way, those who attributed them to the land denied that the sea could have covered these places; they all joined in extolling little known powers of Nature as capable of producing anything whatsoever. And it may be that a third opinion has sufficient in it to be acceptable, that some of the said bodies may be attributed to the land and some to the sea; yet almost everywhere there is deep silence about the doubt of the ancients, except that some mention floods and a succession of years of unknown duration, but only incidentally, and, as it were, in dealing with something else. Thus, in order to satisfy the laws of analysis to the best of my ability, I wove and unravelled the web of this investigation many many times, and examined its individual parts until there seemed to me to be left no further difficulty in the reading of authors, nor in the objections of friends, nor in the inspection of sites, that I had not either solved, or about which I had at least decided, from what I had learned up till now, how far a solution was possible.

The first question was whether the glossopetrae of Malta had been at one time the teeth of sharks; it was immediately obvious that this was similar to the general question of whether bodies resembling marine bodies which are found far from the sea were produced in the sea in past times. But since other bodies are also found on land that resemble those which grow in fresh waters, in the air, and in other fluids, if we grant the earth the power to produce these bodies, we cannot deny it the ability of producing the rest. Thus, it was necessary to extend the investigation to all those bodies dug from the earth that are seen to resemble those bodies that elsewhere are observed growing in fluid; but many other bodies are also found in rocks, endowed with a certain shape; if it be said that they were produced by the power of the place, it is necessary to admit that all the rest were produced by the same power, so at length, I saw that the point had been reached where every solid naturally enclosed within a solid should be examined to determine if it was produced in the same place as it was found, that is, the nature of the place where it is found should be examined and then the nature of the place where it was produced. But indeed, no one will readily determine the place of production who is ignorant of the method of production, and all discussion on the method of production is futile unless we have some certain knowledge of the nature of matter; from this it is clear how many problems must be solved to satisfy one line of inquiry.

The second reason, one which nourishes doubts, seems to me to be that in considering the natural world those things which cannot be determined with certainty are not kept separate from those that can be so determined; as a result, the principal schools of philosophy are reduced to two classes; some indeed are prevented by scruples from putting faith even in demonstrations, for fear that the same error exists in them that they often detect in other declarations; others, on the contrary, would by no means show themselves constrained to hold as certain only those things in which people of sound mind and sound perception could express belief, they being of the opinion that all those things are true that seem to them admirable and ingenious. Indeed, the advocates of experiments have rarely had the restraint either to avoid rejecting entirely even the most certain principles of nature or to avoid considering their own self contrived principles as proved. [*Here Steno may be referring to Descartes, whose methodology he rejected.—RKW*] Thus, to avoid this reef also, I decided to press with all my might in physics for what Seneca often urges strongly regarding moral precepts; he states that the best moral precepts are those which are in common use, widely accepted, and which are jointly proclaimed by all from every school, Peripatetics, Academics, Stoics, and Cynics; and indeed those principles of nature could not but be best that are in common use, widely accepted, and are considered admissible by all from every school, whether those who are eager for novelty in everything or those who are devoted to the teaching of the past.

The following three points constitute Steno's recognition that solids that are found imbedded within solids (such as fossil shark teeth) took on their solidity first, and the surrounding solid, originally liquid, conformed to the imbedded solid's shape as it solidified. They come from pages 161 and 165.

Strata of the earth are related to fluid deposits because:

1. The pulverized matter of the strata could not have been reduced to that form unless, having been mixed with some fluid, it was extracted from that fluid by its own weight and was spread out by the motion of the said superincumbent fluid.
2. The larger bodies contained in these same strata obey for the most part the laws of gravity, not only with respect to the position of any individual body but also to the relative positions of different bodies to each other.
3. The pulverized matter in the strata has so adapted itself to bodies contained in it that it has not only filled the most minute cavities in each contained body but has taken on the smoothness and lustre of the said body on that part of its surface which touches it, even though the coarse nature of the pulverized material seems incapable of such smoothness and lustre.

Sediments then are formed when the contents of a fluid sink under their own weight regardless of whether these contents have been conveyed there from elsewhere or have been secreted gradually from particles of the fluid itself, either in its upper surface or from all the particles of fluid....

The four principles that follow constitute an early form of uniformitarian thinking. The fourth point reflects what is known as Steno's law of superposition, whereas the finalle statement is his insight into the principle of original horizontality: any strata that diverge from horizontal were disturbed subsequent to their formation.

The following can be considered certain about the position of strata:

1. At the time when a given stratum was being formed, there was beneath it another substance that prevented the pulverised materials from sinking further; consequently, when the lowest stratum was being formed, either there was another solid substance underneath it or some fluid existed there which was not only different in nature from the fluid above it but was also heavier than the solid sediment from the fluid above it.
2. When an upper stratum was being formed, the lower stratum had already gained the consistency of a solid.
3. When any given stratum was being formed, it was either encompassed at its edges by another solid substance or it covered the whole globe of the earth. Hence, it follows that wherever bared edges of strata are seen, either a continuation of the same strata must be looked for or another solid substance must be found that kept the material of the strata from being dispersed.
4. When any given stratum was being formed, all the matter resting on it was fluid and, therefore, when the lowest stratum was being formed, none of the upper strata existed.

As far as form is concerned, it is certain that when any given stratum was being produced, its lower surface and its edges corresponded with the surfaces of lower and lateral bodies but that its upper surface was as far as possible parallel to the horizon; all strata, therefore, except the lowest, were bounded by two horizontal planes. Hence, it follows that strata which are either perpendicular to the horizon or inclined to it were at one time parallel to the horizon.

13. John Ray

John Ray brought the same order and systematic study to biological species as Steno did to geology. His deep affection for nature—particularly plants—was likely acquired from his mother, who was an herbalist. Ray was born in 1628 in Black Notley, Essex, England, and he entered Cambridge University at age sixteen. He advanced rapidly, becoming a fellow in 1649 (the year Charles I was beheaded and a constitutional Commonwealth, under Cromwell, established shortly afterward), lecturer in 1651, a junior dean in 1658, and college steward the following two years. In 1660, he was ordained a priest in the Anglican Church, but he continued his investigations in embryology and plant physiology and published his famous *Catalogue of Cambridge Plants* in that same year.

In 1661, however, he was forced to leave Cambridge when he refused to subscribe to the Act of Uniformity, establishing the Book of Common Prayer and unifying the Anglican Church. Under it, every man had to go to church once a week. He was not alone: Fourteen bishops were dismissed, as well. He continued his experiments and collecting, however, taking the opportunity to travel throughout England. In 1667, Ray was elected a fellow of the Royal Society.

His obsession with establishing a consistent and scientific taxonomy extended to all living forms, and he published profusely, not only on plants, but also on insects, fish, birds, and mammals. Seeking a rational—and natural—pattern upon which to base his classification, Ray (a good Baconian investigator) was interested in identifying similarities among disparate forms, particularly in plants. To accomplish this, he refused to focus on a narrow attribute or two as the basis for taxonomy, which he considered wholly artificial, and instead considered the overall structure or general morphological pattern. Thus, he was the first to make the fundamental distinction between monocots and dicots, and inter alia discovered that stems and trunks conduct water. The other great systematist, Carolus Linnaeus (born two years after Ray's death in 1705) classified plants by their reproductive anatomy.

In his philosophy of classification, he saw the organism as a functional unit, and this form-follows-function approach entailed the view that each species is biologically discrete and distinct. Mayr (1982, 256) quotes from his 1686 *History of Plants*, on species:

> no surer criterion for determining species has occurred to me than the distinguishing features that perpetuate themselves in propagation from seed. Thus, no matter what variations occur in the individuals or the species, if they spring from the seed of one and the same plant, they are accidental variations and not such as to distinguish a species.... Animals likewise that differ specifically preserve their distinct species permanently; one species never springs from the seed of another nor vice versa.

Aside from his very popular works of science, his publications supporting natural theology—the belief that nature reveals God's handiwork—were equally influential. Based on sermons he had delivered at Cambridge, *The Wisdom of God Manifested in the Works of the Creation*, published in 1691, followed a year later with *Three Physico-Theological Discourses*, were widely read and discussed.

In all of his publications, we find an abiding interest in adaptation and the causal relationship between the form of a plant or animal part and its function. His studies went beyond living forms to include plant and animal fossils, many of these sent to Ray by admirers and colleagues. Like Steno, Ray did not hesitate to support the idea that fossils were once living forms. He was unwilling, however, to accept any fossils as representing extinct forms, because it was inconceivable that God would allow it.

In 1673, Ray married Margaret Oakley of Launton. His marriage, his subsequent children, and the growing infirmities of age slowed his pace of travel, but not his writing. In 1679, he and his family moved back to his home in Black Notley. He remained there until his death at age seventy-six in 1705.

The Correspondence of John Ray

The following letter is from Ray to Francis Willughby, one of Ray's most famous students at Trinity College. From 1663 to 1666, the two of them toured Europe, winding up at Montpellier, where Steno was studying the year before. During the trip, the two made vast collections that they intended to use in producing complete systematic descriptions of all plants and animals. Willughby took responsibility for animals, but he died in 1672 after having completed only an ornithology and ichthyology. In this letter, Ray describes his ambitious methodology in preparing his masterful catalogue of British plants (published in 1670) and a second one on private and university gardens.

Sir,—I have herewithal sent you one of my books, which you had received a week sooner had not the bookbinder deceived me.... You remember that we lately, out of 'Gerard,' 'Parkinson,' and 'Phytologia Britannica,' made a collection of rare plants, whose places are therein mentioned, and ranked them under the several counties. My intention now is to carry on and perfect that design; to which purpose I am now writing to all my friends and acquaintance who are skillful in Herbary, to request them this next summer each to search diligently his country for plants, and to send me a catalogue of such as they find, together with the places where they grow. In divers counties I have such as are skillful and industrious: for Warwickshire and Nottinghamshire I must beg your assistance, which I hope, and am confident, you will be willing to contribute. After that, partly by my own search, partly by the mentioned assistance, I shall have got as much information and knowledge of the plants of each country as I can (which will require some years), I do design to put forth a complete P. B. [*Phytologia Britannica—RKW*], which I hope to bring into as narrow a compass as this book. First, I shall give the names of all plants that are or shall then be found growing in England in an alphabetical order, together with their synonyma, excepting such as are mentioned in this catalogue, whose synonyma I shall omit, setting down only one name, and referring for the rest to 'Cat. Cant.' I shall also put a full Index Anglicolatinus after the manner of that in this catalogue: then I shall put in the counties, with the several rare plants in them marshalled alphabetically. Instead of putting the particular places to each plant in the first catalogue, I shall only refer to this:—as suppose at Sedum tridactylites alpinum, after I have given the several synonyma, and the English name; instead of adding the place, I will say, vide Carnarvonshire, &c. My second design is to make another catalogue, which I will call 'Horti Angliae.' I intend to write to all the noted gardens, to procure a catalogue of each; Oxford garden and Tradescants I have already. Then I shall out of my own garden, and all these, make up one catalogue. Herein I shall give the synonyma of each plant; and those that are not in my garden, I shall name in what places they are ; as suppose Olea sativa, after I have put down his synonyma and English name, I shall add Tradescants garden, and so of the rest. Into this catalogue I shall not admit any that grow wild in England, lest it swell too big. To this also I shall add a complete Index Anglicolatinus. You have my designs, and I desire your judgment of them.... I am, &c.
Coll. Trin., Feb. 25, 1659.

Experiments Concerning the Motion of the Sap in Trees

1. In Birch trees the sap issues out of the least twiggs of branches, and fibres of roots, in proportion to their bigness.
2. In all trees the gravity promotes the bleeding, so that from a branch or root, that bendes downward, there will issue a great deal more sap, than from another of the same bigness in a more erect posture.
3. Branches and young trees cut quite off when they are full of sap, and held perpendicularly, will bleed, as we experimented in willow, birch, and sycamore: And if you cut off their tops, and invert them, they will bleed also at the little ends. Hence one may conjecture, that the narrowness of the pores is not the sole cause of the ascent of the sap, for, water that hath ascended in the little glass pipes, will not fall out againe by its own gravity, if the pipes be taken out of the water.

The preceding analogy regarding glass pipes refers to a report to the Royal Society by Robert Hooke in 1661. In it, Hooke described capillary action (without so naming it), in which the height of the water in the tube is proportional to its bore.

4. Roots of birch and sycamore cut asunder will bleed both ways, that is, from that part remaining to the tree, and from that part separated, but a great deal faster from the part remaining to the tree. But in a cold snowy day the root of one sycamore, we had bared, bled faster from the part separated, and ten times faster than it did in warm weather before.
5. In birches, the sap does not issue out of the bark, be it never so thick, but as soone as ever you have cut the bark quite through, then it first begins to bleed....

Ray, like fellow naturalists in the seventeenth century, was unsure about the presence of fossils—plants and animals—in earlier strata, and particularly with "formed stones." He also disagreed with naturalist John Woodward, professor of Physic at Gresham College, Cambridge. In his An Essay towards the Natural History of the Earth *(1695), Woodward argued that the biblical flood had covered the entire earth, dissolving all materials on its surface, which then precipitated out by gravity to form the various geological strata. The following selections, from letters to Edward Lhwyd, keeper of the Ashmolean Museum at Oxford, speak to these issues. They all date 1695.*

Further Correspondence of John Ray

[264] I know it is a hard thing to give a good account of the originall of Fossil shels and formed stones, and a satisfactory answer to all objections against either opinion: therefore a man hazards his reputation that is positive and confident on [265] either side. The like may be said of the delineations of plants upon Slate or other stones. I did once embrace a middle way...

[256] I do not see how this disposition of beds can consist with Dr Woodward's hypothesis...I have formerly objected against the Generall Deluge bringing in shels the causes that the Scripture assigns of that Deluge viz. a rain of 40 days, and the breaking up the fountains of the great Deep or bringing the subterraneous waters upon ye superficies of the Earth....

[258] One or two things let me ask you, whether the impressions be all of leaves or parts of leaves smooth & extended, or crumpled & folded up. For if they be all extended & smooth without any folds, plaits or wrinkles it is an unanswerable proof that they never were the impressions of Plants. And seeing the leaves of every plant have two different superficies, whether the two contiguous lamina of slate, between which such leafes of Plants are supposed to have layn, have two different impressions, one of ye superiour, the other of ye inferiour surface of ye leaf.

[259] I should before now have sent you advice of ye receipt of the Box of stones adorned with ye effigies of plants whether engraven or impressed upon them, but that I was loath to trouble you with empty & unnecessary letters. I thank you for ye sight of these figures, by the inspection whereof 1 am inclined to think, that they are the vestigia or impressions of Plants themselves rather then *lusus naturæ*; though I confesse those various & confused figures dispersed through the flakes of the Glocestershire slate would perswade me to be of your opinion & embrace a middle way: though when I consider that we shall make but little advantage of it; but shall be as much puzzled to give an account, how those bodies we acknowledge to have been the spoils of Animals should come to be lodged in the places where now they are found, as if we granted all of them to have been such, I cannot but stagger & remain irresolute....

[260] Such a diversity as we find of figures in one leaf of Fern, & so circumscribed, in an exact similitude to the plants themselves I can hardly think to proceed from any shooting of salts or the like. The instances you give, though they come near, yet doe not come fully up to a parallelism with such figures. Yet on ye other side there follows such a train of consequences, as seem to shock the Scripture-History of ye novity of the World; at least they overthrow the opinion generally received, & not without good reason, among Divines & Philosophers, that since ye first Creation there have been no species of Animals or Vegetables lost, no new ones produced. But whatever may be said for ye Antiquity of the Earth it self & bodies lodged in it, yet that ye race of mankind is new upon ye earth & not older then ye Scripture makes it, may I think by many argumts be almost demonstratively proved....

---

14. THOMAS BURNET

While both Steno and Ray were devout Christians, and saw no conflict between their beliefs and science, they had no particular desire to enter into a headlong confrontation with the discrepancies between the earth as they found it to be and the biblical accounts of Earth's formation. Thomas Burnet, however, burned with the need to reconcile the two without ceding the scientific facts as then understood. He was convinced that God's purpose, as revealed in Genesis, and His plan, as described in Revelation, could be read in the catastrophic events of Earth's geological history. The result was his widely read *Sacred Theory of the Earth (Telluris Theoria Sacra)*, which was published in 1681.

Thomas Burnet was not a scientist. Neither did he make any original observations nor carry out experiments as his contemporaries did. But he read their work and then visited the rugged Alps and saw for himself the "heaps of Stones and Rubbish" and the "multitude of vast Bodies thrown together in confusion." He was unwilling to bend the facts to fit scripture, but he was also unwilling to doubt the truth of scripture. What he did, rather, was to contrive a scientifically logical history of the world that at once accorded with empirical observations and interpreted scripture to support it agreeably.

Born in Yorkshire, England, in 1635, we know that he was accepted at Clare Hall, Cambridge and, in 1654, entered Christ's College as a fellow. Ten years later, he was senior proctor and, treading the tenuous line between mutual Catholic and Protestant suspicions, rose to chaplain-in-ordinary to William III. His attempted conciliation between science and religious doctrine kept him on the edge of heresy and persecution. The following excerpts follow his principle arguments that the deluge could only have come from waters deep in the earth, which arose in violence and wrought utter destruction.

The Sacred Theory of the Earth: Bk. 1, Ch. 1, Introduction

It seems to me very reasonable to believe, that besides the precepts of Religion, which are the principal subject and design of the Books of holy Scripture, there may be providentially conserv'd in them the memory of things and times so remote, as could not be retriev'd, either by History, or by the light of Nature; and yet were of great importance to be known, both for their own excellency, and also to rectifie the knowledge of men in other things consequential to them.

Bk. 1, Ch. 2

This is a short story of the greatest thing that ever yet happened in the world, the greatest revolution and the greatest change in Nature; and if we come to reflect seriously upon it, we shall find it extremely difficult, if not impossible, to give an account of the waters that compos'd this Deluge, whence they came or whither they went. If it had been only the Inundation of a Country, or of a Province, or of the greatest part of a Continent, some proportionable causes perhaps might have been found out; but a Deluge overflowing the whole Earth, the whole Circuit and whole Extent of it, burying all in water, even the greatest Mountains, in any known parts of the Universe, to find water sufficient for this Effect, as it is generally explained and understood, I think is impossible.

Bk. 1, Ch. 6

Let us therefore resume that System of the Ante-diluvian Earth, which we have deduc'd from the Chaos, and which we find to answer St. *Peter's* description, and *Moses* his account of the Deluge. This Earth could not be obnoxious to a Deluge, as the Apostle supposeth it to have been, but by a dissolution; for the Abysse was enclos'd within its bowels. And Moses doth in effect tell us, there was such a dissolution, when he saith, *The fountains of the great Abysse were broken open*. For fountains are broken open no otherwise than by breaking up the ground that covers them: We must therefore here enquire in what order, and from what causes the frame of this exteriour Earth was dissolv'd, and then we shall soon see how, upon that dissolution, the Deluge immediately prevail'd and overflow'd all the parts of it.

15. William Harvey

William Harvey is justly known for his experimental resolution of the flow of blood in the circulatory system, earning him recognition as foremost anatomist of his time. He conducted substantial investigations into the reproductive system, however, as well as embryology, and it is for these contributions—so important as predecessors of evolutionary theory—that we include him here.

Born at home in Kent, in 1578, Harvey received his early education in Canterbury. At age sixteen, he received a scholarship to study medicine at Cambridge. There he earned his bachelor of arts in 1597. He completed his medical education at the University of Padua in 1602 and returned to England. He married Elizabeth Browne, daughter of a physician to Elizabeth I, in 1604, and the same year became a candidate to the Royal College of Physicians. He became a physician to James I in 1618.

At Padua, Harvey's professors included Hieronymus Fabricius, who claimed discovery of "valves" in the veins. His interest piqued, Harvey continued research back in England, announcing discovery of the circulation of blood in 1616 (published in 1628 as *An Anatomical Exercise on the Motion of the Heart and Blood in Animals*). In turn, Harvey was named physician to James's successor, Charles I, in 1630—the year the Plymouth Company expedition sailed for America.

At the beginning of the English Civil War, in 1642, Harvey followed Charles I as he fled with his court to Oxford and was appointed doctor of physic at the university. Here, he established a professional relationship with Boyle, Hooke, Wren, and others. In 1646, he returned to London with Charles, staying there until his death in 1657.

In 1651, Harvey culminated his work on sexual reproduction and embryology with the publication of *On the Generation of Animals*. He was one of the first to recognize that sperm and egg each contributed to formation of the embryo, and that the embryo grows its parts rather than existing preformed in the egg. Both of these went against Aristotelian doctrine and received wisdom, which, although largely dismissed by the scientific community, was still supported by the many in the Anglican Church and in Cromwell's Commonwealth government, which courted biblical literalism. Harvey, therefore, exercised caution when opposing classical teaching.

In all of his research, Harvey was hampered greatly by lacking a microscope. Had he lived longer, he might have shared Robert Hooke's microscope, and thus actually seen the capillaries that he postulated and the embryonic growth he envisioned on the basis of sheer logic.

Anatomical Exercises on the Generation of Animals

In his introduction, Harvey gave faint praise for both Aristotle and Galen. Nonetheless, he and his teacher, Fabricius, found numerous errors in Aristotle's interpretations of animal generation, and Harvey's explanation for circulation overturned Galen's highly respected version. He softens his critique of the ancients here by emphasizing the necessity of reinterpretation derived in part simply through a clearer vocabulary.

On Animal Generation, Introduction

Wherefore, courteous reader, be not displeased with me, if, in illustrating the history of the egg, and in my account of the generation of the chick. I follow a new plan, and occasionally have recourse to unusual language. Think me not eager for vainglorious fame rather than anxious to lay before you the observations that are true, and that are derived immediately from the nature of things. That you may not do me this injustice, I would have you know that I tread in the footsteps of those who have already thrown a light upon this subject, and that, wherever I can, I make use of their words. And foremost of all among the ancients I follow Aristotle; among the moderns, Fabricius of Aquapendente;

the former as my leader, the latter as my informant of the way. For even as they who discover new lands, and first set foot on foreign shores, are wont to give them new names which mostly descend to posterity, so also do the discoverers of things and the earliest writers with perfect propriety give names to their discoveries. And now I seem to hear Galen admonishing us, that we should but agree about the things, and not dispute greatly about the words.

The following brief passage demonstrates Harvey's redefining of terms: spontaneous *now refers to natural abnormalities that are congenital, rather than to the generation of life from nonliving matter.*

On Animal Generation, Ex. 1

Even the creatures that arise spontaneously are called automatic, not because they spring from putrefaction, but because they have their origin from accident, the spontaneous act of nature, and are equivocally engendered, as it is said, proceeding from parents unlike themselves.

On Animal Generation, Ex. 25

For an egg is to be viewed as a conception proceeding from the male and the female, equally endued with the virtue of either, and constituting an unity from which a single animal is engendered....

It appears, consequently, that for one egg there is one soul or vital principle. But whether is this that of the mother, or that of the father, or a mixture of the two?...But how can vital principles be mingled, if the vital principle (as form) be act and substance, which it is, according to Aristotle?...For some eggs are esteemed to be longer, others shorter lived; some engender chickens endowed with the qualities and health of body..., others produce young that are predisposed to disease. Nor is it to be said that this is from any fault of the mother, seeing that the diseases of the father or male parent are transferred to the progeny, although he contributes nothing to the matter of the egg; the procreative or plastic force which renders the egg fruitful alone proceeding from the male; none of its parts contributed by him. For the semen which is emitted by the male during intercourse does by no means enter the uterus of the female, in which the egg is perfected; nor can it, indeed...by any manner or way get into the inner recesses of that organ....

On Animal Generation, Ex. 33

The medical writers with propriety maintain, in opposition to the Aristotelians, that both sexes have the power of acting as efficient causes in the business of generation; inasmuch as the being engendered is a mixture of the two which engender: both form and likeness of the body, and species are mixed, as we see in the fowl. And it does indeed seem consonant with reason to hold that they are the efficient causes of conception whose mixture appears in the thing produced....And, therefore, it is that the male and female by themselves, and separately, are not genetic, but become so united *in coitu*, and made one animal, as it were; whence, from the two as one, is produced and educed that which is the true efficient proximate cause of conception....And they go on to assert that the mixture proceeding immediately from intercourse is deposited in the uterus and forms the rudiments of the conception. That things are very different, however, is made manifest by our preceding history of the egg, which is a true conception.

On Animal Generation, Ex. 45

In the following passage, Harvey makes it clear that the classical concept of preformation—that the entire organism exists in miniature in the generative body (usually the sperm)—was erroneous. Harvey's alternative, epigenesis, was that new parts are added as the embryo develops. We now know this to be wrong in its specifics, but without a microscope to visualize sperm, egg, and zygote, this was an acceptable explanation. Harvey was innovative, however, in recognizing that somehow the embryo developed by accretion.

Now it appears clear from my history that the generation of the chick from the egg is the result of epigenesis, rather than of metamorphosis, and that all its parts are not fashioned simultaneously, but emerge in their due succession and order; it appears, too, that its form proceeds simultaneously with its growth, and its growth with its form; also that the generation of some parts supervenes on others previously existing, from which they become distinct; lastly, that is [sic: its] origin, growth, and consummation are brought about by the method of nutrition; and that at length the fœtus is thus produced....

On Animal Generation, Ex. 50

In the following selection, Harvey follows the traditional belief in a blending of parental traits. This remained conventional wisdom until the rediscovery of Mendel's experiments in the opening years of the twentieth century.

[T]he work of the father and mother is to be discerned both in the body and mental character of the offspring, and in all else that follows or accompanies temperament. In the mule, for instance, the body and disposition, the temper and voice, of both parents (of the horse and the ass, e.g.) are mingled; and so, also, in the hybrid between the wolf and the dog, &c., corresponding traits are conspicuous....

For all the arts are but imitations of nature in one way or another; as our reason or understanding is a derivative from the Divine intelligence, manifested in his works....

Wherefore, according to my opinion, he takes the right and pious view of the matter, who derives all generation from the same eternal and omnipotent Deity, on whose nod the universe itself depends. Nor do I think that we are greatly to dispute about the name by which this first agent is to be called or worshipped; whether it be God, Nature, or the Soul of the universe....

On Animal Generation, Ex. 56

Exercise 56, following, is an insight waiting to happen. It is clear that Harvey recognizes common patterns of embryonic development (insect metamorphosis was a clear exception) and is close to identifying a natural law of development. It would have been incremental for him to have used these observations to infer an evolutionary principle—as Ernst Haeckel did in the nineteenth century—but it is clear that scientific thinking is moving in the right direction!

[On second-month human embryos:] The embryo, when present, was of the length of the little finger-nail, and in shape like a little frog, save that the head was exceedingly large and the extremities very short, like a tadpole in the month of June, when it gets its extremities, loses its tail, and assumes the form of a frog.... The face was the same as that of the embryo of one of the lower animals—the dog or cat, for instance, without lips, the mouth gaping, and extending from ear to ear.

16. Robert Hooke

The tragicomic model used so effectively in Shakespeare's later plays could have found further expression in the life of Robert Hooke. An elusive historical character because so much of his story disappeared from the historical stage after his death, Hooke was a study in contrasts. He was at once a generous colleague whose collaboration with others brought immense credit and personal joy to all, and a jealous and paranoid protector of ideas from perceived plagiarism. These qualities, however, were tandem and not simultaneous.

Hooke was fortunate to bring his genius for invention to an atmosphere of experimentation, at a place where such experimentation was most encouraged, and at a time when public demonstration of experimental results was most propitious. The place was Oxford, where Robert Boyle, Thomas Willis, Christopher Wren, Richard Lower, John Mayow, and others reveled in nonstop experiments. The time was the mid-seventeenth century, when the Royal Society was created and became a venue for scientific demonstrations (1660) and the first coffeehouses, where debate, discussion, and exchange of new ideas prevailed, were established. (Oxford had the first coffeehouses in England: in 1650, the Grand Café opened, followed, in 1654, by the Queen's Lane Coffeehouse. In 1654, St. Michael's Alley opened in London. By 1675, there were over 3,000 coffeehouses in England alone.)

Robert Hooke was born in 1635 on the Isle of Wight and died in London in 1703. An inheritance upon his father's death in 1648 allowed Hooke to enter Westminster School in London where, in addition to learning Hebrew, Greek, and the geometry of Euclid, he became reasonably proficient in playing the organ. In 1653, Hooke secured a chorister position at Christ Church, Oxford. Here, propitiously, began his life's work in science. He entered Wadham College, under the influential leadership of Wilkins, became chemical assistant to Willis, and was hired as an experimental assistant to Boyle. His ingenious invention of Boyle's air pump, and its public demonstration, led to his election as fellow of the Royal Society in 1663 (he and the others had formed the nucleus of the society's charter in 1660). He was also awarded (he never earned it) a master of arts in 1663 and in 1691 received the degree of doctor of physics. Hooke's report on the capillary action of glass tubes, in 1661, has been previously mentioned in connection with Ray's experiments with plants. This tract led to Hooke's appointment as curator of experiments for the society in November of that year.

Hooke's experiments with respiration, along with those by Harvey, Lower, Mayow, Wren, and others, collectively revolutionized science. These demonstrated new and better understanding of animal physiology. They revealed a new and promising use of experimental apparatus in testing hypotheses on both living and dead animals. They directly paved the way to further discoveries (e.g., the discovery of oxygen by Lavoisier and Priestly, demonstration of the nature of the heartbeat by Purkinje and others) in the following centuries.

Beginning in 1663, Hooke's interest was trained on the microscope and what could be revealed through its use. His long and diverse series of observations, published as *Micrographia*, appeared in 1665.

Hooke also contributed significantly to geology. His 1668 *Lectures and Discourses of Earthquakes and Subterranean Eruptions* anticipated James Hutton. Hooke claimed that mountain building was largely due to earthquakes, in addition to heat in Earth's depths, and that this constant building force was balanced by the destructive force of erosion.

Many of the disappointments Hooke experienced, particularly in his later years, were of his own making. With his broad interests and obsessive personality, he quickly moved from one idea and experiment to another, seldom taking the time and patience to carefully work them up and write about them. At times, he became morose and irascible, engaging in bitter disputes with Newton over who, exactly, devised the law of inverse squares, and with Henry Oldenberg and Christian Huygens over the invention of the spring balance and escapement mechanisms for the pocket watch. There is evidence, as well, that he developed the architecture for the dome of St. Paul's Cathedral in London, for which Christopher Wren is given credit. At other times, Hooke generated excellent ideas that were simply impractical. After the great London fire of 1666, he was appointed (with Wren) as an architect to redesign the city. Evidence suggests that he single-handedly surveyed most of the city. His elaborate grid plan for London's reconstruction was rejected in favor of following the existing streets.

As he grew older, infirmities wracked him. He suffered from edema, dizziness, insomnia, and chest pains. It may be ironic that, with all of his work on the heart and circulation, his death in 1703 may have been due in part to cardiovascular complications. It is claimed that his obscurity after death was enhanced by a vengeful Isaac Newton, who as president of the Royal Society, either destroyed or sequestered much of his work. This may have included his official society portrait, which has never been found.

Micrographia

It is the great prerogative of Mankind above other Creatures, that we are not only able to behold the works of Nature, or barely to sustein our lives by them, but we have also the power of considering, comparing, altering, assisting, and improving them to various uses. And as this is the peculiar priviledge of humane Nature in general, so is it capable of being so far advanced by the helps of Art, and Experience, as to make some Men excel others in their Observations, and Deductions, almost as much as they do Beasts. By the addition of such artificial Instruments and methods, there may be, in some manner, a reparation made for the mischiefs, and imperfection, mankind has drawn upon it self, by negligence, and intemperance, and a wilful and superstitious deserting the Prescripts and Rules of Nature, whereby every man, both from a deriv'd corruption, innate and born with him, and from his breeding and converse with men, is very subject to slip into all sorts of errors.

The only way which now remains for us to recover some degree of those former perfections, seems to be, by rectifying the operations of the Sense, the Memory, and Reason, since upon the evidence, the strength, the integrity, and the right correspondence of all these, all the light, by which our actions are to be guided is to be renewed, and all our command over things is to be establisht....

The first thing to be undertaken in this weighty work, is a watchfulness over the failings and an inlargement of the dominion, of the Senses....

The next care to be taken, in respect of the Senses, is a supplying of their infirmities with Instruments, and, as it were, the adding of artificial Organs to the natural; this in one of them has been of late years accomplisht with prodigious benefit to all sorts of useful knowledge, by the invention of Optical Glasses. By the means of Telescopes, there is nothing so far distant but may be represented to our view; and by the help of Microscopes, there is nothing so small, as to escape our inquiry; hence there is a new visible World discovered to the understanding. By this means the Heavens are open'd, and a vast number of new Stars, and new Motions, and new Productions appear in them, to which all the ancient Astronomers were utterly Strangers. By this the Earth it self, which lyes so neer us, under our feet, shews quite a new thing to us, and in every little particle of its matter; we now behold almost as great a variety of Creatures, as we were able before to reckon up in the whole Universe it self.

It seems not improbable, but that by these helps the subtilty of the composition of Bodies, the structure of their parts, the various texture of their matter, the instruments and manner of their inward motions, and all the other possible appearances of things, may come to be more fully discovered; all which the ancient Peripateticks were content to comprehend in two general and (unless further explain'd) useless words of Matter and Form. From whence there may arise many admirable advantages, towards the increase of the Operative, and the Mechanick Knowledge, to which this Age seems so much inclined, because we may perhaps be inabled to discern all the secret workings of Nature, almost in the same manner as we do those that are the productions of Art, and are manag'd by Wheels, and Engines, and Springs, that were devised by humane Wit.

In this kind I here present to the World my imperfect Indeavours; which though they shall prove no other way considerable, yet, I hope, they may be in some measure useful to the main Design of a reformation in Philosophy, if it be only by shewing, that there it not so much requir'd towards it, any strength of Imagination, or exactness of Method, or depth of Contemplation (though the addition of these, where they can be had, must needs produce a much more perfect composure) as a sincere Hand, and a faithful Eye, to examine, and to record, the things themselves as they appear.

Micrographia, Observ. XVIII. Of the Schematisme or Texture of Cork, and of the Cells and Pores of some other such frothy Bodies

In the following passage, Hooke describes his discovery of the cell, and so names it.

I took a good clear piece of Cork, and with a Pen-knife sharpen'd as keen as a Razor, I cut a piece of it off, and thereby left the surface of it exceeding smooth, then examining it very diligently with a *Microscope*, me thought I could perceive it to appear a little porous; but I could not so plainly distinguish them, as to be sure that they were pores, much less what Figure they were of: But judging from the lightness and yielding quality of the Cork, that certainly the texture could not be so curious, but that possibly, if I could use some further diligence, I might find it to be discernable with a *Microscope*, I with the same sharp Penknife, cut off from the former smooth surface an exceeding thin piece of it, and placing it on a black object Plate, because it was it self a white body, and casting the light on it with a deep *plano-convex Glass*, I could exceeding plainly perceive it to be all perforated and porous, much like a Honey-comb, but that the pores of it were not regular; yet it was not unlike a Honey-comb in these particulars.

First, in that it had a very little solid substance, in comparison of the empty cavity that was contain'd between, as does more manifestly appear by the Figure A and B of the *XI. Scheme*, for the *Interstitia*, or walls (as I may so call them) or partitions of those pores were never as thin in proportion to their pores, as those thin films of Wax in a Honey-comb (which enclose and constitute the *sexangular cells*) are to theirs.

Next, in that these pores, or cells, were not very deep, but consisted of a great many little Boxes, separated out of one continued long pore, by certain *Diaphragms*, as is visible by the Figure B, which represents a sight of those pores split the long-ways.

Lectures and Discourses on Earthquakes and Subterraneous Eruptions

In the following selection, Hooke gives one of many explanations of the sources for fossil remains in buried strata. He agrees with Da Vinci and Steno (and others) that ancient shells reflect previous land under water or, when found at higher elevations, reflect upheavals. The last selection reveals his break from convention in accepting species extinction. His suggestion that new varieties within a species have occurred is just a step short of a concept of evolution, although Hooke remained convinced of species fixity.

[291] Eleventhly, that there have been many other Species of Creatures in former Ages, of which we can find none at present; and that 'tis not unlikely also but that there may be divers new kinds now, which have not been from the beginning.

[327]...[T]here may have been divers Species of things wholly destroyed and annihilated, and divers others changed and varied, for since we find that there are some kinds of Animals and Vegetables peculiar to certain places, and not to be found elsewhere; if such a place have been swallowed up, 'tis not improbable but that those Animal Beings may have been destroyed with them....

[T]here may have been divers new varieties generated of the same Species, and that by the change of the Soil on which it was produced;...

Ideas to Think About

1. Compare the nature and focus of the arguments of Steno and Ray by taking their positions regarding the nature and locations of fossils. Compare these with Da Vinci's arguments.
2. Describe the difference between Harvey's explanation for the generation of an animal from the egg and that of earlier writers, going back to Aristotle. What may have led to this difference?
3. In what way might Harvey's "epigenesis" idea have been used to explain the early human embryo?
4. Compare Hooke's arguments on true versus false knowledge to those of Francis Bacon. How does the use of instruments influence this comparison? What can explain Hooke's view of the universe as being like a machine rather than due to mysterious supernatural forces?
5. Does Burnet's argument follow that of the Scholastics, particularly Roger Bacon? How is it different?

For Further Reading

Ingenious Pursuits, Lisa Jardine. New York: Nan A. Talese, Doubleday (1999). This is an historical and literary journey into the (primarily) seventeenth-century world of scientific discovery and invention. This union of technology and imagination created the Western intellectual tradition. It also presaged the Industrial Revolution to follow in the eighteenth century. More than all else, Jardine demonstrates that where science was, art was not far behind! Among the many early scientists treated are Steno, Ray, Harvey, and Hooke.

A Social History of Truth: Civility and Science in Seventeenth-Century England, Steven Shapin. Chicago: The University of Chicago Press (1994). This is a compellingly crafted treatise on how facts are secured in this early community of scientists. Beyond experimentation and demonstration and beyond the origins of research design lay a strong reliance on mutual trust, credible testimony, and the gentlemanly resolution of disagreement. The focus is on Robert Boyle, but pays due homage to his colleague and technical assistant, Robert Hooke. The two shared adjacent quarters and laboratories at University College, Oxford.

UNIT 4

Eighteenth-Century Enlightenment: Classification and Description

> I am not yet so lost in lexicography, as to forget that words are the daughters of earth, and that things are the sons of heaven. Language is only the instrument of science, and words are but the signs of ideas: I wish, however, that the instrument might be less apt to decay, and that signs might be permanent, like the things which they denote.
>
> —SAMUEL JOHNSON, *Preface to his Dictionary, 1755*

There was no little debate during the eighteenth century over species and the words used to designate them. To some, species did not exist outside our naming them—they were daughters of Earth. Natural groupings of animals—the sons of heaven—might or might not conform to our "species." Yet naming them was critically important for communication and shared information.

This new century inherited from the previous both a new vision of reality and a series of new routes for its discovery. Newton had already demonstrated that laws governing the heavens were those that also governed the earth. Harvey and Boyle had collaborated to expand the Baconian method and discovered that knowledge can actually be generated by experiment. Steno and Ray had expanded the concepts of time and change to a world that now had a long material history wrought by strong mechanical change. Nature, from cosmos to the cell, provided its own evidence for regularity, and understanding it lay in the discovery of its laws.

Such discovery, however, demanded a concept of order and predictability in the natural world to provide its legs. The individual sciences gained some of their independent legitimacy in this century, and the first stage in the discrete emergence of any science is an inventory of its phenomenological inclusions and limits. We thus see newer attempts at classification of nature. Unlike earlier taxonomies and descriptions, these were largely devoid of attempts of biblical correspondence or mystical speculation, but rather drew upon discoveries about nature and laws governing it. We also see a continued interest in the investigation into origins—of the earth and of life.

The classification of humans and the scientific study of human varieties marked the beginning—in subject of interest if not in legitimacy as a science—of physical anthropology. Johann Blumenbach added greatly to Buffon's initial attempts to classify human races and clarified much of the controversy over their species' status and their origins.

The Greek concept of the *scala naturae*, utilized in the medieval period to reflect God's original creation of the vast diversity of forms, with humankind at the top, was reintroduced in Bonnet's "Great Chain of Being" as a reflection of the natural hierarchy of living forms, with taxonomies from Linnaeus, Buffon, and Cuvier reflecting this model. In the physical sciences, new theories of Earth's formation, based on mechanistic and not mythic forces, likewise appeared.

As scientific discoveries helped to clarify some of the profound questions about nature, the scientific disciplines began to solidify into the compartments we find familiar: botany, geology, biology, chemistry.

Not surprisingly, what followed was both an empirical and philosophical disconnect between the material and the mystical. The last serious alchemist was Isaac Newton.

17. Charles Bonnet

A Swiss naturalist who turned to philosophy when his eyesight began to fail, Charles Bonnet is best known for advancing the "Great Chain of Being" model of life on Earth. He also conducted experiments that confirmed parthenogenesis in aphids and demonstrated the spontaneous generation of lost parts in the fresh-water hydra. These revelations contributed to his views of species fixity and preformation.

Bonnet was born in 1720 in Geneva, apparently never travelled beyond its borders, and had a long but childless marriage. Partial deafness at age seven led to private tutoring for much of his education. Making law his profession, his passion was always nature, and although he is said to have practiced law, we know only of his work in natural science.

This work was impressive and earned him multiple honors. He was named a fellow of the Royal Society (London) in 1743, subsequently a foreign member of the Royal Swedish Academy and the Imperial Academies in Germany and Russia, and became a member of the Royal Societies of Montpellier and Göttingen.

He disagreed with the theory of epigenesis—the Aristotelian notion that form develops with no predetermined blueprint—and refuted it in 1762 in *Considérations sur les corps Organisés* (Considerations on Organized Bodies). In this he supported preformation theory—the assertion that all living forms were created at the same time, and that all subsequent individuals exist preformed but in miniature in the fertilized egg. Thus, each species is immortal and immutable.

Two years after this, he published the first of a two-volume work, *The Contemplation of Nature*, in which he describes the idea that all beings form an unbroken chain, arranged in a scale from lowest to highest. These were all created at once, but thereafter continued to generate through inherent forces.

Yet, Bonnet also espoused a theory of catastrophism: Periodic catastrophes cause many if not most beings to suddenly die, while the survivors climb the great chain to reach new heights. Humankind, already at the peak, would evolve into angels at the next disaster, plants would become animals, and minerals would become plants.

This modified concept of evolution—in which the next higher form is inherent in the germplasm of the preexisting one—influenced some of Erasmus Darwin's ideas.

The Contemplation of Nature, Vol. 1

[23] Between the lowest and highest degree of corporeal or spiritual perfection, there is an almost infinite number of intermediate degrees. The result of these degrees composes the *universal chain.* This unites all beings, connects all worlds, comprehends all the spheres. One sole being is out of this chain, and that is He that made it.

A thick cloud conceals from our sight the noblest parts of this immense chain, and admits us only to a slight view of some ill-connected links, which are broken, and doubtless greatly differing from the natural order.

We behold its winding course on the surface of our globe, see it pierce into its entrails, penetrate into the abyss of the sea, dart itself into the atmosphere, sink far into the celestial spaces, where we are only able to descry it by the flashes of fire it emits hither and thither.

The Contemplation of Nature, Vol. II

[66] If before the discovery of the polypus [*hydra—RKW*], those who form general rules had been asked their sentiments concerning a being that multiplied by slips and shoots, and that might be grafted, they would undoubtedly have answered, that this being was a plant. But if they had been told that this being lived by prey, which it could seize with a net, devour and digest it, they would then have called this being an *animal plant,* and would have thought they had given a happy definition of it. If they had afterwards learnt, that it possessed a property unknown in the plant, that of being turned inside out like a glove, they would in all probability have judged that this being was neither plant nor animal, and would have placed it in a particular class.

The polypus is not, strictly speaking, an animal-plant; it is still less a being that neither belongs to the class of animals or that of vegetables: it is a real animal, but one that has more relations with the plant than other animals have.

Nature descends by degrees from man to the polypus, from the polypus to the sensitive, from the sensitive to the truffle. The superior species are always connected by some character with the inferior; the latter to more inferior still. We have much contemplated this wonderful chain. *Organized* matter has received an almost infinite number of various modifications, and all are shaded like the colours of a prism. We make points on the representation, trace lines on it, and term that forming *genera* and *classes.* We only perceive the most glaring strokes, and the delicate shadowings escape our sight.

Plants and animals then are only modifications of organized matter. They all partake of a similar [67] essence, and the distinguishing attribute is unknown to us. We thought ourselves acquainted with the principle properties of the animal body, *irritability* came and convinced us of our ignorance; and this new property, concerning which we make so many and such curious experiments, is at present only known to us by some *effects.*

18. Carolus Linnaeus

Whereas John Ray devised a system of classification that revealed patterns in nature, and thus recognized that diversity is far from random, Linnaeus created a taxonomy that organized those patterns in a hierarchical series from closer to more distant sharing of traits. Together they revealed that the growing diversity of living things—knowledge compounded with each returning ship and its holds of new specimens—constituted greater complexity than could have been imagined. For both, classification of static organisms with discrete morphological boundaries reinforced the idea that species were fixed from the moment of their unique creation.

Linnaeus, however, in introducing a series of nested designs—uniqueness at very detailed levels of attributes that combine in more general categories to include an ever enlarging circle of shared traits—revealed patterns of relationships instead of separations. Perhaps God's designs were few and natural law led to variations upon them. Perhaps species (dare we think it?) were not fixed! (Breeders of stock had long demonstrated this, of course, but this was artificial interference in nature's business.)

Contributing to this sense of unease was the growing incredulity that, with the burgeoning numbers of diverse species, the Ark of Noah was becoming unacceptably large. In Ray's time, he estimated that there were "some 10,000 kinds of insects, 1,300 other kinds of animals, and 20,000 species of plants in the world" (Thomson 2005, 68). When Linnaeus was published, these numbers had been exceeded by a magnitude, and today, estimates range from ten to one hundred million. The first task of any emergent science is to establish its phenomenological limits (what is included, what is not). The second is to classify what is included. Linnaeus's binomial taxonomic system did just that, and his categories—although now multiplied by subdivisions—are remarkably the same today.

Born in Stenbrohult, Sweden, in 1707, Linnaeus entered the University of Lund at age twenty to study medicine. He transferred to Uppsala the next year, where he began collecting plants—not an academic diversion, as botanical pharmacology was part of the curriculum. He earned his medical degree in 1735 in the Netherlands, after having completed two botanical expeditions. In that year, he also published his *Systema Naturae*, which became the taxonomist's bible for generations.

He returned to practice medicine and teach in Sweden in 1738 where, at the University of Uppsala, he trained and influenced a generation of naturalists. On far-reaching expeditions to Japan and Southeast Asia, South America, Africa, the Near East, and even North America, his students collected and sent back the plant and animal specimens that enlarged the world of life and knowledge for Europeans. Captain James Cook took one of his students on each of his two world-circling voyages—much as Darwin would later accompany Fitzroy on the *Beagle*.

Despite evidence in Linnaeus's groupings that suggests common descent, he always maintained that God's plan was to establish fixed forms, and that any subsequent changes in species were part of that plan. He was quick to recognize the battleground of competition in nature, however, although this was God's way of weeding out superfluous numbers in order to maintain nature's balance. This was the same natural theology that we find in Steno and Ray.

In addition to bringing order to nature through hierarchical rankings, Linnaeus brought consistency through nomenclature: He had inherited a hodge-podge of undisciplined habits of assigning Latin names, and he insisted on rules. In his tenth and expanded edition of *Systema Naturae*, Linnaeus classified humankind into the categories he established: kingdom, Animalia; class, Vertebrata; order, Primates; genus, Homo; and species, sapiens. These categories, and many of the others he established, are still valid today—although numerous subgroupings have been added. In that same edition he recognized four human races, subspecies of the species *H. sapiens*: Americanus, Asiaticus, Africanus, and Europeanus.

Reflections on the Study of Nature

We see in the following passages the continuing Scholastic concept of the scala naturae, *which remained strong in the eighteenth century even in the face of more progressive and developmental views. Unlike Hooke, Linnaeus here disputes any notion of extinction. The notion of ecological equilibrium maintained by natural checks on population growth anticipates Adam Smith and Malthus.*

[2] The knowledge of one's self is the first step towards wisdom: this was the favourite precept of the wise Solon, and was written in letters of gold on the entrance of the temple of Diana.

A man surely cannot be said to have attained this self-knowledge, unless he has at least made himself acquainted with his origin, and the duties that are incumbent upon him.

Men and all animals increase and multiply in such a manner, that however few at first, their numbers are continually and gradually increasing. If we trace them backwards, from a greater to a lesser number, we at length arrive at one original pair. Now mankind, as well as all other creatures, being formed with such exquisite and wonderful skill, that human wisdom is utterly insufficient to imitate the most simple fibre, vein, or nerve, much less a finger, or other contriving or executive organ; it is perfectly evident, that all these things must originally have been made by an omnipotent and omniscient Being; for "he who formed the ear, shall he not hear? And he who made the eye, shall he not see?"

Moreover, if we consider the generation of Animals, we find that each produces an offspring after its own kind, as well as Plants, Tænias, and Corallines; that all are propagated by their branches, by buds, or by feed; and that from each proceeds a germ of the same nature with its parent; so that all living things, plants, animals, and even mankind themselves, form one "chain of universal Being," from the beginning to the end of the world: in this sense truly may it be said, that there is nothing new under the sun.

[10–11] Quadrupeds, which wander and sport in the fields, convert all other things to their use: by their joint endeavours they purge the earth from putrefying carcases; by their voracious appetites they fix bounds to the number of living creatures; they join in the contracts of love; and, when urged by hunger, unite in pursuit of their prey. Thus, whilst all things are purified, all things are renewed, and an equilibrium is maintained; so that of all the species originally formed by the Deity, not one is destroyed.

[13] Thus we learn, not only from the opinions of moralists and divines, but also from the testimony of nature herself, that this world is destined to the celebration of the Creator's glory, and that man is placed in it to be the publisher and interpreter of the wisdom of God: and indeed he who does not make himself acquainted with God from the consideration of nature, will scarcely acquire knowledge of him from any other source; for "if we have no faith in the things which are seen, how should we believe those things which are not seen?"

[17–18] Nature always proceeds in her accustomed order, for her laws are unchangeable; the omniscient God has instituted them, and they admit of no improvement,

It is so evident that the continent is gradually and continually increasing by the decrease of the waters, that we want no other information of it than what nature gives us: mountains and vallies, petrifactions and the strata of the earth, the depths of the ocean and all the various kinds of stones, proclaim it aloud. As the dry land increases at this day, so it is probable that it has all along gradually extended itself from the beginning: if we therefore enquire into the original appearance of the earth, we shall find reason to conclude, that instead of the present wide-extended regions, one small island only was in the beginning raised above the surface of the waters.

If we trace back the multiplication of all plants and animals, as we did that of mankind, we must stop at one original pair of each species. There must therefore have been in this island a kind of living museum, so furnished with plants and animals, that nothing was wanting of all the present produce of the earth.

Lachesis Lapponica (Travels to Lapland), Vol. 1

[27] The people [*of the Helsingland region*] seemed somewhat larger in stature than in other places, especially the men. I inquired whether the children are kept longer at the breast than is usual with us, and was answered in the affirmative. They are allowed that nourishment more than twice as long as in other places. I have a notion that Adam and Eve were giants, and that mankind from one generation to another, owing to poverty and other causes, have diminished in size. Hence, perhaps, the diminutive stature of the Laplanders....

[28] The clay contained small and delicately smooth white bivalve shells quite entire, as well as some larger brown ones, of which great quantities are to be found near the water side. I am therefore convinced that all these valleys and marshes have formerly been under water, and that the highest hills only then rose above it.

19. George-Louis Leclerc, Comte de Buffon

The strong theological framework within which life's diversity had long been explained—species fixity, separate creation, no extinctions, the *scala naturae*, and the attenuated chronology of the earth—began to crumble in the eighteenth century. Buffon was one of its demolishers. Immersed in the French Enlightenment, a friend of Voltaire, and a member of the French Academy of Sciences in 1734 and its literary sister, the French Academy, in 1753, Buffon was surrounded by literary and philosophical avant-garde. Public debates and essay contests—in Paris, Berlin, and elsewhere—encouraged self-expression and freedom from the constraint of social convention. Buffon's revolutionary biology fit in well.

Born at Montbard, in Côte-d'Or, in 1707, he entered a small Jesuit college in Dijon at age ten, followed by the University of Dijon, and finally the University of Angers. His two passions were natural history and mathematics. He moved to Paris in 1732, where he stayed until his death in 1788. Both his well-argued ideas and his talent in writing gained him rapid popularity and respect. In 1739, he was appointed director of the Parisian *Jardin du Roi*, the equivalent to London's Kew Gardens, and under his direction, it became a major museum and research facility.

Buffon's earliest research was in mathematics, where he first applied calculus to probability theory, but he is most famous for his multivolume *Histoire Naturelle*, published from 1749 to 1785. In thirty-five volumes (translated into English in a reduced nine-volume set), Buffon made meticulous descriptions of each known animal, interlaced with a body of theoretical speculation.

Like Linnaeus and Steno, Buffon recognized that the presence of fossil shells in particular horizontal strata indicated natural processes were involved in Earth's formation. Unlike the others, Buffon reasoned that the time this took was vast—far longer than the accepted biblical reckoning of a 4004 BC origin, made by Archbishop Ussher in the previous century. In biology, Buffon likewise went against convention. Emphasizing the history in natural history, he believed that understanding the past meant examining the whole organism and its relation to its environment. He was less interested, therefore, in formal taxonomy than in descriptive anatomy, and he criticized Linnaeus's binomial system, claiming that species were, after all, convenient human creations for categorizing nature rather than actual biological entities. His view that species were not fixed, but instead were part of a dynamic system of interaction with the environment, introduced a new perspective on nature:

to understand her, one must understand the narrative account of her ever-changing qualities. What is important is the story that explains the process rather than the taxonomy that describes the product.

Histoire Naturelle, Vol. 1

[vol. 1: 13] The changes which the earth has undergone, during the last two or three thousand years, are inconsiderable, when compared with the great revolutions which must have taken place in those ages that immediately succeeded the creation. For, as terrestrial substances could not acquire solidity but by the continued action of gravity, it is easy to demonstrate, that the surface of the earth was at first much softer than it is now; and, consequently, that the same causes which at present produce but slight, and almost imperceptible alterations during the course of many centuries, were then capable of producing very great revolutions in a few years. It appears, indeed, to be an uncontrovertible [*sic*] fact, that the dry land which we now inhabit, and even the summits of the highest mountains, were formerly covered with the waters of the sea; for shells, and other marine bodies, are [14] still found upon the very tops of mountains. It likewise appears, that the waters of the sea have remained for a long track of years upon the surface of the earth; because, in many places, such immense banks of shells have been discovered, that it is impossible so great a multitude of animals could exist at the same time. This circumstance seems likewise to prove, that, although the materials on the surface of the earth were then soft, and, of course, easily disunited, moved, and transported, by the waters; yet these transportations could not be suddenly effected. They must have been gradual and successive, as sea-bodies are sometimes found more than 1000 feet below the surface. Such a thickness of earth or of stone could not be accumulated in a short time. Although it should be supposed, that, at the deluge, all the shells were transported from the bottom of the ocean, and deposited upon the dry land; yet, beside the difficulty of establishing this supposition, it is clear, that, as shells are found incorporated in marble, and in the rocks of the highest mountains, we must likewise suppose, that all these marbles and rocks were formed at the same time, and at the very instant when the deluge took place; and that, before this great revolution, there were neither mountains, nor marbles, nor rocks, nor clays, nor matter of any kind similar to what we are now acquainted with, as they all, with few exceptions, contain shells, [15]and other productions of the ocean. Beside, at the time of the universal deluge, the earth must have acquired a considerable degree of solidity, by the action of gravity for more than sixteen centuries. During the short time the deluge lasted, it is, therefore, impossible, that the waters should have overturned and dissolved the whole surface of the earth, to the greatest depths that mankind have been able to penetrate.

Histoire Naturelle, Vol. 2

[vol. 2: 51] I conceive, then, that the organic particles sent from all parts of the body into the testicles and seminal vessels of the male, and into the ovarium of the female, compose the seminal fluid, which, in either sex, as formerly observed, is a kind of extract from the several parts of the body. These organic particles, instead of uniting and forming an individual similar to that in whose body they are contained, as happens in vegetables, and some imperfect animals, cannot accomplish this end without a mixture of the fluids of both sexes. When this mixture is made, if the organic particles of the male exceed those of the female, the result is a male; and, if those of the female most abound, a female is generated. I mean not that the organic particles of the male or of the female could singly produce [52] individuals: A concurrence or union of both is requisite to accomplish this end. Those small moving bodies, called *spermatic animals*, which, by the assistance of the microscope, are seen in the seminal fluids of all male animals, are, perhaps, organized substances proceeding from the individual which contains them; but, of themselves, they are incapable of expansion, or of becoming animals similar to those in whom they exist. We shall afterwards demonstrate, that there are similar animalcules in the seminal fluids of females, and point out the place where this fluid is to be found....

[59] But the strongest proof of the truth of our present doctrine arises from the resemblance of children to their parents. Sons, in general, resemble their father more than their mothers, and daughters have a greater resemblance to their mothers than their fathers; because, with regard to the general habit of a body, man resembles a man more than a woman, and a woman resembles a woman more than a man. But, as to particular features or habits, children sometimes resemble the father, sometimes the mother, and sometimes both. A child, for example, will have the eyes of the father, and the mouth of the mother, or the colour of the mother and the stature of the father. Of such phaenomena it is impossible to give any explication, unless we admit that both parents have contributed to the formation of the child, and, consequently, that there has been a mixture of two seminal fluids.

Histoire Naturelle, Vol. 3

[vol. 3: 279] Men never admired [280]the apes, till they saw them imitate human actions. It is not, indeed, an easy matter to distinguish some copies from the originals. There are, besides, so few who can clearly perceive the difference between genuine and counterfeit actions, that, to the bulk of mankind, the apes must always excite surprise and humiliation.

The apes, however, are more remarkable for talents than genius. Though they have the art of imitating human actions, they are still brutes, all of which, in various degrees, possess the talent of imitation. This talent, in most animals, is entirely limited to the actions of their own species. But the ape, although he belongs not to the human species, is capable of imitating some of our actions. This power, however, is entirely the effect of his organization. He imitates the actions of men, because his structure has a gross resemblance to the human figure. What originates solely from organization and structure, is thus ignorantly ascribed, by the vulgar, to intelligence and genius. . . .

[292] As every operation of nature is conducted by shades, or slight gradations, a scale may be formed for ascertaining the intrinsic qualities of every animal, by taking, for the first point, the material part of man, and by placing the animals successively at different distances, in proportion as they approach or recede from that point, either in external form, or internal organization. Agreeable to this scale, the monkey, the dog, the elephant, and other quadrupeds, will hold the first rank; the cetaceous animals, [293]who, like the quadrupeds, consist of flesh and blood, and are viviparous, will hold the second; the birds, the third, because they differ more from man than the quadrupeds or cetaceous animals; and, were it not for oisters and polypi, which seem to be the farthest removed from man, the insects would be thrown into the lowest rank of animated beings . . .

[294] It is impossible, therefore, that the brutes have a certain knowledge of the future from an intellectual principle similar to ours. Why, then, ascribe to them, upon such slight grounds, a quality so sublime? Why unnecessarily degrade the human species? Is it not less unreasonable to refer the cause to mechanical laws, established, like the other laws of nature, by the will of the Creator?

[206] Upon the whole, every circumstance concurs in proving, that mankind are not composed of species essentially different from each other; that, on the contrary, there was originally but one species, who, after multiplying and spreading over the whole surface of the earth, have undergone various changes by the influence of climate, food, mode of living, epidemic diseases, and the [207]mixture of dissimilar individuals; that, at first, these changes were not so conspicuous, and produced only individual varieties; that these varieties became afterwards specific, because they were rendered more general, more strongly marked, and more permanent, by the continual action of the same causes; that they are transmitted from generation to generation, as deformities or diseases pass from parents to children; and that, lastly, as they were originally produced by a train of external and accidental causes, and have only been perpetuated by time and the constant operation of these causes, it is probable that they will gradually disappear, or, at least, that they will differ from what they are at present, if the causes which produce them should cease, or if their operation should be varied by other circumstances and combinations.

[405] An individual is a solitary, a detached being, and, has nothing in common with other beings, except that it resembles, or rather differs from them. All the similar individuals which exist on the surface of the earth, are regarded as composing the species of these individuals. It is neither, however, the number nor the collection of similar individuals, but the constant succession and renovation of these individuals, which constitute the species. A being, whose duration was perpetual, would not make a species. Species, then, is an abstract and general term, the meaning of which can only be apprehended by considering Nature in the succession of time, and in the constant destruction and renovation of beings. It is by comparing present individuals with those which are past, that we [406] acquire a clear idea of species; for a comparison of the number or similarity of individuals is an accessory only, and often independent of the first: the ass resembles the horse more than the spaniel does the greyhound; and yet the latter are of the same species, because they produce fertile individuals; but as the horse and ass produce only unfertile and vitiated individuals, they are evidently of different species. . . .

We shall introduce no artificial or arbitrary divisions [in these volumes]. Every species, every succession of individuals, who reproduce and cannot mix, shall be considered and treated separately; and we shall employ no other families, genera, orders, or classes, than what are exhibited by Nature herself.

---

20. Johann Friedrich Blumenbach

Blumenbach is widely considered the father of physical anthropology. Although not the first to look at human variation, his detailed craniometric studies of human skulls from different continents and regions were the first scientific attempts to make sense out of human races. He was well positioned to do so. Born in 1752 in Gotha, Germany, between Frankfurt and Dresden, he studied medicine at the ancient University of Jena, completing his medical degree at Göttingen in 1775.

While there, he took an elective course in natural history under Christopher W. Buttner. Buttner's illustrated lectures on far-flung peoples and countries were captivating. "It was thus I was led to write as the dissertation for my doctorate, *On the natural variety of mankind;* and the further prosecution of this interesting subject laid the foundation of my anthropological collection, which has in process of time become everywhere quite famous for its completeness in its way" (Marx 1840:5).

He graduated with his doctor of medicine degree in 1775, and his thesis was published the following year. It quickly became a standard for human racial studies. He was appointed that year as professor of medicine at Göttingen, where he stayed until his death in 1840.

In 1813, he became a foreign member of the Royal Swedish Academy of Sciences. His famous treatise was translated into English and published by the Anthropological Society of London in 1864, one year after it was founded. This carries some irony, as the Anthropological Society was founded in a split from the Ethnological

Society of London (founded in 1843), not only so it could provide a focus on physical anthropology, but because its founders firmly believed in *polygenism*, the multiple origins of humans. The Ethnological Society believed humans to be a single species with a single origin, an argument strongly supported by Blumenbach's thesis.

On the Natural Varieties of Mankind

[82] Man then alone is destitute of what are called *instincts*, that is, certain congenital faculties for protecting himself from external injury, and for seeking nutritious food, &c. All his instincts are artificial *(kunst-triebe)*, and of the others there are only the smallest traces to be seen. Mankind therefore would be very wretched were it not preserved by the use of *reason*, of which other animals are plainly destitute. I am sure they are only endowed with innate or common and truly material sense (which is not wanting either to man), especially after comparing everything which I have read upon the rational mind of animals with their mode of life and actions, and what perhaps is the most important speculation, and demands most attention, with the phenomena of death, which are very much like both in animals and men. Instinct always remains the same, and is not advanced by cultivation, nor is it smaller or weaker in the young animal than in the adult. Reason, on the contrary, may be compared to a developing germ, which in the process of time, and by the accession of a social life and other external circumstances, is as it were developed, formed, and cultivated....

[84] He is born naked and weaponless, furnished with no instinct, entirely dependent on society and education. This excites the flame of reason by degrees, which at last shows itself capable of happily supplying, by itself, all the defects in which animals seem to have the advantage over men. Man brought up amongst the beasts, destitute of intercourse with man, comes out a beast. The contrary however never occurs to beasts which live with man. Neither the beavers, nor the seals, who live in company, nor the domestic animals who enjoy our familiar society, come out endowed with reason.

Contributions to Natural History, Part 1

[281–82] "Yes, that's the way of the world," says Voltaire; "we can't get any more purple, for the Murex has long since been exterminated. The poor little shell must have been eaten up by some other larger animals." "God forbid," answer the physico-theologians; "it is impossible that Providence can allow of the extinction of a species." Thus says the noble village pastor of Savoy in *Emilie*, "There is no creature in the universe that may not equally be looked upon as the common centre of all the rest." And, says another in addition, "There is no one, so to say, which is not that for all the rest of the creation, which the head of Phidias was for the shield of his artificial Minerva, which could not be removed without the whole of the great work falling to pieces."

"Rather than that," says Linnisus, "let nature create new sorts. Thus not far from Upsala, on the island Sbdra-Gaesskiaeret, a new plant has appeared, the *Peloria*, that is undoubtedly a sort of new creation." "Ah," they answer, "nature is an old hen, which will certainly lay nothing more fresh at this time of day." "Certainly not," decides Haller; "and such errors should be denounced, because they will be eagerly snapped up by the atheists, who will be only too glad to demonstrate the instability of nature as well by the appearance of new species, as by the pretended extermination of old kinds. And this must not be; for if order in the physical world comes to an end, so also will order in the moral world, and at last it is all over with all religion."

If I may presume to put in a word here myself, my opinion is that on all sides too much has been made of the matter. The murex exists up to the present day just as much as in the time of the old Phoenicians and Greeks;—the peloria is a monstrous freak of nature, and no new particular independent species. Nature is made common, but is not exactly an old hen,— and the creation is something more solid than that statue of Minerva,—and it will not go to pieces even if one species of creatures dies out, or another is newly created,—and it is more than merely probable, that both cases have happened before now,—and all this without the slightest danger to order, either in the physical or in the moral world, or for religion in general. For my own part it is exactly in these things that I find the guidance of a higher hand most unmistakable; so that in spite of this recognized instability of nature, the creation continues going on its quiet way; and on that very account it is my opinion that it is well worth the trouble, after such an immense deal has been written upon the pretended unchangeableness of the creation, just once to recollect on the other hand the proofs of the great alterations in it.

[283] Every paving-stone in Göttingen is a proof that species, or rather whole genera, of creatures must have disappeared. Our limestone swarms likewise with numerous kinds of lapidified marine creatures, among which, as far as I know, there is only one single species that so much resembles any one of the present kinds, that it may be considered as the original of it; and this is that particular kind of the Terebratulae in the Mediterranean and Atlantic waters, which from their appearance have received the name of the cock and hen. For one of the two delicate bellied shells rises behind over the other at the junction, and so when it is seen in profile it has some resemblance to a cock which is treading on a hen.

[287] After therefore that organic creation in the Preadamite primitive epoch of our planet had fulfilled its purpose, it was destroyed by a general catastrophe of its surface or shell, which probably lay in ruins some time, until it was put together again, enlivened with a fresh vegetation, and vivified with a new animal creation. In order that it might provide such a harvest, the Creator took care

to allow in general powers of nature to bring forth the new organic kingdoms, similar to those which had fulfilled that object in the primitive world. Only the formative force having to deal with materials, which must of course have been much changed by such a general revolution, was compelled to take a direction differing more or less from the old one in the production of new species.

[294] But no one knows the exact original wild condition of man. There is none, for nature has limited him in no wise, but has created him for every mode of life, for every climate, and every sort of aliment, and has set before him the whole world as his own and given him both organic kingdoms for his aliment. But the consequence of this is that there is no second animal besides him in the creation upon whose *solidum vivum* so endless a quantity of various *stimuli,* and therefore so endless a quantity of concurring causes of degeneration, must needs operate.

[303–4] Consequently I do not see the slightest shadow of reason why I, looking at the matter from a physiological and scientific point of view, should have any doubt whatever that all nations, under all known climates, belong to one and exactly the same common species.

Still, in the same way as we classify races and degenerations of horses and poultry, of pinks and tulips, so also, in addition, must we class the varieties of mankind which exist within their common original stock. Only this, that as all the differences in mankind, however surprising they may be at the first glance, seem, upon a nearer inspection, to run into one another by unnoticed passages and intermediate shades; no other very definite boundaries can be drawn between these varieties, especially if, as is but fair, respect is had not only to one or the other, but also to the peculiarities of a natural system, dependent upon all bodily indications alike. Meanwhile, so far as I have made myself acquainted with the nations of the earth, according to my opinion, they may be most naturally divided into these five principal races:

1. *The Caucasian race.* The Europeans, with the exception of the Lapps, and the rest of the true Finns, and the western Asiatics this side the Obi, the Caspian Sea, and the Ganges along with the people of North Africa. In one word, the inhabitants nearly of the world known to the ancient Greeks and Romans. They are more or less white in colour, with red cheeks, and, according to the European conception of beauty in the countenance and shape of the skull, the most handsome of men.
2. *The Mongolian.* The remaining Asiatics, except the Malays, with the Lapps in Europe, and the Esquimaux in the north of America, from Behring's Straits to Labrador and Greenland. They are for the most part of a wheaten yellow, with scanty, straight, black hair, and have flat faces with laterally projecting cheek-bones, and narrowly slit eyelids.
3. *The Ethiopian.* The rest of the Africans, more or less black, generally with curly hair, jaw-bones projecting forwards, puffy lips, and snub noses.
4. *The American.* The rest of the Americans; generally tan-coloured, or like molten copper, with long straight hair, and broad, but not withal flat face, but with strongly distinctive marks.
5. *The Malay.* The South-sea islanders, or the inhabitants of the fifth part of the world, back again to the East Indies, including the Malays, properly so called. They are generally of brownish colour (from clear mahogany to the very deepest chestnut), with thick black ringleted hair, broad nose, and large mouth.

Each of these five principal races contains besides one or more nations which are distinguished by their more or less striking structure from the rest of those of the same division.

---

21. JAMES HUTTON

James Hutton was born in Edinburgh in 1726—in the seat and at the height of the Scottish Enlightenment. At once the most radical and least reverent of Enlightenment schools, it was peopled by the likes of David Hume and Adam Smith, both of whom Hutton came to know and respect.

Known as the "father of modern geology" for identifying cyclic processes involved in changing Earth's crust and in introducing a uniformitarian view of those processes, Hutton came only late to the study of geology. At age eighteen, he became interested in medicine, took classes at the University of Edinburgh, went to Paris to study at age twenty-one, and took the doctor of medicine degree in 1749 at the University of Leyden. He lived for a while in London, but returned to Scotland in the 1750s to take over the family farms at Berwickshire. His father had died when James was young, but left substantial wealth to the family, allowing James to pursue his interests.

Those interests turned to geology by mid-decade, and he made many trips into the countryside to observe what turned out to be sedimentary strata subjected to uplift by heat and pressure from the earth's core. These unconformities led Hutton to propose a cyclic theory of uplift and erosion, contrasting this "Vulcanist" (or "Plutonist") theory with the currently popular "Neptunist" theory—going back in one form to Steno—which explained rock formation as the result of a catastrophic flood. One effect of his observations was realization of the enormous period of time it had taken for all of these processes to result in today's appearance, and the fact that the processes of the distant past were the same as those operating in his own time.

Though he composed and read (or had read for him) several papers before the Royal Society of Edinburgh in 1785, it was not until 1788 that his work was printed and available. His *Theory of the Earth* went through several drafts until 1795, when he published a two-volume version. This included reviews of the ideas presented by Thomas Burnet and Comte de Buffon.

Unfortunately, Hutton was a writer of turgid prose and had an apparent philosophy that the more words one uses the more convincing can be one's argument. It had, however, a somnolent affect, and it remained for others, principally John Playfair, to convert Hutton's ideas into a more readable form. Playfair's 1802 restatement, *Illustrations of the Huttonian Theory of the Earth*, caught the attention of Charles Lyell and resulted in a broad acceptance and acknowledgment of Hutton's important contributions. Hutton had died in 1797, five years before Playfair's popularization.

Abstract of a Dissertation read in the Royal Society of Edinburgh

The solid parts of the present land appear, in general, to have been composed of the productions of the sea, and of other materials similar to those now found upon the shores.... Hence we are led to conclude, that the greater part of our land, if not the whole, had been produced by operations natural to this globe; but that, in order to make this land a permanent body, resisting the operations of the waters, two things had been required; 1st, The consolidation of masses formed by collections of loose and incoherent materials; *2dly*, The elevation of those consolidated masses from the bottom of the sea, the place where they were collected, to the stations in which they now remain above the level of the ocean.

Theory of the Earth, Vol. I, Part II

[263] Let us now consider how far the other proposition, of strata being elevated by the power of heat above the level of the sea, may be confirmed from the examination of natural appearances.

The strata formed at the bottom of the ocean are necessarily horizontal in their position, or nearly so, and continuous in their horizontal direction or extent. They may change, and gradually assume the nature of each other, so far as concerns the materials of which they are formed; but there cannot be any sudden change, fracture or displacement naturally in the body of a stratum. But, if these strata are cemented by the heat of fusion, and erected with an expansive power acting below, we may expect to find every species of fracture, dislocation and contortion, in those bodies, and every degree of departure from a horizontal towards a vertical position.

[217] Now, if we are to take the written history of man for the rule by which we should judge of the time when the species first began, that period would be but little removed from the present state of things. The Mosaic history places this beginning of man at no great distance; and there has not been found, in natural history, any document by which a high antiquity might be attributed to the human race. But this is not the case with regard to the inferior species of animals, particularly those which inhabit the ocean and its shores. We find in natural history monuments which prove that those animals had long existed; and we thus procure a measure for the computation of a period of time extremely remote, though far from being precisely ascertained.

[304] We have now got to the end of our reasoning; we have no data further to conclude immediately from that which actually is: But we have got enough; we have the satisfaction to find, that in nature there is wisdom, system, and consistency. For having, in the natural history of this earth, seen a succession of worlds, we may from this conclude that there is a system in nature; in like manner as, from seeing revolutions of the planets, it is concluded, that there is a system by which they are intended to continue those revolutions. But if the succession of worlds is established in the system of nature, it is in vain to look for any thing higher in the origin of the earth. The result, therefore, of our present enquiry is, that we find no vestige of a beginning,—no prospect of an end.

Ideas to Think About

1. Compare the concept of a "chain of being" as it is expressed by Bonnet, Linnaeus, and Buffon. How is the relationship among forms, or "links," in this chain differently conceptualized by the three, and why?
2. Compare the ideas of the development of the earth as expressed by Linnaeus, Buffon, and Hutton. Only in the latter do we see recognition of a uniform and continuous process. Why?
3. Among these authors, where do we find the acceptance of extinction, and where is it denied? What might explain the differences?
4. As an explanation for Earth's history, what is the likely source for the Vulcanist idea of Hutton, in opposition to the common Neptunist ideas of Steno and others?
5. Compare Blumenbach, Hutton, and Buffon in contemplating man's place in the natural order.
6. Which of the authors espouses a theory of catastrophism in Earth's history, and why?

For Further Reading

Linnaeus: The Compleat Naturalist, Wilfrid Blunt. Princeton, NJ: Princeton University Press (2002). This is a reissued classic and highly rated for its compelling style and personal anecdotes about the man who gave everything two names and changed the natural world. Short (288 pages), this is a gem of a book. In Sweden, children know him as the "Prince of Flowers." In his obsession with plant reproductive variations, he is also known as the first "botanical pornographer"!

The Man Who found Time: James Hutton and the Discovery of the Earth's Antiquity, Jack Repcheck. New York: Basic Books (2009). This is an engrossing and well-written account of the keen mind but acerbic personality of this important geologist.

Women and Enlightenment in Eighteenth Century Britain, Karen O'Brien. Cambridge: Cambridge University Press (2009). A major change in the eighteenth-century view of the social order was new attention to the place and role of women. We see not only the rise of women authors, but also the treatment of women in the important writings of Malthus, Locke, Hume, and Adam Smith. This is a well-written treatment of the Enlightenment influence on women writers and women as subject matter.

A World on Fire: A Heretic, an Aristocrat, and the Race to Discover Oxygen, Joe Jackson. New York: Viking (2005). We did not discuss either Joseph Priestly or Antoine Lavoisier in our treatment of the Enlightenment, but this engaging book is a relevant insight into the great political upheavals in France and England during the Enlightenment. It gives a great characterization of the heretic, Priestly, who opposed kingly and ecclesiastical power and was forced to flee to America, and the Frenchman, Lavoisier, who was guillotined for being a scientist during the Terror.

UNIT 5

Eighteenth-Century Enlightenment: Evolution and Progress

> If we take in our hand any volume; of divinity or school metaphysics, for instance; let us ask, Does it contain any abstract reasoning concerning quantity or number? No. Does it contain any experimental reasoning, concerning matter of fact and existence? No. Commit it then to the flames: for it can contain nothing but sophistry and illusion.
>
> —DAVID HUME, *Enquiry Concerning Human Understanding*

The Enlightenment sought to expand the confined history of the past into the broadening horizon of immense time. In so doing, it found in the cosmos as well as on the earth itself a systematic and sometimes cyclic process of change. In that change, now circumscribed by natural law, there was a new concept of progress: A human spirit previously doomed by original sin to a repentant life of pious hope and unquestioning faith was now becoming liberated by free will to the pursuit of improvement.

Time, change, and natural law constitute evolution's three-legged stool. The eighteenth century witnessed the initial attempts to construct that stool. All three of these, we must be quick to recognize, were still denied by church doctrine: The earth was still considered quite young, life forms were fixed in the ranked separate creations of the great chain of being, and God—not nature—was still considered the prime mover. These medieval concepts remained strong in numerous theological circles, became allied with important political movements, and thus influenced the conduct of science in the eighteenth century. Natural theology was one of these concepts, not new but newly revived.

Yet, under the Roman Catholic Church, religious doctrine had always been rather clear-cut and not at all uniform in opposition to scientific thought. However, throughout the seventeenth century, schisms within the church, fostered in the previous century by the Reformation, led to vying factions, religious wars (the Thirty Years' War, the English Civil War, and the Glorious Revolution), and opposing church doctrines. At any given time, one could be found on the dangerous side of the political and religious issue. Scientific discovery was not shackled, but often scientific communication became hostage to political pragmatism. The religious component of Enlightenment resistance, then, as Susan Neiman tells us (Neiman 2009, 216–7) was not a movement against religion, but against religious authority "in its day the basis of all political authority, bolstered with webs of superstition and absurdity waiting to be swept away."

What we find in eighteenth-century society—on the Continent, in England, and, of course, in the American Colonies—is a new interest in the industry of applied science combined with two new ideologies: (1) natural and social progress, and (2) personal and social freedom. The two are not unrelated, and we find in both the strong current of interest in morality and justice that dominated philosophical discourse. For many, progress in technology embedded moral progress. Kant claimed that morality emerged intuitively from rationality.

The robust optimism in applied science, derived from the scientific demonstrations of the seventeenth century, led to the Industrial Revolution of the eighteenth century. The equally strong optimism in scientific materialism and its rational basis led to the eighteenth century's label—the Age of Enlightenment. And the push for liberty led to the American Revolution. We find in this century an unabashed love affair with the human spirit and its unshackled reach for a better future—largely through science.

I say "largely," but perhaps I should not have. The profound and disturbing ideas of this age spawned creative energies in prose and poetry. Their authors—through their works—provided still another search for a better future, as well as a thoughtful and sometimes troublesome marriage of social conscience and scientific discovery. Though not included in the selections for this unit and the previous, the further readings appended here, and in Unit 6, include some of this literary harvest.

These, then, are the social, political, and religious contexts in which scientific achievements in the eighteenth century must be measured. These are important in their own light, as well as forerunners for Darwin's ideas.

22. Erasmus Darwin

It was midcentury when Erasmus Darwin—fresh with his doctor of medicine degree—settled in the Midlands on the Birmingham Plateau and became friend and colleague with a curious group of intellectuals who were destined to be creators of the Industrial Revolution. The Lunar Society included men such as Matthew Boulton, Josiah Wedgwood, Joseph Priestly, James Watt, and others. Many of them trained, as did Darwin, at the University of Edinburgh, and others at the University of Glasgow. In these two of the famed four far northern universities, intellectual independence and the new rationalism attracted freethinkers not welcome at either Oxford or Cambridge. Many were deists, if they espoused any religion at all. A compelling account of this society and its participants may be found in *The Lunar Men* by Jenny Uglo (2002).

Born in 1731, in Nottinghamshire, England, Erasmus was the youngest of seven children. He and his nearest brother, John, attended St. John's College in Cambridge together, and both eventually became successful physicians. For his medical education, Edinburgh offered better studies, and in 1753, Erasmus travelled there, accompanied by his eldest brother, Robert, a successful botanist.

A large, rotund man with florid complexion and a persistent stammer, his good humor and vociferous presence filled every room he entered. Erasmus was gregarious and virile in his approach to life, and he embodied Enlightenment strides toward the liberation of sex and women. He wrote a short book on the education of women and once claimed that sex was a good treatment for hypochondria. Practicing what he preached, Erasmus sired, by two wives, twelve children (his fourth, Robert, was Charles Darwin's father) and at least two illegitimate children. He died suddenly—and presumably happy—on April 18, 1802.

Like Buffon, whose works Erasmus read and whose ideas he supported, Erasmus was interested in the breadth of humanities and sciences, and was adept at writing—particularly poetry. His long poem, *The Botanic Garden* (1789) is famous in part for its suggestion that biological forms have evolved over time. This idea becomes more explicit in his *Zoonomia; or, The Laws of Organic Life* (1794–96). In his posthumously printed *The Temple of Nature* (1803), Darwin expresses the evolutionary idea in the lilting verse so typical of the romantic period to follow:

> Organic Life beneath the shoreless waves
> Was born and nurs'd in Ocean's pearly caves;
> First forms minute, unseen by spheric glass,
> Move on the mud, or pierce the watery mass;
> These, as successive generations bloom,
> New powers acquire, and larger limbs assume;
> Whence countless groups of vegetation spring,
> And breathing realms of fin, and feet, and wing.

Zoonomia, Vol. 1

*Charles Darwin never knew his grandfather, Erasmus, and despite his curious attempt at a biography (*The Life of Erasmus Darwin, *published in 1879 and reissued by Cambridge University Press in a larger edition restoring original edits in 2002), Charles seems to have been influenced little by the elder Darwin's evolutionary ideas. "I had previously read the* Zoönomia *of my grandfather" wrote Charles, "in which similar views [to those of Lamarck] are maintained, but without producing any effect on me. Nevertheless it is probable that the hearing rather early in life such views maintained and praised may have favoured my upholding them under a different form in my* Origin of Species. *At this time I admired greatly the* Zoönomia*; but on reading it a second time after an interval of ten or fifteen years, I was much disappointed, the proportion of speculation being so large to the facts given." Darwin did admire Erasmus's advocacy of experiment, and wrote in* The Life of Erasmus Darwin*: "His remarks... on the value of experiments and the use of hypotheses show that he had the true spirit of the philosopher." Although the prodigious work of Buffon was read and praised by the elder Darwin, Charles admitted that he had not read him.*

6. From this account of reproduction it appears, that all animals have a similar origin, viz. from a single living filament; and that the difference of their forms and qualities has arisen only from the different irritabilities and sensibilities, or voluntarities, or associabilities, of this original living filament; and perhaps in some degree from the different forms of the particles of the fluids, by which it has been first stimulated into activity. And that from hence, as Linnæus has conjectured in respect to the vegetable world, it is not impossible, but the great variety of species of animals, which now tenant the earth, may have had their origin from the mixture of a few natural orders. And that those animal and vegetable mules, which could continue their species, have done so, and constitute the numerous families of animals and vegetables which now exist; and that

those mules, which were produced with imperfect organs of generation, perished without reproduction, according to the observation of Aristotle; and are the animals, which we now call mules....

In the following passage, Darwin's "single living filament" is equivalent, we may assume, to today's "cell." Although Robert Hooke first identified the cell by name in 1665, the cell theory—as the basic building block of life—would not be articulated until 1839 by Schwann and Schleiden. The term "appetency" refers to an instinctive inclination or tendency—a concept with teleological implications as Darwin applies it. The term also refers to a chemical affinity, hence his analogy. Azote *is a French term for nitrogen.*

7. All animals therefore, I contend, have a similar cause of their organization, originating from a single living filament, endued indeed with different kinds of irritabilities and sensibilities, or of animal appetencies; which exist in every gland, and in every moving organ of the body, and are as essential to living organization as chemical affinities are to certain combinations of inanimate matter.

If I might be indulged to make a simile in a philosophical work, I should say, that the animal appetencies, are not only perhaps less numerous originally than the chemical affinities; but that like these latter, they change with every new combination; thus vital air and azote, when combined, produce nitrous acid; which now acquires the property of dissolving silver; so with every new additional part to the embryon, as of the throat or lungs, I suppose a new animal appetency to be produced.

In this early formation of the embryon from the irritabilities, sensibilities, and associabilities, and consequent appetencies, the faculty of volition can scarcely be supposed to have had its birth. For about what can the fetus deliberate when it has no choice of objects? But in the more advanced state of the fetus, it evidently possesses volition; as it frequently changes its attitude, though it seems to sleep the greatest part of its time; and afterwards the power of volition contributes to change or alter many parts of the body during its growth to manhood, by our early modes of exertion in the various departments of life. All these faculties then constitute the vis fabricatrix, and the vis conservatrix, as well as the vis medicatrix of nature, so much spoken of, but so little understood by philosophers.

The following section contains the notion of the inheritance of acquired characters, a prelude to Lamarck but common to intellectual thought since the ancient Greeks. The short paragraph beginning "When we consider all these changes" is an explicit refutation of the preformation theory supported by many naturalists as the best explanation for embryonic development. De Maupertuis likewise opposed preformation.

8. When we revolve in our minds, first, the great changes, which we see naturally produced in animals after their nativity, as in the production of the butterfly with painted wings from the crawling caterpillar; or of the respiring frog from the subnatant tadpole; from the feminine boy to the bearded man, and from the infant girl to the lactescent woman; both which changes may be prevented by certain mutilations of the glands necessary to reproduction.

Secondly, when we think over the great changes introduced into various animals by artificial or accidental cultivation, as in horses, which we have exercised for the different purposes of strength or swiftness, in carrying burthens or in running races; or in dogs, which have been cultivated for strength and courage, as the bulldog; or for acuteness of his sense of smell, as the hound and spaniel;... Add to these the great changes of shape and colour, which we daily see produced in similar animals from our domestication of them, as rabbits, or pigeons; or from the difference of climates and even of seasons; thus the sheep of warm climates are covered with hair instead of wool; and the hares and partridges of the latitudes, which are long buried in snow, become white during the winter months; add to these the various changes produced in the forms of mankind, by their early modes of exertion; or by the diseases occasioned by their habits of life; both of which became hereditary, and that through many generations. Those who labour at the anvil, the oar, or the loom, as well as those who carry sedan-chairs, or who have been educated to dance upon the rope, are distinguishable by the shape of their limbs; and the diseases occasioned by intoxication deform the countenance with leprous eruptions, or the body with tumid viscera, or the joints with knots and distortions.

Thirdly, when we enumerate the great changes produced in the species of animals before their nativity; these are such as resemble the form or colour of their parents, which have been altered by the cultivation or accidents above related, and are thus continued to their posterity. Or they are changes produced by the mixture of species as in mules; or changes produced probably by the exuberance of nourishment suppled to the fetus, as in monstrous births with additional limbs; many of these enormities of shape are propagated, and continued as a variety at least, if not as a new species of animal....

When we consider all these changes of animal form, and innumerable others, which may be collected from the books of natural history; we cannot but be convinced, that the fetus or embryon is formed by apposition of new parts, and not by the differentiation of a primordial nest of germes, included one within another, like the cups of a conjurer.

Fourthly, when we revolve in our minds the great similarity of structure which obtains in all the warm blooded animals, as well quadrupeds, birds, and amphibious animals, as in mankind; from the mouse and bat to the elephant and whale; one is led to conclude, that they have alike been produced from a similar living filament. In some this filament in its advance to maturity has acquired hands and fingers, with a fine sense of touch, as in mankind.

Fifthly, from their first rudiment, or primordium, to the termination of their lives, all animals undergo perpetual transformations; which are in part produced by their own exertions in consequence of their desires and aversions, of their pleasures and their pains, or of irritations, or of associations; and many of these acquired forms or propensities are transmitted to their posterity.

The selections that follow—on characteristics reflecting sexual competition, seeding strategies, and protection—are tantamount to the principle of natural selection that Erasmus's grandson devised.

As air and water are supplied to animals in sufficient profusion, the three great objects of desire, which have changed the forms of many animals by their exertions to gratify them, are those of lust, hunger, and security. A great want of one part of the animal world has consisted in the desire of the exclusive possession of the females; and these have acquired weapons to combat each other for this purpose, as the very thick, shield-like, horny skin on the shoulder of the boar is a defence only against animals of his own species, who strike obliquely upwards, nor are his tushes for other purposes, except to defend himself as he is not naturally a carnivorous animal. So the horns of the stag are sharp to offend his adversary, but are branched for the purpose of parrying or receiving the thrusts of horns similar to his own, and have therefore been formed for the purpose of combatting other stags for the exclusive possession of the females;...

The birds, which do not carry food to their young, and do not therefore marry, are armed with spurs for the purpose of fighting for the exclusive possession of the females, as cocks and quails. It is certain that these weapons are not provided for their defence against other adversaries, because the females of these species are without this armour. The final cause of this contest amongst the males seems to be, that the strongest and most active animal should propagate the species, which should thence become improved.

Another great want consists in the means of procuring food, which has diversified the forms of all species of animals.... Some birds have acquired harder beaks to crack nuts, as the parrot. Others have acquired beaks adapted to break the harder feeds, as sparrows. Others for the softer seeds of flowers, or the buds of trees, as the finches. Other birds have acquired long beaks to penetrate the moister soils in search of insects or roots, as woodcocks; and others broad ones to filtrate the water of lakes, and to retain aquatic insects, as ducks. All which seem to have been gradually produced during many generations by the perpetual endeavor of the creatures to supply the want of food, and to have been delivered to their posterity with constant improvement of them for the purposes required.

The third great want amongst animals is that of security, which seems much to have diversified the forms of their bodies and colour of them; these consist in the means of escaping other animals more powerful than themselves. Hence some animals have acquired wings instead of legs, as the smaller birds, for the purpose of escape. Others great length of fin, or of membrane, as the flying fish, and the bat. Other great swiftness of foot, as the hare. Others have acquired hard or armed shells, as the tortoise and the echinus marinus.

In these final selections, Darwin concludes that the earth is "millions of ages old," and that not only have all animals arisen from an original form, but that plants and animals together share a common ancestral "filament." Extinction, he reasons, was likewise part of the long process. In his final reference to Hume, Erasmus concludes that the world itself was "generated, rather than created." There is no evidence that Erasmus ever met Hume, but he knew and respected Hume's famous colleague, Rousseau.

From thus meditating on the great similarity of the structure of the warm-blooded animals, and at the same time of the great changes they undergo both before and after their nativity; and by considering in how minute a portion of time many of the changes of animals above described have been produced; would it be too bold to imagine, that in the great length of time, since the earth began to exist, perhaps millions of ages before the commencement of the history of mankind, would it be too bold to imagine, that all warm-blooded animals have arisen from one living filament, which THE GREAT FIRST CAUSE endued with animality, with the power of acquiring new parts attended with new propensities, directed by irritations, sensations, volitions, and associations; and thus possessing the faculty of continuing to improve by its own inherent activity, and of delivering down those improvements by generation to its posterity, world without end?...

Shall we then say that the vegetable living filament was originally different from that of each tribe of animals above described? And that the productive living filament of each of those tribes was different originally from the other? Or, as the earth and ocean were probably peopled with vegetable productions long before the existence of animals; and many families of these animals long before other families of them, shall we conjecture that one and the same kind of living filament is and has been the cause of all organic life?

If this gradual production of the species and genera of animals be assented to, a contrary circumstance may be supposed to have occurred, namely, that some kinds by the great changes of the elements may have been destroyed. This idea is shewn to our senses by contemplating the petrifactions of shells, and of vegetables, which may be said, like busts and medals, to record the history of remote times. Of the myriads of belemnites, cornua ammonis, and numerous other petrified shells, which are found in the masses of lime-stone, which have been produced by them, none now are ever found in our seas, or in the seas of other parts of the world, according to the observations of many naturalists. Some of whom have imagined, that most of the inhabitants of the sea and earth of very remote times are now extinct;

as they scarcely admit, that a single fossil shell bears a strict similitude to any recent ones, and that the vegetable impressions or petrifactions found in iron-ores, clay, or sandstone, of which there are many of the fern kind, are not similar to any plants of this country, nor accurately correspond with those of other climates, which is an argument countenancing the changes in the forms, both of animals and vegetables, during the progressive structure of the globe, which we inhabit.

The late Mr. David Hume, in his posthumous works, places the powers of generation much above those of our boasted reason; and adds, that reason can only make a machine, as a clock or a ship, but the power of generation makes the maker of the machine; and probably from having observed, that the greatest part of the earth has been formed out of organic recrements; as the immense beds of limestone, chalk, marble, from the shells of fish; and the extensive provinces of clay, sandstone, ironstone, coals, from decomposed vegetables; all which have been first produced by generation, or by the secretions of organic life; he concludes that the world itself might have been generated, rather than created; that is, it might have been gradually produced from very small beginnings, increasing by the activity of its inherent principles, rather than by a sudden evolution of the whole by Almighty fiat.—What a magnificent idea of the infinite power of THE GREAT ARCHITECT! THE CAUSE OF CAUSES! PARENT OF PARENTS! ENS ENTIUM!

23. Jean-Baptiste Lamarck

Lamarck became a celebrated biologist—both in botany and in invertebrate zoology—after an initial career as a soldier, following his father and brothers in a long family tradition. Born with the aristocratic name of Jean-Baptiste Pierre Antoine de Monet, Chevalier de Lamarck in 1744, he served in the French Army from 1761 until 1768, when an injury forced him to leave the military. In Paris, he began to study medicine and botany.

His expertise developed rapidly, and, in 1778, he published a widely reputed book on French plants, *Flore Française*. Through the patronage and friendship of Comte de Buffon, Lamarck became assistant botanist at the royal botanical garden and was appointed chair of botany in 1788. In 1793, with the founding of the Muséum Nationale d'Histoire Naturelle, he was appointed professor of zoology.

Though Lamarck's early beliefs followed the essentialist ideas of his time, his study of invertebrates slowly convinced him that species were not fixed and that the environment strongly influenced an animal's organs and behavior. A lecture in 1800 at the museum was the first that revealed his radical ideas.

Zoological Philosophy

[9] To observe nature, to study her productions in their general and special relationships, and finally to endeavour to grasp the order which she everywhere introduces, as well as her progress, her laws, and the infinitely varied means which she uses to give effect to that order: these are in my opinion the methods of acquiring the only positive knowledge that is open to us,—the only knowledge moreover which can be really useful to us. It is at the same time a means to the most delightful pleasures, and eminently suitable to indemnify us for the inevitable pains of life.

And in the observation of nature what can be more interesting than the study of animals? There is the question of the affinities of their organisation with that of man, there is the question of the power possessed by their habits, modes of life, climates and places of habitation, to modify their organs, functions and characters. There is the examination of the different systems of organisation which are to be observed among them, and which guide us in the determination of the greater or lesser relationships that fix the place of each in the scheme of nature. There is finally the general classification that we make of these animals from considerations of the greater or lesser complexity of their organisation; and this classification may even lead us to a knowledge of the order followed by nature in bringing the various species into existence.

[10–11] After the organization of man had been so well studied, as was the case, it was a mistake to examine that organization for the purposes of an enquiry into the causes of life, of physical and moral sensitiveness, and, in short, of the lofty functions which he possesses. It was first necessary to try to acquire knowledge of the organisation of the other animals. It was necessary to consider the differences which exist among them in this respect, as well as the relationships which are found between their special functions and the organisation with which they are endowed.

[11] These different objects should have been compared with one another and with what is known of man. An examination should have been made of the progression which is disclosed in the complexity of organisation from the simplest animal up to man, where it is the most complex and perfect. The progression should also have been noted in the successive acquisition of the different special organs, and consequently of as many new functions as of new organs obtained. It might then have been perceived how *needs,* at first absent and afterwards gradually increasing in number, have brought about an inclination towards the actions appropriate to their satisfaction; how actions becoming habitual and energetic have occasioned the development of the organs which execute them; how the force which stimulates organic movements can in the most imperfect animals exist outside of them and yet animate them; how that force has been subsequently transported and fixed in the animal itself; and, finally, how it has become the source of sensibility, and last of all of acts of intelligence.

[12] The belief has long been held that there exists a sort of scale or graduated chain among living bodies. Bonnet has developed this view; but he did not prove it by facts derived from their organisation; yet this was necessary especially with regard to animals. He was unable to prove it, since at the time when he lived the means did not exist.

[35] It is not a futile purpose to decide definitely what we mean by the so-called *species* among living bodies, and to enquire if it is true that species are of absolute constancy, as old as nature, and have all existed from the beginning just as we see them to-day; or if, as a result of changes in their environment, albeit extremely slow, they have not in course of time changed their characters and shape....

I shall show in one of the following chapters that every species has derived from the action of the environment in which it has long been placed the *habits* which we find in it. These habits have themselves influenced the parts of every individual in the species, to the extent of modifying those parts and bringing them into relation with the acquired habits.

[36]... [S]pecies have really only a constancy relative to the duration of the conditions in which are placed the individuals composing it; nor that some of these individuals have varied, and constitute races which shade gradually into some other neighbouring species. Hence, naturalists come to arbitrary decisions about individuals observed in various countries and diverse conditions, sometimes calling them varieties and sometimes species. The work connected with the determination of species therefore becomes daily more defective, that is to say, more complicated and confused....

[T]he individuals of the species perpetuate themselves without variation only so long as the conditions of their existence do not vary in essential particulars. Since existing prejudices harmonise well with these successive regenerations of like individuals, it has been imagined that every species is invariable and as old as nature, and that it was specially created by the Supreme Author of all existing things.

[37] I do not mean that existing animals form a very simple series, regularly graded throughout; but I do mean that they form a branching series, irregularly graded and free from discontinuity, or at least once free from it. For it is alleged that there is now occasional discontinuity, owing to some species having been lost. It follows that the species terminating each branch of the general series are connected on one side at least with other neighbouring species which merge into them....

[78–79; italics in original] After having produced aquatic animals of all ranks and having caused extensive variations in them by the different environments provided by the waters, nature led them little by little to the habit of living in the air, first by the water's edge and afterwards on all the dry parts of the globe. These animals have in course of time been profoundly altered by such novel conditions; which so greatly influenced their habits and organs that the regular gradation which they should have exhibited in complexity of organisation is often scarcely recognisable.

These results which I have long studied, and shall definitely prove, lead me to state the following zoological principle, the truth of which appears to me beyond question.

Progress in complexity of organisation exhibits anomalies here and there in the general series of animals, due to the influence of environment and of acquired habits.

24. William Paley

Much as Lamarck closed out the eighteenth century and entered the nineteenth with an immensely effective and logical explanation of nature's complexity without requiring a creator, so William Paley spanned this transition with a passionate and popular explanation that demanded one. One of Christianity's most eloquent and devout messengers, Paley continued an intellectual tradition in theology going back beyond John Ray, insisting that observant eyes and open minds could not avoid seeing divine purpose in nature's intricate designs and could only explain those designs as the product of an omnipotent designer.

Paley was born in Peterborough (Cambridgeshire)—seventy-eight miles due north of London—in 1743. He studied for the Anglican priesthood at Christ's College, Cambridge, graduating in 1763. He became a fellow in 1766 and tutor in 1768. He closely followed John Locke in teaching moral philosophy, and although Locke fundamentally agreed that nature followed intricate design, he was troubled by "monstrosities" that seemed to depart from this. Paley also corresponded with David Hume—the quintessential atheist of the Scottish Enlightenment, who argued against design by using the same monstrosities. The two agreed, however, on the issue of an inherent human moral sense (they rejected it).

Paley rose steadily through the Anglican ranks: He was made vicar of Dalston in 1780 and archdeacon of Carlisle in 1782. In 1785, he published *The Principles of Moral and Political Philosophy*, which the following year was incorporated into the examinations at Cambridge. In 1794, his *View of the Evidences of Christianity* was also added and remained there until the 1920s. *Natural Theology* was published in 1802. Charles Darwin, at Cambridge, read all three, stating that the logic of *Principles* and *Natural Theology* "gave me as much delight as did Euclid."

Natural Theology

[1–2] In crossing a heath, suppose I pitched my foot against a *stone*, and were asked how the stone came to be there; I might possibly answer, that, for any thing I knew to the contrary, it had lain there for ever: nor would it perhaps be very easy to show the absurdity of this answer. But suppose I had found a *watch* upon the ground, and it should be inquired how the watch happened to be in that place; I should hardly think of the answer which I had before given, that, for any thing I knew, the watch might have always been there. Yet why should not this answer serve for the watch as well as for the stone? why is it not as admissible in the second case, as in the first? For this reason, and for no other, viz. that, when we come to inspect the watch, we perceive (what we could not discover in the stone) that its several parts are framed and put together for a purpose, *e. g.* that they are so formed and adjusted as to produce motion, and that motion so regulated as to point out the hour of the day; that, if the different parts had been differently shaped from what they are, of a different size from what they are, or placed after any other manner, or in any other order, than that in which they are placed, either no motion at all would have been carried on in the machine, or none which would have answered the use that is now served by it....

[3] This mechanism being observed (it requires indeed an examination of the instrument, and perhaps some previous knowledge of the subject, to perceive and understand it; but being once, as we have said, observed and understood), the inference, we think, is inevitable, that the watch must have had a maker: that there must have existed, at some time, and at some place or other, an artificer or artificers who formed it for the purpose which we find it actually to answer; who comprehended its construction, and designed its use.

[55] Every observation which was made in our first chapter, concerning the watch, may be repeated with strict propriety, concerning the eye; concerning animals; concerning plants; concerning, indeed, all the organized parts of the works of nature.

25. Thomas Malthus

Thomas Malthus was born in Westcott, County Surrey, bordering Greater London to the north and Kent on the east, in 1766. One of eight children born to a family of independent means, Thomas entered Jesus College, Cambridge, in 1784. After graduating with honors in mathematics, he studied for the master of arts and received it in 1791. He became a fellow there in 1793.

He took orders in the Anglican Church in 1797, becoming a curate the following year. It was in this year, 1798, that he published his influential *Essay on the Principle of Population*—a book that saw six revised editions between then and 1826. It all began as a short response to both his father's views and those of Rousseau, who was a friend of his father, that man's social and moral future would be progressively improved and result in greater happiness. Similar optimism had been expressed by William Godwin, the popular novelist and political economist of the late eighteenth century, and the Marquis de Condorcet, the French political scientist with liberal views on women's rights, economics, and education.

In responding to these thinkers, Malthus instead was struck by the ever-present and growing poverty he saw in society, a poverty that was the direct result of social progress. The paper expanded into a book. In it, Malthus argued that population expands more rapidly than the food supply, leading to overpopulation, misery, and despair. The two checks on population—positive checks such as famine and disease that increase the death rate, and preventative checks such as abortion and birth control that decrease the birth rate—are not equally effective, and misery trumps common sense. Only in later editions, and in response to criticisms, did Malthus become more optimistic in stressing the possibility of moral restraint in reproduction.

Malthus's suggestion that the experience of animal husbandry might be successfully applied to human reproduction anticipated the eugenics of Francis Galton in the following century—although the idea was first suggested by Plato centuries before.

Malthus's influence spread across the social as well as the biological sciences. Modern economic and social theory derives in part from debates stimulated by his writing, and both Charles Darwin and Alfred Russell Wallace acknowledged his influence on ideas of natural selection.

An Essay on the Principle of Population

[13–14] I say, that the power of population is indefinitely greater than the power in the earth to produce subsistence for man.

Population, when unchecked, increases in a geometrical ratio. Subsistence increases only in an arithmetical ratio. A slight acquaintance with numbers will shew the immensity of the first power in comparison of the second.

By that law of our nature which makes food necessary to the life of man, the effects of these two unequal powers must be kept equal.

This implies a strong and constantly operating check on population from the difficulty of subsistence. This difficulty must fall somewhere and must necessarily be severely felt by a large portion of mankind.

[16] This natural inequality of the two powers of population and of production in the earth, and that great law of our nature which must constantly keep their effects equal, form the great difficulty that to me appears insurmountable in the way to the perfectibility of society. All other arguments are of slight and subordinate consideration in comparison of this. I see no way by which man can escape from the weight of this law which pervades all animated nature.

[99–100] Notwithstanding, then, the institution of the poor laws in England, I think it will be allowed that considering the state of the lower classes altogether, both in the towns and in the country, the distresses which they suffer from the want of proper and sufficient food, from hard labour and unwholesome habitations, must operate as a constant check to incipient population.

To these two great checks to population, in all long occupied countries, which I have called the preventive and the positive checks, may be added vicious customs with respect to women, great cities, unwholesome manufactures, luxury, pestilence, and war.

All these checks may be fairly resolved into misery and vice.

And that these are the true causes of the slow increase of population in all the states of modern Europe, will appear sufficiently evident from the comparatively rapid increase that has invariably taken place whenever these causes have been in any considerable degree removed.

---

26. Adam Smith

Born in Kirkcaldy, Scotland, probably in 1723, Adam Smith could look south across the Firth of Forth to Edinburgh, one of the two centers of the Scottish Enlightenment. Smith chose to study at the other, entering the University of Glasgow at age fourteen. He excelled there, and after three years was awarded a scholarship to study at Balliol College, Oxford. This was not a happy experience, and he returned to Glasgow before his scholarship expired.

At Glasgow, he studied moral philosophy and became chair of that department in 1751. In that year, he also met David Hume, who had been teaching at the University of Edinburgh, and they became lifelong friends and intellectual colleagues. Smith was also greatly influenced by other free thinkers and progressive intellectuals at Glasgow. Among the several student societies at the university was the Anderson Club, whose members included Smith, professor of medicine Joseph Black, physicist and inventor John Robison, James Watt, and the club's namesake, John Anderson, professor of natural philosophy (Uglo 2002, 33f.).

Smith published *The Theory of Moral Sentiments* in 1759, a work so popular that he attracted students from other schools. He subsequently shifted his interest to political economy; left the university in 1763; and, in three years of travel through Europe, he came to know Voltaire, Benjamin Franklin, Turgot, d'Alembert, and others. From these influences, he developed his ideas concerning labor as capital and the source of a country's wealth. At the beginning of the American Revolution—in 1776—he published his famous *Wealth of Nations*, a resolute argument that unrestrained competition, self-interest, and a free market was the best assurance for a just achievement of social and personal success.

He thus argued against control of wealth by a government run by the upper classes. In this, he followed the arguments of Rousseau, Hume, and the American founders. In the following passage, Smith echoes the findings of Malthus and illustrates how intellectual sentiment at the close of the century was to influence natural philosophy and the ideas of Charles Darwin.

An Inquiry into the Nature and Causes of the Wealth of Nations, Bk. 1, Ch. 8

It deserves to be remarked, perhaps, that it is in the progressive state, while the society is advancing to the further acquisition, rather than when it has acquired its full complement of riches, that the condition of the labouring poor, of the great body of the people, seems to be the happiest and the most comfortable. It is hard in the stationary, and miserable in the declining state. The progressive state is, in reality, the cheerful and the hearty state to all the different orders of the society; the stationary is dull; the declining melancholy.

Poverty, though it no doubt discourages, does not always prevent marriage. It seems even to be favourable to generation. A half-starved Highland woman frequently bears more than twenty children, while a pampered fine lady is often incapable of bearing any, and is generally exhausted by two or three. Barrenness, so frequent among women of fashion, is very rare among those of inferior stock. Luxury in the fair sex, while it inflames perhaps the passion for enjoyment, seems always to weaken, and frequently to destroy altogether, the powers of generation....

In some places one half the children born die before they are four years of age; in many places before they are seven; and in almost all places before they are nine or ten. This great mortality, however, will everywhere be found chiefly among the children of the common people, who cannot afford to tend them with the same care as those of better station.....

Every species of animals naturally multiplies in proportion to the means of their subsistence, and no species can ever multiply beyond it. But in civilised society it is only among the inferior ranks of people that the scantiness of subsistence can set limits to the further multiplication of the human species; and it can do so in no other way than by destroying a great part of the children which their fruitful marriages produce.

The liberal reward of labour, by enabling them to provide better for their children, and consequently to bring up a greater number, naturally tends to widen and extend those limits. It deserves to be remarked, too, that it necessarily does this as nearly as possible in the proportion which the demand for labour requires. If this demand is continually increasing, the reward of labour must necessarily encourage in such a manner the marriage and multiplication of labourers, as may enable them to supply that continually increasing demand by a continually increasing population. If the reward should at any time be less than what was requisite for this purpose, the deficiency of hands would soon raise it; and if it should at any time be more, their excessive multiplication would soon lower it to this necessary rate. The market would be so much understocked with labour in the one case, and so much overstocked in the other, as would soon force back its price to that proper rate which the circumstances of the society required. It is in this manner that the demand for men, like that for any other commodity, necessarily regulates the production of men, quickens it when it goes on too slowly, and stops it when it advances too fast.

Ideas to Think About

1. It is interesting to critically compare two potentially conflicting ideas that become prominent during the Enlightenment and persist in the nineteenth and twentieth centuries: the idea of common origin in a linear sequence of progress, and the idea of discontinuous variation radiating outward from a species. Examine these ideas as they emerge in authors from Unit 4.
2. The same ideas are seen in the writings of Erasmus Darwin and Lamarck. How do they differ and how are they similar?
3. Erasmus Darwin describes two quite distinct processes that account for new species: One is the idea of the common "living filament," and the other is hybrids of interspecies crosses. Can you find historical precedent for each of these ideas?
4. Describe the similarities and differences in Erasmus Darwin's and Lamarck's notions of the function of the environment and of acquired characteristics in species transformation.
5. What is Erasmus's rationale for arguing against a created world?
6. What arguments does Lamarck use to place humankind within the natural order and as part of the same natural forces for change? Do you see any precedent for this in earlier writers we have discussed?
7. Both Erasmus and Paley make use of analogy: Erasmus, in arguing for evolutionary change in species (his paragraph 7), and Paley, in arguing for intelligent design. Compare these two uses and critically analyze their internal coherence and logic.
8. Both Malthus and Adam Smith agree on the relationship between food supply and population growth and likewise agree on the nature of poverty and mortality. Do they agree on the nature of the solution?

For Further Reading

Erasmus Darwin: A Life of Unequalled Achievement, Desmond King-Hele. London: Giles de la Mare (1999). King-Hele brings to vibrant life this larger-than-life hero of the industrial and scientific revolution. This is a totally engaging and unpretentious biography of a blustery and engaging, but also unpretentious physician, engineer, poet, inventor, and scientist. King-Hele deftly surrounds Erasmus with famous and not-so-famous friends who influenced him and whom he influenced, including Ben Franklin and Samuel Taylor Coleridge.

Being Shelley, Ann Wroe. London: Vintage Books (2008). As social introspection necessarily followed the industrial revolution, so personal introspection into life's meaning followed the intellectual agnosticism of the accompanying science. Percy Bysshe Shelley epitomizes this soulful and lonely effort. Ann Wroe's tender, caressing, and lyrical account of a poet "into which earthly life keeps intruding" matches the sometimes haunting and often cryptic lyricism of Shelley himself.

A Newton Among Poets: Shelley's Use of Science in Prometheus Unbound, Carl Grabo. Chapel Hill: The University of North Carolina Press (1930). This classic treatise began as notes to a new edition of Shelley's epic poem, but the science in Shelley's intellectual life grew the treatise into a (small) book-sized manuscript. It is good that it did: Grabo's intercalation of Shelley's obsession with science into this elliptical poem is insightful, particularly as it reveals Erasmus Darwin's influence on the poet.\

UNIT 6

Early Nineteenth Century: The Road to Darwin

> Amid all the revolutions of the globe the economy of Nature has been uniform, and her laws are the only things that have resisted the general movement. The rivers and the rocks, the seas and the continents have been changed in all their parts; but the laws which direct those changes, and the rules to which they are subject, have remained invariably the same.
>
> —John Playfair

The great prosperity that followed the Enlightenment and its Industrial Revolution included the transformation of Enlightenment appeals for freedom and justice into action. By midcentury, Britain had repealed capital punishment for theft (1830s), abolished slavery (1833) and public hangings (1868), and made significant strides in electoral reform (the Great Reform Act of 1832). In France, stung by memories of the revolution, social progress was delayed for much of the century, but distrust of the old wealth-encrusted social hierarchies was still prominent in social thought. Meanwhile, a rising middle class across the European horizon sought its own identity—resisting the old aristocratic mentality but fearing a loss of its own growing status in the hierarchy.

Competing theories of the good society and its reflection in nature invoked science in their support. Science had been politicized during the Enlightenment, but the emerging romanticism of the Victorian nineteenth century altered the terms of this as it reacted negatively to the more strident mechanistic and materialistic view of nature this new century had inherited. Nature, in all of its diversity and seemingly unconnected adaptations, the romanticists insisted, demonstrated a common and consistent pattern. It was this pattern that gave nature its compelling power and beauty. It was this pattern, in its harmonious elegance, that revealed a divine plan. Furthermore, the Enlightenment had separated man from nature and fragmented both nature (the reduction of the whole to its parts) and human understanding (the false distinction between intellect and spirit, the pragmatic and the sublime). It had furthermore invited arrogance: Human achievement requires control of nature, and sometimes its destruction.

Nowhere was this duel between the two sides of nature—and of man—better expressed than in Mary Shelley's *Frankenstein* (1818), often hailed as the first novel in the genre of modern science fiction. Her more dystopian *The Last Man*, where the modern achievements of the Industrial Revolution failed a dead planet, appeared in 1826.

The pendulum of romanticism swung back at midcentury when Comte and positivism dominated intellectual thought. Furthermore, the romantic movement in science did not itself displace Enlightenment values or ideology in the century's first half. The faith in the Enlightenment model of science, industry, and society maintained a strong hold on what was becoming a science of professionals in culture and a culture of professionalism in science. A truly scientific community was emerging (the word *scientist* first appears in 1840); The British Association for the Advancement of Science, formed in 1831, was a reaction against the Royal Society's dominance by "aristocratic amateurs" (Bowler 2003, 107). The American Association for the Advancement of Science was formed in 1848.

It was in this divisive intellectual climate that the continuing question over the role of the supernatural in earth's formation dominated scientific discussion. As the conventional religious model—a young earth populated by specially created life forms in which extinction did not occur—was challenged by new scientific observations and interpretations in geology and paleontology, the challenge intensified. Serious efforts to reconcile what could be salvaged from biblical accounts with what was becoming accepted in the scientific accounts appeared in numerous forms. A prominent one was the establishment of the *Bridgewater Treatises* in 1829. In that year, the Rev. Francis Henry Egerton, last Earl of Bridgewater, died and bequeathed a sum of £8,000 to establish an annual publication series under control of the Royal Society of London. The society was to select, each year, a person to write a treatise "on the Power, Wisdom and Goodness of God as manifested in the Creation illustrating such work by all reasonable arguments. . . ." In 1885, Adam Lord Gifford bequeathed ten times this amount for a similar lecture series, on natural theology as a science, to the four Scottish universities.

The authors selected for this unit represent the debate mentioned previously. They thus extend the views and arguments of Bonnet, Lamarck, Buffon, and Hutton, which were treated earlier. They also extend the strong typological thinking—the ghost of Aristotle haunting, still—that was to continue characterizing science for the next century (and our own, to some extent), and which was a major reason for the opposition to Darwin by philosophers and fellow scientists, as we shall see.

Cuvier and Owen were conservatives, politically but not religiously. To them, all species were stable; but they also equated transformationism with liberal political views such as democracy, and thus were opposed to evolutionary ideas. Adam Sedgwick (to appear in Units 7 and 10) and William Buckland were religious conservatives and thus supported geological interpretations that did not contradict creation and the flood.

As we now enter Darwin's century, these selections also identify many paleontologists and geologists who influenced Charles Darwin and a few who were personally known to him. We will also find that professional conflicts, strongly aligned as they often were with political or religious ideology, often strained personal ties and friendships.

27. Georges Cuvier

Georges Cuvier was born in 1769 in Montbéliard, France, on the eastern border near Basel, Switzerland. His probing curiosity, analytic mind, and flawless memory led him to excel in school. At the gymnasium between ages ten and twelve, he discovered Buffon's multivolume *Histoire Naturelle* and devoured it, rereading it several times. At age fourteen, he entered the Caroline Academy in Stuttgart, where he quickly picked up German. Cuvier impressed the naturalists with whom he came in contact, and in 1795 was invited to Paris, where he became assistant to Jean-Claude Mertrud, chair of comparative anatomy at the *Jardin des Plantes,* which, in its earlier incarnation as the *Jardin du Roi* (1739), was under the directorship of the Comte de Buffon.

Cuvier was also elected to the French Academy of Sciences that year (1795), and in the following year, at age twenty-seven, he read his first scientific paper, *Mémoires sur les espèces d'éléphants vivants et fossils.* This was a significant paper. Through a careful comparative anatomy of the African and Indian elephants and the fossil remains of the mammoth, Cuvier was able to demonstrate that each was a distinct species. Not only did this reveal that there were two living species of elephant, but that the fossil form was an extinct species. He later also established that the American mastodon was another extinct species. This paper helped to remove all doubt that extinction was a fact of the past.

In a second paper that same year, he further confirmed the evidence for extinction in the fossil record. There was a report of an ancient fossil from Paraguay that he named "megatherium." Through comparative anatomy with living forms, he concluded that the megatherium was an extinct sloth—a giant ground-dweller related to the much smaller tree-climbing sloth of South America. After Charles Darwin discovered his own megatherium near Buenos Aires in 1832, he used this evidence to argue his theory of descent with modification.

Cuvier's recognition that fossils in general represent extinct forms rather than unmodified early examples of living forms led to his belief that extinctions represented a series of catastrophes, and that such sudden cataclysms were largely responsible for Earth's history. In this, he differed from Buffon and others who argued that geological changes were slow and systematic results of processes still going on at the time. This "catastrophism" theory pitted geologists in its favor against the "uniformitarian" geologists such as Charles Lyell.

In addition, Cuvier insisted that, just as extinction occurred suddenly, so also did new species arrive suddenly. Fossils not only disappear quickly, but spontaneously appear in the strata as well. Hence, Cuvier rejected Lamarck's "species transmutation" and any other ideas of transitional change. This abrupt appearance and subsequent continuation without change was quite supportive of the claims of creationists opposed to Darwinism later in the nineteenth century and throughout the twentieth.

In his prodigious body of published work, Cuvier became justly famous during his lifetime. He not only established the modern field of paleontology, but also transformed the methodology of comparative anatomy—for living and nonliving forms—that underlay studies of species for generations following.

In recognition of his scientific contributions in 1819, Cuvier was made a peer, becoming Baron Cuvier. He died in Paris of cholera on May 13, 1832. Charles Darwin, at this time, was in Rio de Janeiro, shortly preparing to set sail on the *Beagle* to Paraguay and Buenos Aires and his great megatherium discovery.

Essay on the Theory of the Earth

In the following selections, Cuvier's catastrophism, and opposition to uniformitarianism, is clear—the latter explicit on his page 24. In the selection from page 12, his discussion of the continuing changes in species through time suggests a possible evolutionary transformation, but this he explicitly denies. On page 102, he claims that the lack of transitional forms rules out species transformation. The final quotation, taken from his Natural History of Fishes, *details the methodology by which he denies any lineage of forms through the ages.*

[10] Thus the sea, previous to the deposition of the horizontal strata, had formed others, which, by the operation of problematical causes, were broken, raised, and overturned in a thousand ways; and, as several of those inclined strata which it had formed at more remote periods, rise higher than the horizontal strata which have succeeded them, and which surround them, the causes by which the inclination of these beds was effected, had also made them project above the level of the sea, and formed islands of them, or at least shoals and inequalities; and this must have happened, whether they had been raised by one extremity, or whether the depression of the opposite extremity had made the waters subside. This is the second result, not less clear, nor less satisfactorily demonstrated, than the first, to every one who will take the trouble of examining the monuments on which it is established.

Proofs that such revolutions have been numerous.

But it is not to this subversion of the ancient [11] strata, nor to this retreat of the sea after the formation of the new strata, that the revolutions and changes which have given rise to the present state of the Earth are limited.

When we institute a more detailed comparison between the various strata and those remains of animals which they contain, we presently perceive, that this ancient sea has not always deposited mineral substances of the same kind, nor remains of animals of the same species; and that each of its deposits has not extended over the whole surface which it covered. There has existed a succession of variations; the former of which alone have been more or less general, while the others appear to have been much less so....

Amidst these variations in the nature of the general fluid, it is evident, that the animals which lived in it could not remain the same. Their [12] species, and even their genera, changed with the strata; and, although the same species occasionally recur at small distances, it may be announced as a general truth, that the shells of the ancient strata have forms peculiar to themselves; that they gradually disappear, so as no longer to be seen at all in the recent strata, and still less in the presently existing ocean, in which their corresponding species are never discovered, and where several, even of their genera, do not occur:...

There has, therefore, been a succession of variations in the economy of organic nature, which has been occasioned by those of the fluid in which the animals lived, or which has at least corresponded with them; and these variations have gradually conducted the classes of aquatic animals to their present state, till, at length, at the time when the sea retired from our continents for the last time, its inhabitants did not differ much from those which are found in it at the present day.

[13] We say for the *last* time, because, if we examine with still greater care those remains of organised bodies, we discover, in the midst of even the oldest strata of marine formation, other strata replete with animal or vegetable remains of terrestrial or fresh-water productions; and, amongst the more recent strata, or, in other words, those that are nearest the surface, there are some in which land animals are buried under heaps of marine productions. Thus, the various catastrophes which have disturbed the strata, have not only caused the different parts of our continents to rise by degrees from the bosom of the waves, and diminished the extent of the basin of the ocean, but have also given rise to numerous shiftings of this basin....

[14] *Proofs that these revolutions have been sudden.*

It is of much importance to remark, that these repeated irruptions and retreats of the sea have neither all been slow nor gradual; on the contrary, most of the catastrophes which have occasioned them have been sudden; and this is especially easy to be proved, with regard to the last of these catastrophes, that which, by a twofold motion, has inundated, and afterwards laid dry, our present continents, or at least a part of the land which forms them at the present day.

[15] The breaking to pieces, the raising up and overturning of the older strata, leave no doubt upon the mind that they have been reduced to the state in which we now see them, by the action of sudden and violent causes; and even the force of the motions excited in the mass of waters, is still attested by the heaps of debris and rounded pebbles which are in many places interposed between the solid strata. Life, therefore, has often been disturbed on this earth by terrible events.

[24] [It has] long been considered possible to explain the more ancient revolutions on its surface by means of these still existing causes.... But we shall presently see, that unfortunately the case is different in physical history:—the thread of operations is here broken; the march of Nature is changed; and none of the agents which she now employs, would have been sufficient for the production of her ancient works.

[38] During a long time, two events or epochs only, the Creation and the Deluge, were admitted as comprehending the changes which have been operated upon the globe;

and all the efforts of geologists were directed to account for the present existing state of things, by imagining a certain original state, afterwards modified by the deluge, of which also, as to its causes, its operations, and its effects, each entertained his own theory.

Thus, according to one, [*Cuvier refers to Burnet's* Sacred Theory of the Earth] the earth was at first invested with an uniform light crust, which covered the abyss of the sea; and which being broken up for the production of the deluge, formed the mountains by its fragments....

[39] It is evident, that, even while confined within the limits prescribed by the Book of Genesis, naturalists might still have a pretty wide range: they soon found themselves, however, in too narrow bounds if and when they had succeeded in converting the six days of creation into so many indefinite periods, the lapse of ages no longer forming an obstacle to their views, their systems took a flight proportioned to the periods which they could then dispose of at pleasure.

[102] *Proofs that the Extinct Species of Quadrupeds are not varieties of the presently existing Species.*

I now proceed to the consideration of another objection, one, in fact, which has already been urged against me.

Why may not the presently existing races of land quadrupeds, it has been asked, be modifications of those ancient races which we find in a fossil state; which modifications may have been produced by local circumstances and change of [103] climate; and carried to the extreme difference, which they now present, during a long succession of ages?

Yet to these persons an answer may be given from their own system. If the species have changed by degrees, we ought to find traces of these gradual modifications. Thus, between the palæotheria and our present species, we should be able to discover some intermediate forms; and yet no such discovery has ever been made.

[295] This much is certain, that we are now at least in the midst of a fourth succession of land animals,—that, after the age of reptiles, the age of palaeotheria, the age of mammoths, and that of mastodons and megatheria, has come the age in [296] which the human species, aided by some domestic animals, peaceably governs and fertilizes the earth, and that it is only in the deposits formed since the commencement of this age, in alluvial matters, peat-bogs, and recent concretions, that bones are found in the fossil state, which belong all of them to known and still living animals.

Memoirs of Baron Cuvier

[126] Let it not be imagined, because we place one genus or one family before another, that we consider them as more perfect, or superior to another in the series of beings. He only could pretend to do this, who would pursue the chimerical project of ranging beings in one single line,—a project which we have long renounced. The more progress we have made in the study [127] of nature, the more we are convinced that this is one of the falsest ideas that has ever resulted from the pursuit of natural history; the more we have been convinced of the necessity of considering each being, each group of beings, by itself, and the part it plays by its properties and organisation, and not to make abstraction of any of its affinities, or any of the links which attach it, either to the beings nearest to it, or the most distant from it. Once placed in this point of view, difficulties vanish, all arranges itself for the naturalist: but systematic methods only embrace the nearest affinities; and by placing a being only between two others, they will always be wrong. The true method is, to view each being in the midst of all others: it shows all the radiations by which it is more or less closely linked with that immense network which constitutes organised nature; and it is this only which can give us that great idea of nature, which is true, and worthy of her and her Author; but ten or twenty rays often would not suffice to express these innumerable affinities....

We shall therefore approach to each other those whom nature has approximated, without feeling obliged to put into our groups the beings she has not placed there; and making no scruple, after [128] having demonstrated, for example, all the species which will admit of being arranged in a well-defined genus, all the genera which may be placed in a well-defined family, to leave out one or several isolated species or genera, which are not attached to others in a natural manner; preferring the honest avowal of these irregularities, if we may be allowed to call them so, to those errors which must arise from leaving these species, and anomalous genera, in a series, the characters of which they do not embrace.

28. Geoffroy Saint-Hilaire

Étienne Geoffroy Saint-Hilaire was born in Étampes—just south of Paris—in 1772. Although his father was a procurator, and then a judge, Geoffroy initially planned for the priesthood. He became a canon of the church at fifteen and entered the College of Navarre, University of Paris, at eighteen. After earning a bachelor of law in 1790, his interest in science took him to the nearby College of Cardinal Lemoine. His strong intellect was widely recognized and he rose quickly, first as a subcurator at the Botanic Garden and then, in 1793, professor of zoology at the Museum of Natural History, which had just been organized as part of the Botanic Garden, as Buffon had envisioned. Geoffroy's colleague, responsible for invertebrates, was Jean-Baptiste Lamarck.

In 1795, upon recommendation by the naturalist Abbé Alexandre Henri Tessier, Geoffroy invited Georges Cuvier, a young zoologist, to join the museum. Thus was established a long friendship and professional rivalry that reached its height in three famous debates over biological function and structure in 1830.

It had long been Geoffroy's conviction that function was the result of form, an idea introduced by his predecessor Buffon and a view opposed by Lamarck. Cuvier, a functionalist, insisted (as did Lamarck) that form was the result of function. Whereas Geoffroy believed that an underlying unified body plan is common to all animals, with differences resulting from changes in the basic pattern responsive to environment, Cuvier believed that the organism is a functionally integrated whole.

Geoffroy's initial application of the common body plan was to four-limbed animals. In 1807, he extended this to all vertebrates.

When Geoffroy extended his "unity of plan" idea to include invertebrates, he was trespassing on Cuvier's domain, and the conflict became public. It was not just in the form-versus-function relationship that the two differed, it was also the nature of organic change that derived from it: Geoffroy supported an evolutionist interpretation of change in form through time. Cuvier was adamant that species were fixed, and that replacement could only come through extinction and new creation. Geoffroy was a deist, believing in a God who created nature and let its laws operate without further divine interference; Cuvier was a theist who believed in a God actively involved in His creation.

Geoffroy had two insights that saw later development as rich biological theories. The first of these was the reality of natural selection. We find this, for example, in his 1831 report on the influence of the environment in changing animal forms:

> The external world is all-powerful in alteration of the form of organized [organic] bodies.... these [modifications] are inherited, and they influence all the rest of the organization of the animal, because if these modifications lead to injurious effects, the animals which exhibit them perish and are replaced by others of a somewhat different form, a form changed so as to be adapted to the new environment.

The second insight reflects upon Geoffroy's focus on embryology, which paralleled his interest in comparative anatomy. While he was one of the first to hypothesize *recapitulation* theory—that embryonic stages of higher animals reflect lower forms to which they are related—his idea of a uniform body plan for all animals gave his embryological studies a new importance. Here was evidence that a common plan finds its diversification in embryonic developmental control. Without the molecular evidence, this is surprisingly close to the new evo-devo:

> Nature constantly uses the same materials, and is ingenious only in varying their forms. As if in fact she were constrained by initial givens, we see her always tending to make the same elements appear, in the same number, in the same circumstances, and with the same connections. (Saint-Hilaire, "Considérations sur les pieces de la tête osseuse des animaux vertébrés, et particulièrement sur celles du crâne des Oiseaux," *Annales, Musée d'Histoire Naturelle* 10 [1807]: 342. [*This reference is cited in the same work as the passages that follow.—RKW*])

Let us, finally, put these materials in historical context: Saint-Hilaire and Cuvier lived during the French Revolution, the Reign of Terror, and the Napoleonic era. All three periods of turmoil affected their careers. On the July after Geoffroy entered the College of Navarre, the Bastille was stormed in the city and the revolution began. Just before the massacres of the Terror began in 1792, Geoffroy rescued the priest and scientist, Abbé Haüy, from prison, and it was with Haüy's intervention that Geoffroy gained his appointment to the Museum of Natural History. Then, in 1798, he was chosen to join Napoleon's scientific expedition to Egypt, where his study of the Nile lungfish contributed to his body plan ideas. Napoleon, who returned to France in 1799 to be crowned first consul (later emperor), awarded Geoffroy the Cross of the Legion of Honor for his services.

It was the reorganization of the Botanical Garden following the revolution that led to Cuvier's move to Paris. Then, when Napoleon became first consul, he put himself as head of the National Institute. Hearing of Cuvier's energetic work there, Napoleon appointed him as one of six inspector-generals to establish lyceums throughout France. Cuvier founded those at Nice, Bordeaux, and Marseilles, all of which subsequently became royal colleges.

Anatomical Philosophy

[32]A new era, whose date is fixed by the publication of this book, begins under other auspices. If it is not a new path that is open, at least the field of organization is illuminated by a new principle, that *of connections:* a principle of high philosophical interest, since it finally admits us to the full and complete enjoyment, without the slightest exception in practice, of that other principle fundamental to natural philosophy, that all animals having a spinal cord lodged in a bony case are made on the same model. The forecast to which this truth leads us, that is, the presentiment that we will always find, in every family, all the organic materials that we have perceived in another: that is what I have included in my work under the designation of *Theory of Analogues.*

Memoirs sur l'organisation des insectes

It was in 1820, in three reports to the Academy of Sciences, that Geoffroy insisted that the insects share a basic "skeletal form" with the vertebrates, thus placing arthropods and vertebrates, two of Cuvier's original four "embranchements," together in a single class. What follows is an excerpt from the 1824 publication that united the three reports.

[57–58] From these facts it is to be concluded that insects are *vertebrates*, and if everything can be reduced to a vertebra, it is in the insects that the proposition appears in all its clarity. In the last analysis, we arrive at this result: every animal lives within or outside its vertebral column. We have in fact from now on this great characteristic to differentiate the old vertebrates from the new, who will take their position after the others.

Saint-Hilaire's colleague Étienne Serres, who with Geoffroy and Friedrich Meckel developed embryogenetic "recapitulation," puts the issue succinctly in the following statement (quoted in Le Guyader, *p. 103):*

First we see the transitory form of higher embryos assume fugitively and in passing the organic and permanent attributes of lower animals; further, the permanent organization of the latter sketches in its successive degrees of perfection all the embryonic phases of the one among them that most closely approaches the last of the vertebrates in such a way that, for them as well, the various features of their zoology are only in some way the graduated scale of their organogenesis.

29. RICHARD OWEN

Richard Owen was a brilliant, but personally obnoxious, comparative anatomist and paleontologist who made numerous contributions to biology in both discoveries and, more importantly, in ideas. He knew and followed the methodology of Cuvier in using comparative studies to identify species relationships, but extended his ideas to recognize—and coin the term—"homologies."

Born in Lancaster, England, in 1804, his early education was not promising and he enlisted for a short period in the Royal Navy. He became interested in medicine and worked for a local surgeon before enrolling in medical school at the University of Edinburgh at age twenty. His real interest narrowed to anatomy there, and he quit the university and enrolled in the private Barclay School (which Charles Darwin later attended). There he thrived. Upon the headmaster's recommendation, Owen moved to London and apprenticed with surgeon John Abernathy, who was also president of the Royal College of Surgeons. In 1826, Owen gained membership and licensing.

The Royal College had acquired thirteen thousand animal and human anatomical specimens—the Hunterian Collection—and Owen was given the assistant curatorship. This necessitated a long and patient cataloguing of each one, which he completed in 1830 with the resultant highly specialized knowledge of vertebrate anatomy. He travelled to Paris in 1830, where he met Cuvier, and attended the famous 1831 debate between Cuvier and Geoffroy Saint-Hilaire. Upon his return, Owen was appointed Hunterian Lecturer in Comparative Anatomy and gave his first lecture in 1837. Darwin, having returned from his voyage on the *HMS Beagle*, attended several of these. Owen was given the fossil vertebrates Darwin sent back to study, and the two became professional colleagues. The two became enemies after Darwin's *Origin* was published (Owen's review appears in Unit 10), due, Darwin wrote, more from jealousy than from specific disagreement.

Owen's scientific philosophy was antimaterialist; he followed the dualists in claiming a separate existence of the soul and the "Vital Principle" as the essence of the body, neither of which were materialistic. This principle gave to all living forms a common fundamental body plan, the *archetype*, from which diversification occurred by adaptation. This transcendental anatomy, as Owen developed it, is startlingly similar to the segmentary embryological development we see today in the new science of evolutionary development (evo-devo), to be discussed in a Unit 15. However, Owen's "archetype," far from being a common ancestor, is the "idea" in the mind of the creator. Indeed, Owen compares it to Plato's ἰδέα. With this philosophy, Owen marks himself as an essentialist in the true Aristotelian tradition.

On the Archetype and Homologies of the Vertebrate Skeleton

[1] When the structure of organized beings began to be investigated, the parts, as they were observed, were described under names or phrases suggested by their forms, proportions, relative position, or likeness to some familiar object. Much of the nomenclature of human anatomy has thus arisen, especially that of the osseous system, which, with the rest of man's frame, was studied originally from an insulated point of view, and irrespective of any other animal structure or any common type.

So when the exigences of the veterinary surgeon, or the desire of the naturalist to penetrate beneath the superficial characters of his favourite class, led them to anatomise the lower animals, they, in like manner, seldom glanced beyond their immediate subject, and often gave arbitrary names to the parts which they detected.

Owen coined the term homology *to counter the explanation of Cuvier and others that similar structures reflect similar adaptations or functions. His "special homology" refers to specific structures, whereas the term* general homology *refers to the similarity of overall body plan or structural principle. Owen defines homology (1843) as "the same organ in different animals under any variety of form and function." The term today refers to similarity through common descent.*

[72] It must surely appear a most remarkable circumstance to one acquainted only with the osteology of the human frame, that so many bones should be, by the common consent of comparative anatomists, determinable in the skull of every animal down to the lowest osseous fish. This fact alone, so significant of the unity of plan pervading the vertebrate structure, has afforded me, at least, a huge ground of hope and much encouragement to perseverance in the reconsideration of those points on which a difference of opinion has prevailed; and in the re-investigation of what

is truly constant and essential in characters determinative of special homologies. . . .

Upon this point the anatomical world is at present divided, lacking the required demonstration. The majority of existing authors on comparative anatomy have tacitly abandoned, or with Cuvier and M. Agassiz, have [73] directly opposed the idea of 'special homology' being included in a higher law of uniformity of type.

Yet the attempt to explain, by the Cuvierian principles, the facts of special homology on the hypothesis of the subserviency of the parts so determined to similar ends in different animals,—to say that the same or answerable bones occur in them because they have to perform similar functions—involve many difficulties, and are opposed by numerous phaenomena. We may admit that the multiplied points of ossification in the skull of the human foetus facilitate, and were designed to facilitate, childbirth; yet something more than such a final purpose lies beneath the fact, that most of those osseous centres represent permanently distinct bones in the cold-blooded vertebrates.

In the following conclusion to his work, Owen misses the opportunity to see natural selection as the "adaptive force," preferring the more conventional "intelligent designer" explanation. Without the archetype-mysticism, this comes close to an evo-devo explanation. The lengthy quote Owen inserts at the end comes from John Barclay, "Inquiry into the Opinions, Ancient and Modern, Concerning Life and Organization" (1822). Barclay was a Scottish anatomist and lecturer at the Royal College of Surgeons in London; he died in 1826.

[172] The extent to which the operation of the polarizing or vegetative-repetition-force is so subdued in the organization of a specific animal form becomes the index of the grade of such species, and is directly as its ascent in the scale of being. The lineaments of the common archetype are obscured in the same degree: but even in Man, where the specific organizing force has exerted its highest power in controlling the tendency to type and in modifying each part in adaptive subserviency to, or combination of power with, another part, the extent to which the vegetative repetition of segments and the archetypal features are traceable indicates the degree in which the general polarizing force may have operated in the arrangement of the parts of the developing frame: and it is not without interest or devoid of significance that such evidence should be mainly manifested in the system of organs in whose tissue the inorganic earthy salts most predominate.

With regard to the 'adaptive force,' whatever may be the expressions by which its nature and relations, when better understood, may be attempted to be explained, its effects must ever impress the rightly constituted mind with the conviction, that in every species "ends are obtained and the interests of the animal promoted, in a way that indicates superior design, intelligence and foresight; but a design, intelligence and foresight in which the judgment and reflection of the animal never were concerned; and which, therefore, with Virgil, and with other studious observers of nature, we must ascribe to the Sovereign of the universe, in whom we live, and move, and have our being."

30. Charles Lyell

Perhaps no scientist had more influence on Charles Darwin than did Charles Lyell. Lyell's first volume of *Principles of Geology*, published in 1830, was one of the books Darwin took with him on the *Beagle* in 1831, and Darwin—already schooled in geological fieldwork by Adam Sedgwick the year before—found Lyell's ideas most provocative. Darwin, alas, had no such immediate influence on Lyell, who remained skeptical of natural selection for some time after Darwin's publication.

Charles Lyell was born in 1797 near Dundee, Scotland, surrounded by grand geological formations. His father fostered Charles's interest in natural history, having some accomplishments in botany himself. In 1816, he entered Exeter College at Oxford and came under the influence of the geologist William Buckland, lecturer and fellow of Oxford's Corpus Christi College. Buckland was on the side of the catastrophists; however, Lyell soon became convinced that the same forces at work today were alone responsible for all previous geologic events—the principle of uniformitarianism, suggested by Hutton. This principle, elaborated and supported by evidence in his *Principles of Geology*, was perhaps his greatest contribution to science, but he also helped to demonstrate that only deep time could explain the geological events that had shaped the earth.

Lyell received his undergraduate degree at Exeter, with a second class in classics, in 1819, and his master of arts degree in 1821. In 1823, he was elected joint secretary of the Geological Society of London. Though he took law as his profession, and practiced briefly, his failing eyesight turned him back to geology in 1827. In the 1830s, as his volumes of *Principles of Geology* were being published, Lyell was appointed professor of geology at Kings College, London. In London he met and married (in 1832) Mary Horner, whose father was also involved with the Geological Society.

Lyell received numerous honors during his life. He was made fellow of the Royal Society, awarded its Copley Medal in 1858, and received the Wollaston Medal of the Geological Society of London in 1866. He was knighted and made a baronet. He died in 1875, two years after his wife's death, and was buried in Westminster Abbey.

The quotation by Playfair that opens this unit is cited at the beginning of the first volume of *Principles of Geology*. Playfair was professor of natural philosophy at the University of Edinburgh and a close friend of James Hutton. He popularized and rendered more accessible much of Hutton's work. The quote comes from Playfair's *Illustrations of the Huttonian Theory of the Earth* (1802). Lyell took Hutton's uniformitarian theory and expanded it with the intense attention to detail and use of abundant evidence for which Lyell is justly famous.

Principles of Geology, Vol. I

[85] The establishment, from time to time, of numerous points of identification, drew at length from geologists a reluctant admission, that there was more correspondence between the physical constitution of the globe, and more uniformity in the laws re-[86]gulating the changes of its surface, from the most remote eras to the present, than they at first imagined.

[311] Those geologists who are not averse to presume that the course of Nature has been uniform from the earliest ages, and that causes now in action have produced the former changes of the earth's surface, will consult the ancient strata for instruction in regard to the reproductive effects of tides and currents. It will be enough for them to perceive clearly that great effects now annually result from the operations of these agents, in the inaccessible depths of lakes, seas, and the ocean; and they will then search the ancient lacustrine and marine strata for manifestations of analogous effects in times past. Nor will it be necessary for them to resort to very ancient monuments; for in certain regions where there are active volcanos, and where violent earthquakes prevail, we may examine submarine formations many thousand feet in thickness, belonging to our own era, or, at least, to the era of contemporary races of organic beings.

[271] But it is undeniable, that the great majority of the older commentators have held the deluge, according to the brief account of the event given by Moses, to have consisted of a rise of waters over *the whole earth*, by which the summits of the loftiest mountains on the globe were submerged. Many have indulged in speculations concerning the instruments employed to bring about the grand cataclysm; and there has been a great division of opinion as to the effects which it might be expected to have produced on the surface of the earth.... [According to Dr. Buckland] the deluge is represented as a violent and transient rush of waters which tore up the soil to a great depth, excavated valleys, gave rise to immense beds of shingle, carried fragments of rock and gravel from one point to another, and, during its advance and retreat, strewed the valleys, and even the tops of many hills, with alluvium....

[I]n the narrative of Moses there are no terms employed that indicate the impetuous rushing of the waters, either as they rose or when they re-[272]treated, upon the restraining of the rain and the passing of a wind over the earth. On the contrary, the olive-branch, brought back by the dove, seems as clear an indication to us that the vegetation was not destroyed, as it was then to Noah that the dry land was about to appear.

[273] For our own part, we have always considered the flood, if we are required to admit its universality in the strictest sense of the term, as a preternatural event far beyond the reach of philosophical inquiry, whether as to the secondary causes employed to produce it, or the effects most likely to result from it.

In the following passage, Lyell denies any continuity from species to species in the fossil record and goes on to deny any great antiquity to the human species. Following Darwin's publication, he came to accept species transformation, and in his The Antiquity of Man *(1863) he accepted evidence of man's antiquity as well.*

[153] It is, therefore, clear, that there is no foundation in geological facts, for the popular theory of the successive development of the animal and vegetable world, from the simplest to the most perfect forms; and we shall now proceed to consider another question, whether the recent origin of man lends any support to the same doctrine, or how far the influence of man may be considered as such a deviation from the analogy of the order of things previously established, as to weaken our confidence in the uniformity of the course of nature. We need not dwell on the proofs of the low antiquity of our species, for it is not controverted by any geologist; indeed, the real difficulty which we experience consists in tracing back the signs of man's existence on the earth to that comparatively modern period when species, now his contemporaries, began to predominate. If there be a difference of opinion respecting the occurrence in certain deposits of the remains of man and his works, it is always in reference to strata confessedly of the most modern order; and [154] it is never pretended that our race co-existed with assemblages of animals and plants, of which *all the species* are extinct.

Principles of Geology, Vol. II

A good part of volume II is taken up in discussing the various transmutation theories and in demonstrating that a single creation for each species is more valid. Here is one such passage.

[23] It is by no means improbable that when the series of species of certain genera is very full, they may be found to differ less widely from each other, than do the mere varieties or races of certain species. If such a fact could be established, it would by no means overthrow our confidence in the reality of species, although it would certainly diminish the chance of our obtaining certainty in our results.

It is almost necessary, indeed, to suppose, that varieties will differ in some cases, more decidedly than some species, if we admit that there is a graduated scale of being, and assume that the following laws prevail in the economy of the animate creation:—first, that the organization of individuals is capable of being modified to a limited extent by the force of external causes; secondly, that these modifications are, to a certain extent, transmissible to their offspring; thirdly, that there are fixed limits beyond which the descendants from common parents can never deviate from a certain type; fourthly, that each species springs from one original stock, and can never be permanently confounded, by intermixing with the progeny of any other stock; fifthly, that each species shall endure for a considerable period of time. Now if we assume, for the present, these rules hypothetically, let us see what consequences may naturally be expected to result.

We must suppose, that when the Author of Nature creates an animal or plant, all the possible circumstances in which its descendants are destined to live are foreseen, and that an [24] organization is conferred upon it which will enable the species to perpetuate itself and survive under all the varying circumstances to which it must be inevitably exposed. Now the range of variation of circumstances will differ essentially in almost every case. Let us take for example any one of the most influential conditions of existence, such as temperature. In some extensive districts near the equator, the thermometer might never vary throughout several thousand centuries for more than 20° Fahrenheit; so that if a plant or animal be provided with an organization fitting it to endure such a range, it may continue on the globe for that immense period, although every individual might be liable at once to be cut off by the least possible excess of heat or cold beyond the determinate quantity. But if a species be placed in one of the temperate zones, and have a constitution conferred on it capable of supporting a similar range of temperature only, it will inevitably perish before a single year has passed away.

Principles of Geology, Vol. III

In this final selection, Lyell reveals his unique methodology: Begin with the most recent geological evidence, for which a reasonable chronology is known, and work backward. Though there was little alternative, given the lack of chronometry, to this approach, it is no longer used. It is an approach that puts severe limits on adducing radically different causation: It assumes a steady-state world with uniform forces acting in uniform intensity, in which actual catastrophic events are disallowed.

[2] In the first volume we enumerated many prepossessions which biased the minds of the earlier inquirers, and checked an impartial desire of arriving at truth. But of all the causes to which we alluded, no one contributed so powerfully to give rise to a false method of philosophizing as the entire unconsciousness of the first geologists of the extent of their own ignorance respecting the operations of the existing agents of change.

They imagined themselves sufficiently acquainted with the mutations now in progress in the animate and inanimate world, to entitle them at once to affirm, whether the solution of certain problems in geology could ever be derived from the observation of the actual economy of nature, and having decided that they could not, they felt themselves at liberty to indulge their imaginations, in guessing at what *might be,* rather than in inquiring *what is;*...

It appeared to them more philosophical to speculate on the possibilities of the past, than patiently to explore the realities of the present, and having invented theories under the influence of such maxims, they were consistently unwilling to test their validity by the criterion of their accordance with the ordinary operations of nature.

31. Robert Chambers

The 1844 publication of *Vestiges of the Natural History of Creation* caused an immediate stir in Britain, creating waves that washed far afield, onto the continent and across to the New World within just a few years. It was audacious: It presumed to explain the formation of the solar system, the earth, and all of life on it, including humankind. It was irreverent: It supported a clearly materialistic evolution of life, suggesting that God's role was passive. Most intriguingly, it was anonymous: No author's name appeared on the title page or cover.

Forty years later—in 1884—the author's identity was revealed to be Robert Chambers, a very successful publisher. This appeared on the title page of the twelfth edition. Robert had been dead for thirteen years. He had known that the large (four hundred pages) volume would cause controversy and feared the negative impact on his business if it were known that he was the author. But his strong interest in science, self-taught through very wide reading, and his desire to provoke formal scientific discussion on species transformation through natural law, led him to write the book and arrange for clandestine publication. He achieved his purpose: Twenty thousand copies were sold in the first decade, and it gained notoriety across its readership. Anglicans and scientists condemned the book. Some, such as Quakers and Unitarians, found it appealing. Naturalists were quick to point out its wild claims were without factual basis.

Born in Scotland, Robert Chambers and his brother William began as booksellers on the streets of Edinburgh, eventually forming the successful British publishing house of W. & R. Chambers. Robert acquired an excellent writing ability, and the brothers wrote and produced a weekly magazine, *Vestiges.*

Charles Darwin read *Vestiges* and found it "capitally-written" but factually flawed. He also undoubtedly saw in this some of the derision he himself would later endure for his own evolutionary ideas.

Vestiges of the Natural History of Creation, Ch. 12

[148] In pursuing the progress of the development of both plants and animals upon the globe, we have seen an advance in both cases, along the line leading to the higher forms of organization. Amongst plants, we have first sea-weeds,

afterwards land plants; and amongst these the simpler (cellular and cryptogamic) before the more complex. In the department of zoology, we see zoophytes, radiata, mollusca, articulate, existing for ages before there were any higher forms. The first step forward gives fishes, the humblest class of the vertebrata; and, moreover, the earliest fishes partake of the character of the next lowest sub-kingdom, the articulata. Afterwards come land animals, of which the first are reptiles, universally allowed to be the type next in advance from fishes, and to be connected with these by the links of an insensible gradation. From reptiles we advance to birds, and thence to mammalia, which are commenced by marsupialia, acknowledgedly low forms in their class. That there is thus a progress of some kind, the most superficial glance at the geological [149] history is sufficient to convince us. Indeed the doctrine of the gradation of animal forms has received a remarkable support from the discoveries of this science, as several types formerly wanting to a completion of the series have been found in a fossil state.*

[150] In examining the fossils of the lower marine creation, with a reference to the kind of rock in connexion, with which they are found, it is observed that some strata are attended by a much greater abund- [151] ance of both species and individuals than others. They abound most in calcareous rocks, which is precisely what might be expected, since lime is necessary for the formation of the shells of the mollusks and articulata, and the hard substance of the crinoidea and corals; next in the carboniferous series; next in the tertiary; next in the *new* red sandstone; next in slates; and lastly, least of all, in the primary rocks. This may have been the case without regard to the origination of new species, but more probably it was otherwise; or why, for instance, should the polypiferous zoophyta be found almost exclusively in the limestones? There are, indeed, abundant appearances as if, throughout all the changes of the surface, the various kinds of organic life invariably *pressed in,* immediately on the specially suitable conditions arising, so that no place which could support any form of organic being might be left for any length of time unoccupied. Nor is it less remarkable how various species are withdrawn from the earth, when the proper conditions for their particular existence are changed. The trilobite, of which fifty species existed during the earlier formations, [152] was extirpated before the secondary had commenced, and appeared no more. The ammonite does not appear above the chalk. The species, and even genera of all the early radiata and mollusks were exchanged for others long ago. Not one species of any creature which flourished before the tertiary (Ehrenberg's infusoria excepted) now exists; and of the mammalia which arose during that series, many forms are altogether gone, while of others we have now only kindred species. Thus to find not only frequent additions to the previously existing forms, but frequent withdrawals of forms which had apparently become inappropriate—a constant shifting as well as advance—is a fact calculated very forcibly to arrest attention....

That God created animated beings, as well as the terraqueous theatre of their being, is a fact so powerfully evidenced, and so universally received, that I at once take it for granted. But in the particulars of this so highly supported idea, we surely here see cause for some re-consideration. It may now be inquired,—In what way was the creation [153] of animated beings effected? The ordinary notion may, I think, be not unjustly described as this,— that the Almighty author produced the progenitors of all existing species by some sort of personal or immediate exertion. But how does this notion comport with what we have seen of the gradual advance of species, from the humblest to the highest? How can we suppose an immediate exertion of this creative power at one time to produce zoophytes, another time to add a few marine mollusks, another to bring in one or two conchifers *[a taxonomic subdivision of mollusks—RKW]*, again to produce crustaceous fishes, again perfect fishes, and so on to the end? This would surely be to take a very mean view of the Creative Power—to, in short, anthropomorphize it, or reduce it to some such character as that borne by the ordinary proceedings of mankind. And yet this would be unavoidable; for that the organic creation was thus progressive through a long space of time, rests on evidence which nothing can overturn or gainsay. Some other idea must then be come to with regard to *the mode* in which the Divine Author proceeded in the organic creation. Let us seek in the history of the earth's formation for a new suggestion on this point. We have seen powerful evidence, that the construction of this [154] globe and its associates, and inferentially that of all the other globes of space, was the result, not of any immediate or personal exertion on the part of the Deity, but of natural laws which are expressions of his will. What is to hinder our supposing that the organic creation is also a result of natural laws, which are in like manner an expression of his will?

[156] To a reasonable mind the Divine attributes must appear, not diminished or reduced in any way, by supposing a creation by law, but infinitely exalted. It is the narrowest of all views of the Deity, and characteristic of a humble class of intellects, to suppose him acting constantly in particular ways for particular occasions. It, for one thing, greatly detracts from his foresight, the most undeniable of all the attributes of Omnipotence. It lowers him towards the level of our own humble intellects....

[160] It is also to be observed, that the thing to be accounted for is not merely the origination of organic being upon this little planet, third of a series which is but one of hundreds of thousands [161] of series, the whole of which again form but one portion of an apparently infinite globe-peopled space, where all seems analogous. We have to suppose, that every one of these numberless globes is either a theatre of organic being, or in the way of becoming so.

*Intervals in the series were numerous in the department of the pachydermata; many of these gaps are now filled up from the extinct genera found in the tertiary formation. [*These filled gaps refer to the contributions of Cuvier.—RKW*]

This is a conclusion which every addition to our knowledge makes only the more irresistible.

Vestiges of the Natural History of Creation, Ch. 14

[197] These facts clearly shew how all the various organic forms of our world are bound up in one—how a fundamental unity pervades and embraces them all, collecting them, from the humblest lichen up to the highest mammifer, in one system, the whole creation of which must have depended upon one law or decree of the Almighty, though it did not all come forth at one time. After what we have seen, the idea of a separate exertion for each must appear totally inadmissible. The single fact [198] of abortive or rudimentary organs condemns it; for these, on such a supposition, could be regarded in no other light than as blemishes or blunders—the thing of all others most irreconcilable with that idea of Almighty Perfection which a general view of nature so irresistibly conveys. On the other hand, when the organic creation is admitted to have been effected by a general law, we see nothing in these abortive parts but harmless peculiarities of development, and interesting evidences of the manner in which the Divine Author has been pleased to work....

It is only in recent times that physiologists have observed that each animal passes, in the course of its germinal history, through a series of changes resembling the *permanent forms* of the various orders of animals inferior to it in the scale. Thus, for instance, an insect, standing at the head of the articulated animals, is, in the larva state, a true annelid, or worm, the annelida being the lowest in the same class. The embryo of a crab resembles the perfect animal of the inferior order myriapoda [*a subphylum of Arthropoda, including centipedes—RKW*], and passes through all the forms [199] of transition which characterize the intermediate tribes of Crustacea....His first form is that which is permanent in the animalcule. His organization gradually passes through conditions generally resembling a fish, a reptile, a bird, and the lower mammalia, before it attains its specific maturity. At one of the last stages of his foetal career, he exhibits an intermaxillary bone, which is characteristic of the perfect ape; this is suppressed, and he may then be said to take leave of the simial [*sic: simian*] type, and become a true human creature. Even, as we shall see, the varieties of his race are represented in the progressive development of an individual of the highest, before we see the adult Caucasian, the highest point yet attained in the animal scale.

Chambers's reference to the intermaxillary bone reflects Johann W. Goethe's 1784 proof that humans have such a bone, but it is fused with the maxilla by birth. This demonstrated humankind's position with the higher primates, instead of being quite distinct from them biologically. Chambers's final sentence above demonstrates the racism common in the nineteenth century.

[203] The whole train of animated beings, from the simplest and oldest up to the highest and most recent, are, then, to be regarded as a series of *advances of the principle of development,* which have depended upon external physical circumstances, to which the resulting animals are appropriate. I contemplate the whole phenomena as having been in the first place arranged in the counsels of Divine Wisdom, to take place, not only upon this sphere, but upon all the others in space, under necessary modifications, and as being carried on, from first to last, here and elsewhere, under [204] immediate favour of the creative will or energy.

[212] It has been seen that, in the reproduction of the higher animals, the new being passes through stages in which it is successively fish-like and reptile-like. But the resemblance is not to the adult fish or the adult reptile, but to the fish and reptile at a certain point in their foetal progress; this holds true with regard to the vascular, nervous, and other systems alike. It may be illustrated by a simple diagram. The foetus of all the four classes may be supposed to advance in an identical condition to the point A. The fish there diverges and passes along a line apart, and peculiar to itself, to its mature state at F. The reptile, bird, and mammal, go on together to C, where the reptile diverges in like manner, and advances by itself to R. The bird diverges at D, and goes on to B. The mammal then goes forward in a straight line to the highest point of organization at M.

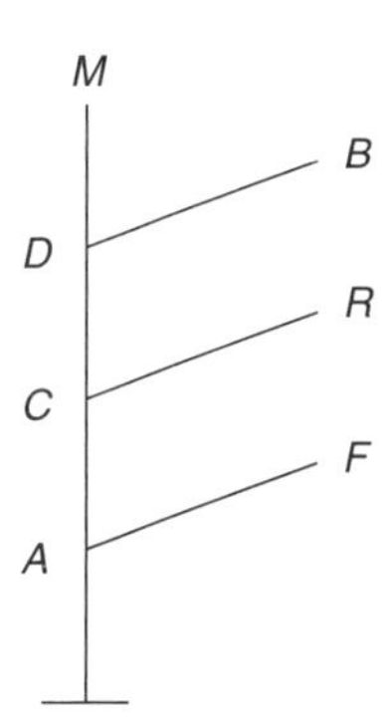

[213] Limiting ourselves at present to the outline afforded by this diagram, it is apparent that the only thing required for an advance from one type to another in the generative process is that, for example, the fish embryo should not diverge at A, but go on to C before it diverges, in which case the progeny will be, not a fish, but a reptile. To protract the *straightforward part of the gestation over a small space*—and from species to species the space would be small indeed—is all that is necessary.

Ideas to Think About

1. Examine the arguments, and evaluate the evidence on which they are based, that led Cuvier to oppose the idea of uniform past and present causes of geological events. Compare these arguments with Buffon's and Lyell's reasoning in favor of uniformity.

2. How does Cuvier defend his conclusion that the succession of animal forms in the fossil record does not represent gradual modification?
3. How does Geoffroy Saint-Hilaire defend his contrary conclusion that animal forms have evolved? In this disagreement with Cuvier, how does the relationship between form and function, as envisioned by the two men, influence their positions? You may also wish to compare these ideas with the similar ones of Buffon, particularly on the influence of nature on species variation.
4. In what ways does Geoffroy's "body plan" concept differ from or agree with Owen's "unity of plan"?
5. Examine the claim of Chambers regarding the progressive evolution of higher forms in the dual framework of geology and embryology that he used to support it.
6. What role does a deity play in the interpretations of Cuvier, Geoffroy, Owen, Lyell, and Chambers? In which cases do you think that belief determines explanation, rather than observation determining belief?

FOR FURTHER READING

The Watch on the Heath: Science and Religion Before Darwin, Keith Thomson. London: Harper Collins (2005). This marvelously written book brings together history and philosophy in a revealing examination of the often uneasy but sometimes rewarding relationship between theological concerns with the nature of life and the universe and the growing attempts of science to render it objectively knowable. It spans the seventeenth through the nineteenth centuries, with carefully woven backtracks and anticipations of today's creationism. A delightfully informative read.

The Measure of God, Larry Witham. London: Harper Collins (2005). This is the story of the Gifford Lectures, endowed in 1887 by Scottish jurist Adam Gifford in a bequest to the four great universities in Scotland to "promote and diffuse the study of Natural Theology in the widest sense of the term—in other words, the knowledge of God." The annual series—continuing in the present—was intended to explore the reconciliation of science and religion. This well-written book examines past lectures and lecturers representing history, the humanities, and nearly all of the sciences.

The Heyday of Natural History, Lynn Barber Cardiff. New York: Doubleday (1984). This is an excellent treatment of Victorian science and those who championed it: Buckland, Lyell, Huxley, Darwin, and most of the others.

UNIT 7

The Age of Darwin, I: The Beagle

> [M]y success as a man of science, whatever this may have amounted to, has been determined, as far as I can judge, by complex and diversified mental qualities and conditions.... With such moderate abilities as I possess, it is truly surprising that I should have influenced to a great extent the belief of scientific men on some important points.
>
> —CHARLES DARWIN, *Autobiography*

These next four units, now that we have arrived at his threshold in time, focus on the most central and enigmatic figure in evolution: Charles Robert Darwin. We are all products of our times, and a deep reading of the extended family of the Darwins and Wedgwoods shows the numerous threads—through both lineal and collateral kin—that weave themselves into the fabric of this person.

32. CHARLES DARWIN

We begin with Erasmus Darwin, Charles's paternal grandfather, whose poetic and scientific mind helped to foster the Enlightenment and the Industrial Revolution. His Midland chums included Josiah Wedgwood 1st, one of the most successful of the pottery dynasty. Josiah Wedgwood married Sarah Wedgwood, a distant cousin, in 1754. Erasmus's son, Robert Waring Darwin (1766–1848), married Josiah's daughter Susannah (1765–1817) in 1796.

It was to become a close-knit family: Susanna's brother Josiah 2nd was Darwin's favorite uncle "Jos" (1769–1843)—and his father-in-law as well, as Charles married Jos's third child, and his maternal parallel cousin, Emma, in 1839. Josiah 2nd's second child, Josiah 3rd (1795–1880), married Charles Darwin's older sister, Caroline (1800–1888). Josiah 2nd's first child, Charlotte, married the Rev. Charles Langton. After her death in 1862, Langton married Charles's oldest sister, Emily Catherine, a spinster in her sixties at the time. If this appears confusing (and there are other Wedgwood-Darwin alliances still later), such marriages were not at all uncommon—particularly among the wealthy, as both families were.

Darwin was quick to acknowledge and appreciate this wealth. "I have had ample leisure from not having to earn my own bread," he wrote in his autobiography. The Wedgwood dynasty began in Burselm (halfway between Birmingham and Manchester), a center of pottery making in the Midlands. The Wedgwood family operated six of the forty-two pot works there, and pioneered in numerous innovations in style, manufacture, and chemistry as they built their fortune. Darwin's father was an esteemed and successful physician as well as an astute businessman. Robert Darwin and Susan Wedgwood grew up together and always knew that they would someday marry. Both families—wealthy, deep in public esteem, genteel in social circles, faithful in all of their obligations to God and country—were thus an integral part of the new social order while carrying on the traditions of the old.

Charles Darwin was born in Shrewsbury, Shropshire, on February 12, 1809, the fourth of six children. Four were girls—Marianne and Susan Elizabeth, both older, and Caroline and Catherine. Charles's brother Erasmus "Ras" was five years older, and the two were particularly close. Both were fascinated by nature when still young—Charles with collecting natural objects and Ras with chemical experiments.

While Charles's fondest childhood memories were these extracurricular activities in science, formal schooling left him wanting. His seven years at school in Shrewsbury, he claimed, were basically wasted time. "The school as a means of education to me was simply a blank," he wrote in his *Autobiography* (p. 32). His lack of interest was of concern to his teachers and his family. "When I left the school I was for my age neither high nor low in it," he wrote, "and I believe that I was considered by all my masters and by my father as a very ordinary boy, rather below the common standard in intellect."

Hoping that Charles's interest in biology would find translation in medicine, his physician-father sent him to the University of Edinburgh in 1825, where brother Ras was already completing his first year. Charles, however, was distressed by clinical visits to ailing patients and bored with the lectures. When Ras left the university after Darwin's first year, Charles expanded his circle of friends to include many in the natural sciences. Chief among these was Robert Grant, who gave up his medical practice to study marine biology—the nearby Firth of Forth provided an ideal source of study—and became a great supporter of evolutionary ideas. He admired Geoffroy

Saint-Hilaire and had cited Darwin's grandfather Erasmus in his dissertation.

Grant and Darwin became good friends, despite an age difference of several years. "One day, when we were walking together," Darwin recalled, "he burst forth in high admiration of Lamarck and his views on evolution. I listened in silent astonishment,..." (*Autobiography*, 38). Being surrounded by influences of scientific freethinkers and political radicals no doubt influenced Darwin's materialist views, although he denied reflecting on it at the time. "Nevertheless," he wrote, "it is probable that the hearing rather early in life such views maintained and praised may have favoured my upholding them under a different form in my 'Origin of Species.'"

Once again, however, Darwin's growing interests were fed not by academic studies but by associations and activities outside the university. Darwin's father lamented this lack of academic focus, as he had in Shrewsbury, and reluctantly accepted that Charles was not destined to become a physician. What, then? The only other esteemed and acceptable vocation for a member of the upper class was in the clergy. With what was a final effort to protect his son from "turning into an idle sporting man" and disgracing the family name, Robert Darwin convinced Charles to study for the clergy. Bidding farewell to Edinburgh, Charles entered Christ's College, Cambridge in 1828.

There, he continued his interest in collecting, particularly beetles, influenced by his second cousin, William Darwin Fox, a clergyman and entomologist. Fox introduced Charles to John Stevens Henslow, a naturalist who had graduated in 1818 and later become professor of mineralogy at Cambridge. Also in 1818, Adam Sedgwick became Woodwardian Professor of Geology at Cambridge. Henslow became a lifelong friend and supporter of Charles Darwin, while Sedgwick gave Charles his first field training in geology. Henslow's daughter married Joseph Dalton Hooker, director of the famous Kew Gardens in London for twenty years. Hooker was to become Darwin's closest friend as he labored over the notes from the *Beagle* and began writing the *Origin of Species*.

This tight circle of friends and colleagues provided the crucible for Darwin's career in science, as well as for his personal conflict with religious dogma—for most of these friends were also devout in their beliefs, some of them buoyed by a literalist interpretation of scripture. Their support of his science on the *Beagle,* however, and their careful study, back in England, of the specimens he sent home from South America, were instrumental not only in advancing knowledge but in promoting the reputation of Darwin as a scientist.

He took his exams for the bachelor's degree in 1831—scoring tenth out of 178—and was presumably on his way to a divinity degree. The Beagle voyage fatefully interrupted this trajectory. Once again, Darwin's passion lay outside academic education and outside professional preparation, and he was again a source of disappointment for his father. First medicine, and now the clergy, were cast aside. Charles, however, was content as he closed out his school years. "Upon the whole the three years which I spent at Cambridge were the most joyful in my happy life; for I was then in excellent health, and almost always in high spirits," Charles wrote later. What happened next could not have been anticipated by either Charles or his father. As usual, the road he would follow was chosen for him by circumstance, not by the careful and deliberative choice which Victorian custom came to identify with the educated gentleman.

The Correspondence of Charles Darwin, Volume 1: 1821–1836

Letter 105—Henslow, J. S., to Darwin, C. R., 24 Aug. 1831

In August of 1831, Charles Darwin accompanied Adam Sedgwick for a geological trip to North Wales. It was on this trip that Charles learned many of the field techniques and observations that he would find so useful on his coming voyage. Upon returning to Shrewsbury, he found the following letter from Henslow, dated August 24, 1831.

MY DEAR DARWIN,

Before I enter upon the immediate business of this letter, let us console together upon the loss of our inestimable friend poor Ramsay, of whose death you have undoubtedly heard long before this.

I will not now dwell upon this painful subject, as I shall hope to see you shortly, fully expecting that you will eagerly catch at the offer which is likely to be made you of a trip to Tierra del Fuego, and home by the East Indies. I have been asked by Peacock, who will read and forward this to you from London, to recommend him a Naturalist as companion to Captain Fitz-Roy, employed by Government to survey the southern extremity of America. I have stated that I consider you to be the best qualified person I know of who is likely to undertake such a situation. I state this not in the supposition of your being a *finished* naturalist, but as amply qualified for collecting, observing, and noting, anything worthy to be noted in Natural History. Peacock has the appointment at his disposal, and if he cannot find a man willing to take the office, the opportunity will probably be lost. Captain Fitz-Roy wants a man (I understand) more as a companion than a mere collector, and would not take any one, however good a naturalist, who was not recommended to him likewise as a *gentleman.* Particulars of salary, &c., I know nothing. The voyage is to last two years, and if you take plenty of books with you, anything you please may be done. You will have ample opportunities at command. In short, I suppose there never was a finer chance for a man of zeal and spirit; Captain Fitz-Roy is a young man. What I wish you to do is instantly to come and consult with Peacock (at No. 7 Suffolk Street, Pall Mall East, or else at the University Club), and learn further particulars. Don't put on any modest doubts or fears about your disqualifications, for I assure you I think you are the

very man they are in search of; so conceive yourself to be tapped on the shoulder by your bum-bailiff and affectionate friend,
J. S. HENSLOW.
The expedition is to sail on 25th September (at earliest), so there is no time to be lost.

Enclosed with letter 105 was the actual offer from George Peacock, who was a math tutor at Cambridge. Beaufort, the hydrologist at the admiralty, had been asked by Capt. FitzRoy to find a suitable companion for the Beagle *voyage. Beaufort turned to his friend Peacock, who in turn asked Henslow. Peacock had discretionary power to make the appointment—subject to the captain's personal okay. In his letter to Darwin, Peacock wrote, "I received Henslow's letter last night too late to forward it to you by the post; a circumstance which I do not regret, as it has given me an opportunity of seeing Captain Beaufort at the Admiralty (the Hydrographer), and of stating to him the offer which I have to make to you. He entirely approves of it, and you may consider the situation as at your absolute disposal. I trust that you will accept it, as it is an opportunity which should not be lost, and I look forward with great interest to the benefit which our collections of Natural History may receive from your labours" (Francis Darwin,* Life and Letters, *192). Darwin had approached his father about Henslow's tentative offer, but his father strongly objected, giving several reasons, and stating, "If you can find any man of common sense who advises you to go I will give my consent." An obedient son, Darwin immediately wrote the following response to Henslow.*

Letter 107—Darwin, C. R., to Henslow, J. S., 30 Aug. 1831

Shrewsbury, Tuesday [August 30, 1831].
MY DEAR SIR,

Mr. Peacock's letter arrived on Saturday, and I received it late yesterday evening. As far as my own mind is concerned, I should, I think *certainly,* most gladly have accepted the opportunity which you so kindly have offered me. But my father, although he does not decidedly refuse me, gives such strong advice against going, that I should not be comfortable if I did not follow it.

My father's objections are these: the unfitting me to settle down as a Clergyman, my little habit of seafaring, *the shortness of the time,* and the chance of my not suiting Captain Fitz-Roy. It is certainly a very serious objection, the very short time for all my preparations, as not only body but mind wants making up for such an undertaking. But if it had not been for my father I would have taken all risks. What was the reason that a Naturalist was not long ago fixed upon? I am very much obliged for the trouble you have had about it; there certainly could not have been a better opportunity....

Yours most sincerely,
My dear Sir,
CH. DARWIN.

I have written to Mr. Peacock, and I mentioned that I have asked you to send one line in the chance of his not getting my letter. I have also asked him to communicate with Captain Fitz-Roy. Even if I was to go, my father disliking would take away all energy, and I should want a good stock of that. Again I must thank you, it adds a little to the heavy but pleasant load of gratitude which I owe to you.

Letter 110—Darwin, C. R. to Darwin, R. W., 31 Aug. 1831

The morning following, Charles rode to Maer (about twenty-five miles northeast of Shrewsbury) to prepare for a hunting trip with his Uncle Jos at the Wedgewood estate, Maer Hall. Darwin mentioned the offer, and that his father had encouraged him to get a second opinion from his brother-in-law. Josiah encouraged him to accept and offered to write to Robert. After lengthy discussion, Darwin wrote his father the following letter, attaching one from Josiah pleading on Darwin's behalf. The two of them then immediately rode back to Darwin's home, The Mount, to allow an oral argument on Charles's behalf.

[Maer] August 31 [1831].
MY DEAR FATHER,

I am afraid I am going to make you again very uncomfortable. But, upon consideration, I think you will excuse me once again, stating my opinions on the offer of the voyage. My excuse and reason is the different way all the Wedgwoods view the subject from what you and my sisters do.

I have given Uncle Jos what I fervently trust is an accurate and full list of your objections, and he is kind enough to give his opinions on all. The list and his answers will be enclosed. But may I beg of you one favour, it will be doing me the greatest kindness, if you will send me a decided answer, yes or no? If the latter, I should be most ungrateful if I did not implicitly yield to your better judgment, and to the kindest indulgence you have shown me all through my life; and you may rely upon it I will never mention the subject again. If your answer should be yes; I will go directly to Henslow and consult deliberately with him, and then come to Shrewsbury.

The danger appears to me and all the Wedgwoods not great. The expense can not be serious, and the time I do not think, anyhow, would be more thrown away than if I stayed at home. But pray do not consider that I am so bent on going that I would for one *single moment* hesitate, if you thought that after a short period you should continue uncomfortable.

I must again state I cannot think it would unfit me hereafter for a steady life. I do hope this letter will not give you much uneasiness. I send it by the car to-morrow morning; if you make up your mind directly will you send me an answer on the following day by the same means? If this letter should not find you at home, I hope you will answer as soon as you conveniently can.

I do not know what to say about Uncle Jos' kindness; I never can forget how he interests himself about me.

Believe me, my dear father,
Your affectionate son,
CHARLES DARWIN.

[Here follows the list of objections which are referred to in the following letter:—

(1) Disreputable to my character as a Clergyman hereafter.
(2) A wild scheme.
(3) That they must have offered to many others before me the place of Naturalist.
(4) And from its not being accepted there must be some serious objection to the vessel or expedition.
(5) That I should never settle down to a steady life hereafter.
(6) That my accommodations would be most uncomfortable.
(7) That you [*i.e.* Dr. Darwin] should consider it as again changing my profession.
(8) That it would be a useless undertaking.]

Letter 109—Wedgwood, Josiah II, to Darwin, R. W., 31 Aug. 1831

Maer, August 31, 1831.

[Read this last.]*

MY DEAR DOCTOR,

I feel the responsibility of your application to me on the offer that has been made to Charles as being weighty, but as you have desired Charles to consult me, I cannot refuse to give the result of such consideration as I have been able to [give?] it.

Charles has put down what he conceives to be your principal objections, and I think the best course I can take will be to state what occurs to me upon each of them.

1. I should not think that it would be in any degree disreputable to his character as a Clergyman. I should on the contrary think the offer honourable to him; and the pursuit of Natural History, though certainly not professional, is very suitable to a clergyman.

2. I hardly know how to meet this objection, but he would have definite objects upon which to employ himself, and might acquire and strengthen habits of application, and I should think would be as likely to do so as in any way in which he is likely to pass the next two years at home.

3. The notion did not occur to me in reading the letters; and on reading them again with that object in my mind I see no ground for it.

4. I cannot conceive that the Admiralty would send out a bad vessel on such a service. As to objections to the expedition, they will differ in each man's case, and nothing would, I think, be inferred in Charles's case, if it were known that others had objected.

5. You are a much better judge of Charles's character than I can be. If on comparing this mode of spending the next two years with the way in which he will probably spend them, if he does not accept this offer, you think him more likely to be rendered unsteady and unable to settle, it is undoubtedly a weighty objection. Is it not the case that sailors are prone to settle in domestic and quiet habits?

*[In C. Darwin's writing.]

6. I can form no opinion on this further than that if appointed by the Admiralty he will have a claim to be as well accommodated as the vessel will allow.

7. If I saw Charles now absorbed in professional studies I should probably think it would not be advisable to interrupt them; but this is not, and, I think, will not be the case with him. His present pursuit of knowledge is in the same track as he would have to follow in the expedition.

8. The undertaking would be useless as regards his profession, but looking upon him as a man of enlarged curiosity, it affords him such an opportunity of seeing men and things as happens to few.

You will bear in mind that I have had very little time for consideration, and that you and Charles are the persons who must decide.

I am,
My dear Doctor,
Affectionately yours,
JOSIAH WEDGWOOD.

Letter 114—Darwin, C. R., to Henslow, J. S., 2 Sept. 1831

Out of deep respect for Josiah's reasoning, and in deference to his son's obvious yearning to go, Charles's father gave his consent. Thrilled, and fearful that he may be too late, Charles sent a hasty letter rejecting his earlier refusal and left for Cambridge at three o'clock in the morning, September 2, on an express coach. From the inn there, much fatigued from his journey, he dispatched the following terse note to Henslow.

Cambridge, Red Lion, 1831.
MY DEAR SIR,
I am just arrived; you will guess the reason. My father has changed his mind. I trust the place is not given away.
I am very much fatigued, and am going to bed.
I dare say you have not yet got my second letter.
How soon shall I come to you in the morning? Send a verbal answer.
Good night,
Yours,
C. DARWIN.

This was to be FitzRoy's second journey to South America aboard the Beagle *for the purpose of surveying the coast, and the need for a naturalist on this scientific voyage was obvious. However, it was to be a five-year voyage on a small ship—a 24-ton brig only ninety feet in length—and FitzRoy was understandably anxious to chose a congenial and compatible companion. Henslow secured an appointment for Charles with the captain on September 5th, and Darwin hurried there. As fortune would have it, the two got on well in this and subsequent meetings. FitzRoy, a devout biblical literalist, was in hopes that Darwin would find geological support for Genesis, and was encouraged that Charles—who at that time fully accepted the proper Victorian position that the flood had actually happened—was at least sympathetic to the prospect. In the brief autobiographical chapter of Francis Darwin's* Life and Letters, *Charles reflects on FitzRoy long after the voyage.*

33. Francis Darwin

Life and Letters of Charles Darwin

[60] Fitz-Roy's character was a singular one, with very many noble features: he was devoted to his duty, generous to a fault, bold, determined, and indomitably energetic, and an ardent friend to all under his sway. He would undertake any sort of trouble to assist those whom he thought deserved assistance. He was a handsome man, strikingly like a gentleman, with highly courteous manners,...

Fitz-Roy's temper was a most unfortunate one. It was usually worst in the early morning, and with his eagle eye he could generally detect something amiss about the ship, and was then unsparing in his blame. He was very kind to me, but was a man very difficult to live with on the intimate terms which necessarily followed from our messing by ourselves in the same cabin. We had several quarrels; for instance, early in the voyage at Bahia, in Brazil, he defended and praised [61] slavery, which I abominated, and told me that he had just visited a great slave-owner, who had called up many of his slaves and asked them whether they were happy, and whether they wished to be free, and all answered "No. " I then asked him, perhaps with a sneer, whether he thought that the answer of slaves in the presence of their master was worth anything? This made him excessively angry, and he said that as I doubted his word we could not live any longer together.

34. Charles Darwin

The Correspondence of Charles Darwin, Volume 1: 1821–1836

Letter 171—Darwin, C. R., to Henslow, J. S., 16 June 1832

The Beagle *left Devonport, Plymouth, two days after Christmas in 1831, arriving back at Falmouth, on England's far southwestern peninsula, on the second of October 1836. Despite his great excitement over the coming scientific opportunities, Darwin knew he was new at this entire game. He had only a preliminary knowledge of nature's trove and of the layered rocks that had captured its earlier forms. He was an experienced insect collector, but was unsure of his own taxonomic skills. He needed constant reaffirmation from Henslow, back in Cambridge, to whom he sent his collections in periodic shipments. Was he doing well? Were his notes and labels accurate? Were his ideas reasonable? In his first letter to Henslow, he mixes the wonder and amazement of a small child with the self-doubt of a novice scientist.*

Rio de Janeiro.
May 18th. 1832
My dear Henslow.—

I have delayed writing to you till this period as I was determined to have a fair trial of the voyage. I have so many things to write about, that my head is as full of oddly assorted ideas, as a bottle on the table is with animals.—You being my chief Lord of the Admiralty, must excuse this letter being full of my's & I's.—After our two attempts to put to sea in spite of the S.W.ly gales, the time at Plymouth passed away very unpleasantly.—I would have written, only I had nothing to say, excepting what had better be left unsaid: so that I only wrote to Shrewsbury.—At length we started on ye 27th of December with a prosperous wind, which has lasted during our whole voyage:—The two little peeps at seasick misery gave me but a faint idea of what I was going to undergo.—Till arriving at Teneriffe (we did not touch at Madeira) I was scarcely out of my hammock & really suffered more than you could well imagine from such a cause.—At Santa Cruz, whilst looking amongst the clouds for the Peak & repeating to myself Humboldt's sublime descriptions, it was announced we must perform 12 days strict quarantine.—We had made a short passage so "Up Jib" & away for St Jago.—You will say all this sounds very bad, & so it was: but from that to the present time it has been nearly one scene of continual enjoyment.—A net over the stern kept me at full work, till we arrived at St Jago: here we spent three most delightful weeks.—The geology was preeminently interesting & I believe quite new: there are some facts on a large scale of upraised coast (which is an excellent epoch for all the Volcanic rocks to [be] dated from) that would interest Mr. Lyell.—One great source of perplexity to me is an utter ignorance whether I note the right facts & whether they are of sufficient importance to interest others.—

On the coast I collected many marine animals chiefly gasteropodous (I think some new).—I examined pretty accurately a Caryophyllea & if my eyes were not bewitched former descriptions have not the slightest resemblance to the animal.... Geology & the invertebrate animals will be my chief object of pursuit through the whole voyage.... I find my life on board, when we are in blue water most delightful; so very comfortable & quiet: it is almost impossible to be idle, & that for me is saying a good deal.—Nobody could possibly be better fitted out in every respect for collecting than I am: many cooks have not spoiled the broth this time;... Touching at the Abrolhos [*an island on the coastal shelf 65 km from the Brazilian mainland.—RKW*], we arrived here on April 4th; when amongst others I received your most kind letter: you may rely on it, during the evening, I thought of the many most happy hours I have spent with you in Cambridge.—I am now living at Botofogo, a village about a league from the city [*Rio de Janeiro.—RKW*], & shall be able to remain a month longer.—The Beagle has gone back to Bahia, & will pick me up on its return.—There is a most important error in the longitude of S America, to settle which this second trip has been undertaken.... A few days after arriving I started on an expedition of 150 miles to Rio Macaò, which lasted 18 days.—Here I first saw a Tropical forest in all its sublime grandeur.—Nothing, but the reality can give any idea, how wonderful, how magnificent the scene is.... I never experienced such intense delight.—I formerly admired Humboldt, I now almost adore him; he alone gives any notion, of the feelings which are raised in the mind on first entering the Tropics....

June 16th.—I have determined not to send a box till we arrive at Monte Video.—it is too great a loss of time both for Carpenters & myself to pack up whilst in harbor.—I am afraid when I do send it, you will be disappointed, not having skins of birds & but very few plants, & geological specimens small: the rest of the things in bulk make very little show....

We sail for Monte Video at the end of this month (June) so that I shall have been here nearly 3 months.—this has been very lucky for me.—as it will be some considerable period before we again cross the Tropic.—I am sometimes afraid I shall never be able to hold out for the whole voyage. I believe 5 years is the shortest period it will consume.—The mind requires a little case-hardening, before it can calmly look at such an interval of separation from all friends.—Remember me most kindly to M^{rs}. Henslow & the <t>wo Signoritas....

Excuse this almost unintelligible letter & believe me dear Henslow—with the warmest feelings of respect & friendship
Yours affectionately
Chas Darwin

The letter Darwin received from Henslow when he arrived in Brazil was a solicitous one—he had already been informed of Darwin's sickness at sea and imagined him disconsolate. He tries here to raise his spirits and encourages him to stick with the voyage as long as he can.

Letter 157—Henslow, J. S., to Darwin, C. R., 6 Feb. 1832

Cambridge
6 Feby 1832
My dear Darwin,
As tomorrow is the first Tuesday in Feby. I select today (my Birthday) for keeping my promise, virtually made by your asking me to write to you on that day. I heard of your adventurous departure thro' Mr Yarrel who was told of it by Capt. King—You had a stout heart to resist the inclination which must necessarily have come over you not to go on whilst you were in such a wretched state of sickness as you are described to have suffered. I trust however that it left you soon afterwards & that you are now an experienced sailor—....

I sometimes blame myself for having hinted to you rather plainly little pieces of advice, lest you should have thought me troublesome—but I am sure your good heart will ascribe my suggestions to the right motive of my being anxious for your happiness—which cannot be enjoyed in this troublesome world without daily restraint & submission to mortifications sometimes trifling sometimes grievous—always when patiently taken refreshing to the spirit—I feel the more anxious for you as I have been so mainly instrumental in your adopting the plans you have—& should your time pass unhappily shall never cease to regret my having recommended you to take the step you have of devoting yourself to the cause of science—Much therefore as [I] should like to see you return laden with the spoils of the Worl<d> yet if you do not find yourself *content* I should much rather see you sooner than I hope to do—You must by the time you get this have had ample experience how far you are qualified to cope with difficulties & whether you can rise superior to them—whether you can enjoy yourself amidst them, & rejoice over them—If you have met with success hitherto be assured that you may go on safely & securely <to> the end—but if you have failed, then don't try any more—but come away—You will only be heaping up greater troubles—I shall endeavor to keep up our correspondence as you may be pleased to direct me how I am to succeed in getting my letters to you, & shall always write to you as freely as I can on the subject of your enterprize, judging from what I can learn from your letters may be the state of your wishes—
Believe me y^{rs} ever affectionately & sincerely J S. Henslow

Letter 196—Henslow, J. S., to Darwin, C. R., 15 & 21 Jan. 1833

When the next crate from South America arrived, Henslow sought to encourage Darwin that his specimens were enthusiastically received. The following letter, although dated 1832, was actually written in 1833—an error most of us make in the early days of any January!

Cambridge
15 Jany 1832
My dear Darwin,

I shall begin a letter to you lest something or other should persuade me to defer it till it becomes too late for the next packet—Wood & I had intended writing by the Decr. packet, but just as was about to do so your letter arrived stating that a Box was on its road, so I thought I had better delay till I had seen its contents. It is now here & every thing has travelled well....

Your account of the Tropical forest is delightful, I can't help envying you—So far from being disappointed with the Box—I think you have done wonders—as I know you do not confine yourself to collecting, but are careful to describe—Most of the plants are very desirable to *me*...

Every individual specimen once arrived here becomes an object of great interest, & tho' you were to send home 10 times as much as you do, yet when you arrive you will often think & wish how you might & had have sent home 100 times as much! things which seemed such rubbish—but now so valuable—However no one can possibly say you have not been active—& that your box is not capital. I shall not wait for Sedgwicks return before I send this but must give you an account of the Geol^{c}. spec^{s}. in the next—I shall now forward this with the vol. of the Dict. Class. to your Brother & wish you a continuance of good success. I

have no fears of your being tired of the expedition whilst you continue to meet with such as you have hitherto, & hope your spirits will not fail you in those dull moments which must occasionally intervene, during the progress of so long an undertaking. Downes & other friends have begged me to remember them to you most kindly & affectionately & M[rs] Henslow adds her best wishes—Mine you well know are ever with you & I need not add that you s[d] believe me
Most affect[ly]. & sincerely y[rs].
J S Henslow

Narrative of the Surveying Voyages of His Majesty's Ships Adventure and Beagle

Darwin took Charles Lyell's volume I of Principles of Geology *on the voyage. Volume II was awaiting him in Montevideo, Uruguay, in early November 1832. He received Volume III in Valparaiso, Chile, in July 1834. He devoured them all. Deep time, gradual change, and fossil connections with the present all slowly replaced the young earth, catastrophism, and fixity of species, as he encountered evidence on the Patagonian coast. The fossils, in particular, in their similarity to current forms, intrigued him. One of his most revealing discoveries, leading to the realization that extinct and living forms bear connections, is recorded in the following journal passage.*

[208] On the south side of the harbor [*St. Julien, 1834.—RKW*], a cliff of about ninety feet in height intersects a plain constituted of the formations above described; and its surface is strewed over with recent marine shells.... In one spot this earthy matter filled up a hollow, or gully, worn quite through the gravel, and in this mass a group of large bones was embedded. The animal to which they belonged, must have lived, as in the case at Bahia Blanca, at a period long subsequent to the existence of the shells now inhabiting the coast. We may feel sure of this, because the formation of the lower terrace or plain, must necessarily have been posterior to those above it, and on the surface of the two higher ones, sea-shells of recent species are scattered. From the small physical change, which the last one hundred feet elevation of the continent could have produced, the climate, as well as the general condition of Patagonia, probably was nearly the same, at the time when the animal was embedded, as it now is. This conclusion is moreover supported by the identity of the shells belonging to the two ages.... I had no idea at the time, to what kind of animal these remains belonged. The puzzle, however, was soon solved when Mr. Owen examined them; for he considers that they formed part of an animal allied to the guanaco or llama, but fully as large as the true camel. As all the existing members of the family of Camelidæ are inhabitants of the most sterile countries, so may we suppose was this extinct kind.

[209] The most important result of this discovery, is the confirmation of the law that existing animals have a close relation in form with extinct species. As the guanaco is the characteristic quadruped of Patagonia, and the vicuna of the snow-clad summits of the Cordillera, so in bygone days, this gigantic species of the same family must have been conspicuous on the southern plains.

Journal of Researches into the Natural History and Geology of the Countries Visited

[173] This wonderful relationship in the same continent between the dead and the living, will, I do not doubt, hereafter throw more light on the appearance of organic beings on our earth, and their disappearance from it, than any other class of facts.

The Zoology of the Voyage of H.M.S. Beagle. Part III: Birds

As important as the revelation that earlier forms show morphological relationship to living forms was to Darwin, this confirmation of the law of succession (recognized by Cuvier) did not immediately drive Darwin's imagination to abandon immutability. It was an important first step. The decisive observations would come later in the Galapagos. In retrospect, we can appreciate the critical sequential nature of Darwin's explorations: Novel ideas seldom appear in sui generis flashes; rather, they build by accretion from experiences tucked away in the mind's recesses. The emergent idea of transmutation through ancient time had to await another emergent idea: transmutation across space as adaptation in isolation. Darwin was helped in recognizing this when he observed separate mockingbird species (genus Mimus) in the islands.

[63] There are five large islands in this Archipelago, and several smaller ones. I fortunately happened to observe, that the specimens which I collected in the two first islands we visited, differed from each other, and this made me pay particular attention to their collection. I found that all in Charles Island belonged to *M. trifasciatus*; all in Albemarle Island to *M. parvulus*, and all in Chatham and James's Islands to *M. melanotus*. I do not rest this fact solely on my own observation, but several specimens were brought home in the Beagle, and they were found, according to their species, to have come from the islands as above named. Charles Island is distant fifty miles from Chatham Island, and thirty-two from Albemarle Island. This latter is only ten miles from James Island, yet the many specimens procured from both belonged respectively to different species. James and Chatham, which possess the same species, are seventy miles apart, but Indefatigable Island is situated between them, which perhaps, has afforded a means of communication.

Darwin's Ornithological Notes (1836)

The following is Darwin's brief discussion of two mockingbird specimens (M. thenca), specimen numbers 3306 and 3307, found on Charles and Chatham Islands, respectively. Brackets are Barlow's notes.

[262] I have specimens from four of the larger Islands; the two above enumerated, and (3349: female. Albermarle Isd.) & (3350: male: James Isd).—The specimens from Chatham & Albermarle Isd appear to be the same; but the other two are different. In each Isld. each kind is *exclusively* found: habits of all are indistinguishable. When I recollect, the fact that the form of the body, shape of scales & general size, the Spaniards can at once pronounce, from which Island any Tortoise may have been brought. When I see these Islands in sight of each other, & [but *del.*] possessed of but a scanty stock of animals, tenanted by these birds, but slightly differing in structure & filling the same place in Nature, I must suspect they are only varieties. The only fact of a similar kind of which I am aware, is the constant asserted difference—between the wolf-like Fox of East & West Falkland Islds.—If there is the slightest foundation for these remarks the zoology of Archipelagoes—will be well worth examining; for such facts [would *inserted*] undermine the stability of Species.

Journal of Researches into the Natural History and Geology

It is chapter XVII of Darwin's Journal of Researches into the Natural History and Geology of the Countries Visited during the Voyage of H.M.S. Beagle..., *later published as The Voyage of the Beagle, that covers the Galapagos. Darwin spent five weeks collecting among these sixteen main islands—remote by some six hundred miles from Ecuador. The primary interest in this chapter has traditionally been the revelations provided by the several species of finches, diversified through adaptive radiation and portrayed by the variety of beak sizes and shapes. Though myth and misinformation surround the more famous interpretations of this unit and the notebooks relating to it, it will be appropriate to quote from it here, and, in our following unit, comment on its significance.*

[372] *September 15th.*—This archipelago consists of ten principal islands, of which five exceed the others in size. They are situated under the Equator, and between five and six hundred miles westward of the coast of America. They are all formed of [373] volcanic rocks; a few fragments of granite curiously glazed and altered by the heat, can hardly be considered as an exception. Some of the craters, surmounting the larger islands, are of immense size, and they rise to a height of between three and four thousand feet. Their flanks are studded by innumerable smaller orifices. I scarcely hesitate to affirm, that there must be in the whole archipelago at least two thousand craters. These consist either of lava and scoriæ, or of finely-stratified, sandstone-like tuff. Most of the latter are beautifully symmetrical; they owe their origin to eruptions of volcanic mud without any lava: it is a remarkable circumstance that every one of the twenty-eight tuff-craters which were examined, had their southern sides either much lower than the other sides, or quite broken down and removed. As all these craters apparently have been formed when standing in the sea, and as the waves from the trade wind and the swell from the open Pacific here unite their forces on the southern coasts of all the islands, this singular uniformity in the broken state of the craters, composed of the soft and yielding tuff, is easily explained.

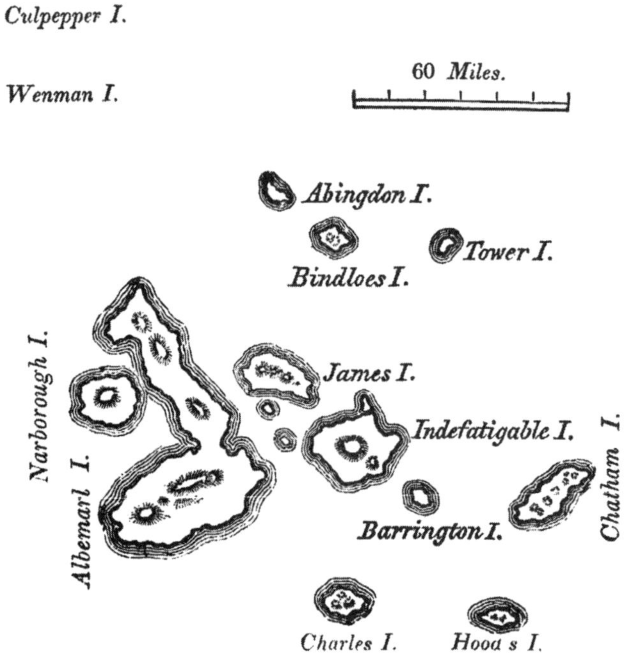

Culpepper I.
60 Miles.
Wenman I.
Abingdon I.
Tower I.
Bindloes I.
James I.
Narborough I.
Indefatigable I.
Chatham I.
Albemarl I.
Barrington I.
Charles I.
Hood's I.

Considering that these islands are placed directly under the equator, the climate is far from being excessively hot; this seems chiefly caused by the singularly low temperature of the surrounding water, brought here by the great southern Polar current. Excepting during one short season, very little rain falls, and even then it is irregular; but the clouds generally hang low. Hence, whilst the lower parts of the islands are very sterile, the upper parts, at a height of a thousand feet and upwards, possess a damp climate and a tolerably luxuriant vegetation. This is especially the case on the windward sides of the islands, which first receive and condense the moisture from the atmosphere.

In the morning (17th) we landed on Chatham Island, which, like the others, rises with a tame and rounded outline, broken here and there by scattered hillocks, the remains of former craters. Nothing could be less inviting than the first appearance. A broken field of black basaltic lava, thrown into the most rugged waves, and crossed by great fissures, is every where covered by stunted, sun-burnt brushwood, which shows little signs of life. The dry and parched surface, being heated by the noonday sun, gave to the air a close and sultry feeling, like that from [374] a stove: we fancied even that the bushes smelt unpleasantly. Although I diligently tried to collect as many plants as possible, I succeeded in getting very few; and such wretched-looking little weeds would have better become an arctic than an equatorial Flora. The brushwood appears, from a short distance, as leafless as our trees during winter; and it was some time before I discovered that not only almost every plant was now in full leaf, but that the greater number were in flower. The commonest bush is one of the Euphorbiaceæ: an acacia and a great odd-looking cactus are the only trees which afford any shade. After the season of heavy rains, the islands are said to appear for a short time partially green....

[377] The natural history of these islands is eminently curious, and well deserves attention. Most of the organic productions are aboriginal creations, found nowhere else; there is even a difference between the inhabitants of the different islands; yet all show a marked relationship with those of America, though separated from that continent by an open space of ocean, between 500 and 600 miles in width. The archipelago is a little world within itself, or rather a satellite attached to America, whence it has derived a few stray colonists, and has received the general character of its indigenous productions. Considering the small size of these islands, we feel the more astonished at the number of their aboriginal beings, and at their confined range. Seeing every [378] height crowned with its crater, and the boundaries of most of the lava-streams still distinct, we are led to believe that within a period, geologically recent, the unbroken ocean was here spread out. Hence, both in space and time, we seem to be brought somewhat near to that great fact—that mystery of mysteries—the first appearance of new beings on this earth....

Of land-birds I obtained twenty-six kinds, all peculiar to the group and found nowhere else, with the exception of one lark-like finch from North America (Dolichonyx oryzivorus), which ranges on that continent as far north as 54°, and generally frequents marshes. The other twenty-five birds consist, firstly, of a hawk, curiously intermediate in structure between a Buzzard and the American group of carrion-feeding Polybori; and with these latter birds it agrees most closely in every habit and even tone of voice. Secondly, there are two owls, representing the short-eared and white barn-owls of Europe. Thirdly, a wren, three tyrant fly-catchers (two of them species of Pyrocephalus, one or both of which would be ranked by some ornithologists as only varieties), and a dove—all analogous to, but distinct from, American species. Fourthly, a swallow, which though [379] differing from the *Progne purpurea* of both Americas, only in being rather duller coloured, smaller, and slenderer, is considered by Mr. Gould as specifically distinct. Fifthly, there are three species of mocking-thrush—a form highly characteristic of America. The remaining land-birds form a most singular group of finches, related to each other in the structure of their beaks, short tails, form of body, and plumage: there are thirteen species, which Mr. Gould has divided into four sub-groups. All these species are peculiar to this archipelago; and so is the whole group, with the exception of one species of the subgroup Cactornis, lately brought from Bow island, in the Low Archipelago. Of Cactornis, the two species may be often seen climbing about the flowers of the great cactus-trees; but all the other species of this group of finches, mingled together in flocks, feed on the dry and sterile ground of the lower districts. The males of all, or certainly of the greater number, are jet black; and the females (with perhaps one or two exceptions) are brown. The most curious fact is the perfect gradation in the size of the beaks in the different species of Geospiza, from one as large as that of a hawfinch to that of a chaffinch, and (if Mr. Gould is right in including his sub-group, Certhidea, in the main [380] group), even to that of a warbler. The largest beak in the genus Geospiza is shown in Fig. 1, and the smallest in Fig. 3; but instead of there being only one intermediate species, with a beak of the size shown in Fig. 2, there are no less than six species with insensibly graduated beaks. The beak of the sub-group Certhidea, is shown in Fig. 4. The beak of Cactornis is somewhat like that of a starling; and that of the fourth sub-group, Camarhynchus, is slightly parrot-shaped. Seeing this gradation and diversity of structure in one small, intimately related group of birds, one might really fancy that from an original paucity of birds in this archipelago, one species had been taken and modified for different ends. In a like manner it might be fancied that a bird originally a buzzard, had been induced here to undertake the office of the carrion-feeding Polybori of the American continent.

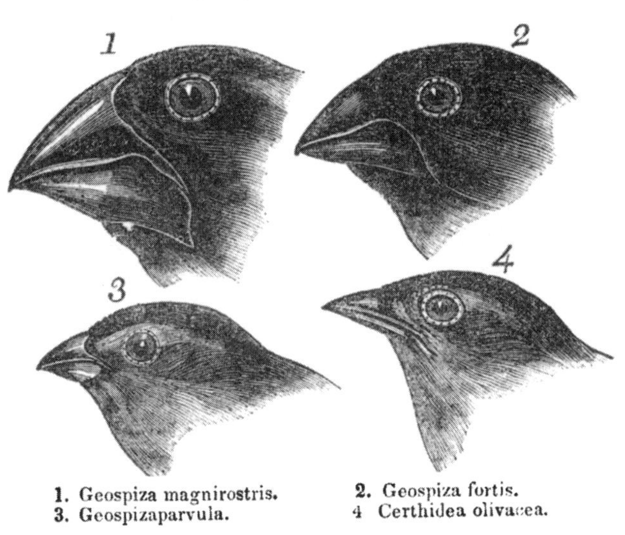

1. Geospiza magnirostris. **2.** Geospiza fortis.
3. Geospizaparvula. **4** Certhidea olivacea.

[393] I have not as yet noticed by far the most remarkable feature in the natural history of this archipelago; it is, that the different [394] islands to a considerable extent are inhabited by a different set of beings. My attention was first called to this fact by the Vice-Governor, Mr. Lawson, declaring that the tortoises differed from the different islands, and that he could with certainty tell from which island any one was brought. I did not for some time pay sufficient attention to this statement, and I had already partially mingled together the collections from two of the islands. I never dreamed that islands, about fifty or sixty miles apart, and most of them in sight of each other, formed of precisely the same rocks, placed under a quite similar climate, rising to a nearly equal height, would have been differently tenanted; but we shall soon see that this is the case. It is the fate of most voyagers, no sooner to discover what is most interesting in any locality, than they are hurried from it; but I ought, perhaps, to be thankful that I obtained sufficient materials to establish this most remarkable fact in the distribution of organic beings.

[397] The distribution of the tenants of this archipelago would not be nearly so wonderful, if, for instance, one

island had a mocking-thrush, and a second island some other quite distinct genus;—if one island had its genus of lizard, and a second island another distinct genus, or none whatever;—or if the different islands were inhabited, not by representative species of the same genera of plants, but by totally different genera, as does to a certain extent hold good; for, to give one instance, a large berry-bearing tree at James Island has no representative species in Charles Island. But it is the circumstance, that several of the islands possess their own species of the tortoise, mocking-thrush, finches, and numerous plants, these species having the same general habits, occupying analogous situations, and obviously filling the same place in the natural economy of this archipelago, that strikes me with wonder. It may be suspected that some of these representative species, at least in the case of the tortoise and of some of the birds, may hereafter prove to be only well-marked races; but this would be of equally great interest to the philosophical naturalist. I have said that most of the islands are in sight of each other: I may specify that Charles Island is fifty miles from the nearest part of Chatham Island, and thirty-three miles from the nearest part of Albemarle Island. Chatham Island is sixty miles from the nearest part of James Island, but there are two intermediate islands between them which were not visited by me. James Island is only ten miles from the nearest part of Albemarle Island, but the two points where the collections were made are thirty-two miles apart. I must repeat, that neither the nature of the soil, nor height of the land, nor the climate, nor the general character of the associated beings, and therefore their action one on another, can differ much in the different islands.

Narrative of the Surveying Voyages of His Majesty's Ships Adventure and Beagle

It should be noted that the book on the voyage was written and published long after Darwin's return to England, and after he had consulted with specialists who had been working on the collections he sent back. It was Gould, for example, who pointed out the correct taxonomy of the finches. Indeed, in the first edition of this work, the only mention of the finch population was a brief one.

[461] The following brief list will give an idea of their kinds. 1st. A buzzard, having many of the characters of Polyborus or Caracara; and in its habits not to be distinguished from that peculiar South American genus; 2d. Two owls; 3d. Three species of tyrant-flycatchers—a form strictly American. One of these appears identical with a common kind (*Muscicapa coronata*? Lath.), which has a very wide range, from La Plata throughout Brazil to Mexico; 4th. A sylvicola, an American form, and especially common in the northern division of the continent; 5th. Three species of mocking-birds, a genus common to both Americas; 6th. A finch, with a stiff tail and a long claw to its hinder toe, closely allied to a North American genus; 7th. A swallow belonging to the American division of that genus; 8th. A dove, like, but distinct from, the Chilian species; 9th. A group of finches, of which Mr. Gould considers there are thirteen species; and these he has distributed into four new sub-genera. These birds are the most singular of [462] any in the archipelago. They all agree in many points; namely, in a peculiar structure of their bill, short tails, general form, and in their plumage. The females are gray or brown, but the old cocks jet-black. All the species, excepting two, feed in flocks on the ground, and have very similar habits. It is very remarkable that a nearly perfect gradation of structure in this one group can be traced in the form of the beak, from one exceeding in dimensions that of the largest grosbeak, to another differing but little from that of a warbler.

Ideas to Think About

1. Populating the highly respectable Darwin genealogy were both physicians and theologians. When it was obvious that Charles Darwin desired neither, against his father's preferences, what social-intellectual contexts prevented his father from insisting that Darwin stick with one of these?
2. From the invitation to join the *Beagle* until halfway into the voyage, Darwin expresses doubt about his adequacy and competence to fulfill his mission. What led to this lack of self-confidence? How was this overcome?
3. What particular discoveries, and when, led Darwin to slowly abandon his belief in species fixity?
4. Distinguish, in the context of voyage chronology and particular discoveries, evidence favoring descent with modification (evolution over time) and evidence favoring the evolution of diversity (evolution over space).

For Further Reading

Darwin and the Beagle, Alan Moorehead. New York: Crescent Books (1969). This is arguably the best popular treatment of Darwin's voyage. Profusely illustrated and entertainingly written, but quite accurate, the book is hard to put down short of completing it! An excellent chronology of the voyage is appended.

Reef Madness, David Dobbs. New York: Pantheon Books (2005). Anyone who seeks to truly appreciate the deep divisions that accompanied Victorian science must read this excellent book. Science in this time sought to rectify misunderstandings of the natural world and reconcile the ideological conflicts about how that understanding is best achieved. This debate within science is almost always overshadowed by the more dramatic science-theology conflict. This delightful and poignant treatment of the two-generation conflict between Louis and Alexander Agassiz and Charles Darwin reflects the methodological war over the proper conduct of science, which in many ways continues today.

UNIT 8

The Age of Darwin, II: After the Voyage

> If we choose to let conjecture run wild, then animals, our fellow brethren in pain, disease, death, suffering and famine—our slaves in the most laborious works, our companions in our amusements—they may partake from our origin in one common ancestor—we may be all netted together.
>
> —Charles Darwin, *Notebook B (1837–38)*

The idea of transmutation was there long before Darwin discovered its sufficient mechanism. As his predecessors have revealed, the problem of species transformation was discussed and debated a generation before Darwin. The value of his long journey lay in the abundant evidence for the sundry faces of transmutation and his growing conviction of its validity. Here he found the fossils that strongly urged the common descent of current forms. Here he identified the strength of isolation in enhancing variation. Here he discovered the relationship of spatial distance, in addition to distance in time, to degrees of species separation. On this journey, he found innumerable instances of specific adaptations to environmental variation and change. But it was not here that he found his famous mechanism of natural selection. This he discovered after the journey.

It is our good fortune that Darwin kept ample and well-organized notes during and after the voyage—notes to which he referred in recapitulating details of the voyage itself, as well as details of his own evolving ideas. We see in these and in his correspondence the methodological approach he took in establishing his case, the care with which he sought and evaluated the scientific opinions of others, and the underlying caution—fueled by anticipated hostility to his theory—that guided his actions.

This critical period of twenty-three years, from the arrival of the *Beagle* at Falmouth on October 2, 1836, to the actual publication of the *Origin* on November 24, 1859 (Darwin wrote the *Introduction* on October 1, 1859), is represented in the readings in this unit. This was a period mixing delight with pain: Darwin married his cousin, Emma Wedgwood, and settled into a life of tranquil domesticity, but suffered great physical discomfort for the rest of his life from a mysterious and undiagnosed chronic condition.

Darwin spent months in Cambridge and especially in London following the voyage, awaiting the Beagle's short journey from Falmouth and then unpacking his crates, seeing specialists who would examine his specimens, writing in his notebooks, and preparing manuscripts for presentation and publication. The pace was frenetic. From his brother Erasmus's lodgings in London, he wrote to his sister, Caroline, on October 24, "These four days since leaving Cambridge have entirely been passed in calling on various naturalist people, but my plans only become more perplexed instead of any clearer.... I do not think mortal man ever talked more than I have during the last three days." To cousin William Darwin Fox, on November 6, he complained, "the busiest time of the whole voyage has been tranquility itself to this last month."

These early months were critical to the development of Darwin's ideas, particularly the confirmation of his field interpretations by scientists at Cambridge and London. He first met Richard Owen, for example, at Charles Lyell's home in the very month Darwin arrived back in England. Owen, anatomist at the Royal College of Surgeons, was the principal scientist who studied the South American fossils and confirmed Darwin's basic identification of the ancient llama—a new species (*Macrauchenia patachonica*). Lyell supported Darwin's interpretations of the recent geological history of the Chilean coast and of the nonvolcanic origins of coral reefs. This latter reversed Lyell's previous beliefs: in a May 1837 letter to Sir J. F. W. Herschel (a polymath scientist who made important contributions to botany, and who Darwin met at the Cape of Good Hope on his return voyage), Lyell wrote, "I am very full of Darwin's new theory of Coral Islands,... I must give up my volcanic crater theory for ever, though it costs me a pang at first, for it accounted for so much." Ornithologist John Gould excitedly confirmed Darwin's discovery of the new lesser rhea from lower Patagonia, naming it *Rhea darwinii*, much to Charles's delight. Equally important was Gould's identification of the several mockingbird species in the Galapagos. Hence, Darwin's stay in London was a productive one—so much so that in March 1837 Charles moved from his brother's place at No. 40 Great Marlborough Street to No. 36, in order to have more space and solitude to write and continue his research.

Darwin encountered a far different London in the 1830s than he had last visited, before the voyage. The London Charles now settled in had added a million souls—numbering about 2.3 million—and had more than doubled its urban space to over ninety acres. When Darwin left on his voyage, coach services provided travelers access to rural villages in the surrounding countryside. The year he returned, rail service connected London to Greenwich. When he had left, Westminster Bridge had been lit by gas lamps, but when he returned, all of London was illuminated.

This was all as much bad news as good: Lighted factories could employ more people for longer hours; rail service made immigration easier; and poverty increased, as did the stench of waste that the city was unequipped to handle. The unbelievable density of humanity—over four hundred people per acre in Greater London—brought the rampant disease, increased mortality, and accelerated reproduction so starkly described by Adam Smith and enumerated by Thomas Malthus. Pollution regularly hid the daytime sun. London in midcentury was the largest city in the world, and perhaps the least desirable to live in. Darwin's contemporary chronicler of urban depravity and misery, Charles Dickens, decried this "shameful testimony to future ages, how civilization and barbarism walked this boastful island together."

Darwin shared Dickens's opinion. He wrote often, while living there, of "this odious dirty smokey town." When he married Emma in 1839 and moved from his digs in Soho on Great Marlborough Street to Upper Gower Street some blocks away, they became reclusive and Emma was occasionally ill, although their conjugality gave Charles a far more favorable opinion of the city. Nevertheless, partly due to Charles's illness, they happily moved to the countryside in Downe in 1842.

This may well have been propitious. On Saturday, September 2, 1854, an epidemic of cholera began in London, originating at the water pump at 40 Broad Street, mere blocks from Great Marlborough. (For a vivid account, see Johnson 2006.) As an ironic aside, the ornithologist John Gould—who classified Darwin's mockingbirds and finches—resided on Broad Street and regularly drank from the pump, but "declined a glass on that Saturday, complaining that it had a repulsive smell" (Johnson 31). He survived.

---

35. NORA BARLOW

The Autobiography of Charles Darwin 1809–1882

Darwin's famous "Red Notebook"—begun on the voyage after leaving the Galapagos, but written predominantly in the early months after his return to England—is the first of several notebooks in which he ruminates theoretically over his experiences and on his conversations with fellow scientists. It is in the post-voyage portion of the Red Notebook *that we see Darwin's emerging questions regarding the nature of species. The following passage from his autobiography helps to clarify the original notes in the* Red Notebook, *which follow this.*

[118] During the voyage of the *Beagle* I had been deeply impressed by discovering in the Pampean formation great fossil animals covered with armour like that on the existing armadillos; secondly, by the manner in which closely allied animals replace one another in proceeding southwards over the Continent; and thirdly, by the South American character of most of the productions of the Galapagos archipelago, and more especially by the manner in which they differ slightly on each island of the group; none of these islands appearing to be very ancient in a geological sense.

It was evident that such facts as these, as well as many [119] others, could be explained on the supposition that species gradually become modified; and the subject haunted me. But it was equally evident that neither the action of the surrounding conditions, nor the will of the organisms (especially in the case of plants), could account for the innumerable cases in which organisms of every kind are beautifully adapted to their habits of life,—for instance, a woodpecker or tree-frog to climb trees, or a seed for dispersal by hooks or plumes. I had always been much struck by such adaptations, and until these could be explained it seemed to me almost useless to endeavour to prove by indirect evidence that species have been modified.

36. Sandra Herbert

The Red Notebook of Charles Darwin

[129] Should urge that extinct Llama [*found at San Julián, Patagonia, and described by Richard Owen.—RKW*] owed its death not to change of circumstances; reversed argument. knowing it to be a desert.—Tempted to believe animals created for a definite time:—not extinguished by change of circumstances:

[130] The same kind of relation that common ostrich [rhea] bears to (Petisse. ...) [This is the lesser rhea.]: extinct Guanaco [Llama] to recent: in former case position, in latter time. (or changes consequent on lapse) being the relation.—As in first cases distinct species inosculate, so must we believe ancient ones: not gradual change or degeneration. from circumstances: if one species does change into another it must be per saltum—or species may perish. = This representation of species important, each its own limit & represented.

37. Gavin de Beer

Darwin's Notebooks on Transmutation of Species. Part I. First notebook [B]

Inosculation here refers to an abrupt relationship rather than a gradual one: inosculate = *joining, as in grafting; from* osculate = *to kiss.*

Notebook B, *Darwin's second theoretical notebook, follows the themes of the* Red Notebook, *but they are more sophisticated and betray a period of serious reflection and critical thinking. It was begun in July 1837. By this time, Darwin had become convinced of the transmutation of species and focused on elaborating its theoretical implications from this point on, although he shared these thoughts with few of his colleagues. He knew they opposed such fancies. In a journal entry for 1837, Darwin writes:*

> In July opened the first note book on 'transmutation of Species'—Had been greatly struck from about month of previous March on character of S. American fossils—& species on Galapagos Archipelago. These facts origin (especially latter) of all my views. (Herbert, 22, n6).

The following passages help to provide details of his reasoning.

[19] Every successive animal is branching upwards different types of organisation improving as Owen says simplest coming in and most perfect and others occasionally dying out; for instance, secondary terebratula may [20] have propagated recent terebratula, but Megatherium nothing.

We may look at Megatherium, Armadillos and Sloths as all offsprings of some still older type. Some of the branches dying out.—

With this tendency to change (and to multiplication when isolated) requires [21] deaths of species to keep numbers of forms equable. But is there any reason for supposing number of forms equable: This being due to subdivisions and amount of differences, so forms would be about equally numerous.—

Changes not result of will of animals, but law of adaptation as much as acid and alkali.

Organized beings represent a tree, *irregularly branched;* some branches far more branched,—hence genera.—As many terminal buds dying, as new ones generated. [22] There is nothing stranger in death of species, than individuals....

[35] Is the shortness of life of *species* in certain orders connected with gaps in the *series of connection?* if starting from same epoch certainly. The absolute end of certain forms from considering S. America (*independent of external causes*) does appear very probable:—Mem.: Horse, Llama, etc. etc.

[36] I think. Case must be that one generation then should have as many living as now. To do this and to have many species in same genus (as is), *requires* extinction.

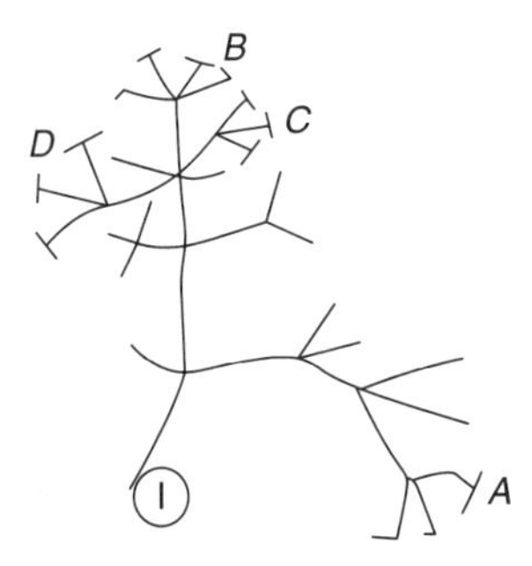

Thus between A and B immens[e] gap of relation, [between] C and B the finest gradation, [between] B and D rather greater distinction. Thus genera would be formed,—bearing relation [37] to ancient types,—with several extinct forms, for if each species as ancient (I) is capable of making 13 recent forms. Twelve of the contemporaries must have left no offspring at all, so as to keep number of species constant.—

With respect to extinction we can easily see that variety of ostrich Petise may not be well adapted, and thus perish out, or on other hand like Orpheus being favourable, [38] many might be produced. This requires principle that the permanent varieties, produced by confined breeding and changing circumstances are continued and produce according to the adaptation of such circumstances, and therefore that death of species is a consequence (contrary to what would appear from America) [39] of non-adaptation of circumstances.

The largeness of present genera renders it probable that many contemporary [genera] would have left scarcely any type of their existence in the present world.—Or we may suppose only each species in each generation only breeds, *like* individuals in a country not rapidly increasing.—[40]

If we thus go very far back to look to the source of the Mammalian type of organization, it is extremely improbable that any of the successors of his relations shall now exist.—....

[42] Hence if this is true that the *greater the groups* the *greater the gaps* (or *solutions* of *continuous structure*) between them.—for instance, there would be great gap between birds and mammalia, still greater between [43] vertebrate and articulate, still greater between animals and plants.—

But yet besides affinities from three elements, from the infinite variations, and all coming from one stock and obeying one law, they may approach—some birds may approach animals and some of the vertebrate invertebrate.—Such or few on each side will yet present some anomaly and bearing [44] stamp of some great main type, and the gradation will be sudden.—

Heaven know[s] whether this agrees with Nature: *Cuidado!*

---

38. Nora Barlow

The Autobiography of Charles Darwin 1809–1882

By the end of 1837, Darwin had worked out the foundational elements to a theory of evolution. He understood that transmutation occurred both through time (the fossils) and across space (isolation). He had demonstrated that species represented functional adaptations to their environments "through the economy of nature." He recognized that extinction had regularly taken place, opening niches for new species. Two elements remained undisclosed: the source of variation itself, and the mechanism by which variations led to new forms. The first of these was never resolved by Darwin nor by his fellow scientists and had to await the rediscovery of Gregor Mendel's principles in the following century. Fortunately, knowledge of the source of variations was not critical to the second element. Darwin's insight followed two courses in discovering the principle of natural selection: first, through selective breeding, and second, through natural mortality. The breeding came from his pigeons and other domesticates with which he experimented at Down House. The mortality evidence came from reading Malthus.

Both of these revelations, prophetically, emancipated him from current typological thinking and led him to a focus on populations of individuals. Eventually he was able to connect the dots between the variations of individuals in a population and the variations of populations in a species.

[119] After my return to England it appeared to me that by following the example of Lyell in Geology, and by collecting all facts which bore in any way on the variation of animals and plants under domestication and nature, some light might perhaps be thrown on the whole subject. My first note-book was opened in July 1837. I worked on true Baconian principles, and without any theory collected facts on a wholesale scale, more especially with respect to domesticated productions, by printed enquiries, by conversation with skilful breeders and gardeners, and by extensive reading. When I see the list of books of all kinds which I read and abstracted, including whole series of Journals and Transactions, I am surprised at my industry. I soon perceived that selection was the keystone of man's success in making useful races of animals and plants. But how selection could be applied to organisms living in a [120] state of nature remained for some time a mystery to me.

In October 1838, that is, fifteen months after I had begun my systematic enquiry, I happened to read for amusement Malthus on *Population*, and being well prepared to appreciate the struggle for existence which everywhere goes on from long-continued observation of the habits of animals and plants, it at once struck me that under these circumstances favourable variations would tend to be preserved, and unfavourable ones to be destroyed. The result of this would be the formation of new species. Here, then, I had at last got a theory by which to work; but I was so anxious to avoid prejudice, that I determined not for some time to write even the briefest sketch of it. In June 1842 I first allowed myself the satisfaction of writing a very brief abstract of my theory in pencil in 35 pages; and this was enlarged during the summer of 1844 into one of 230 pages, which I had fairly copied out and still possess.

But at that time I overlooked one problem of great importance; and it is astonishing to me, except on the principle of Columbus and his egg, how I could have overlooked it and its solution. This problem is the tendency in organic beings descended from the same stock to diverge in character as they become modified. That they have diverged greatly is obvious from the manner in which species of all kinds can be classed under genera, genera under families, families under sub-orders, and so forth; and I can remember the very spot in the road, whilst in my carriage, when to my joy the [121] solution occurred to me; and this was long after I had come to Down. The solution, as I believe, is that the modified offspring of all dominant and increasing forms tend to become adapted to many and highly diversified places in the economy of nature.

---

39. Charles Darwin

The Correspondence of Charles Darwin Volume 6: 1856–1857

Letter 1866—Darwin, C. R. to Lyell, Charles, 3 May 1856

Down Bromley Kent
May 3rd.

My dear Lyell
It was very very good of you to write me so long & very interesting a letter; but I wish you had mentioned whether

you see any further into your very odd case of the vertical divisions of the lava-streams.—. . . .

With respect to your suggestion of a sketch of my view; I hardly know what to think, but will reflect on it; but it goes against my prejudices. To give a fair sketch would be absolutely impossible, for every proposition requires such an array of facts. If I were to do anything it could only refer to the main agency of change, selection,—& perhaps point out a very few of the leading features which countenance such a view, & some few of the main difficulties. But I do not know what to think: I rather hate the idea of writing for priority, yet I certainly sh[d]. be vexed if any one were to publish my doctrines before me.—Anyhow I thank you heartily for your sympathy. I shall be in London next week, & I will call on you on Thursday morning for one hour precisely so as not to lose much of your time & my own: but will you let me this one time come as early as 9 oclock, for I have much which I must do, & the morning is my strongest time.

Farewell | My dear old Patron | Yours | C. Darwin

Letter 1870—Darwin, C. R., to Hooker, J. D., 9 May 1856

Lastly, & of course especially, about myself; I very much want advice & *truthful* consolation if you can give it. I had good talk with Lyell about my species work, & he urges me strongly to publish something. I am fixed against any periodical or Journal, as I positively will *not* expose myself to an Editor or Council allowing a publication for which they might be abused.

If I publish anything it must be a *very thin* & little volume, giving a sketch of my views & difficulties; but it is really dreadfully unphilosophical to give a resumé, without exact references, of an unpublished work. But Lyell seemed to think I might do this, at the suggestion of friends, & on the ground which I might state that I had been at work for 18 years, & yet could not publish for several years, & especially as I could point out difficulties which seemed to me to require especial investigation. Now what think you? I sh[d]. be really grateful for advice. I thought of giving up a couple of months & writing such a sketch, & trying to keep my judgment open whether or no to publish it when completed. It will be simply impossible for me to give exact references; anything important I sh[d]. state on authority of the author generally; & instead of giving all the facts on which I ground any opinion, I could give by memory only one or two. In Preface I would state that the work could not be considered strictly scientific, but a mere sketch or outline of future work in which full references &c sh[d]. be given.—Eheu, eheu, I believe I sh[d]. sneer at any-one else doing this, & my only comfort is, that I *truly* never dreamed of it, till Lyell suggested it, & seems deliberately to think it adviseable.

I am in a peck of troubles & do pray forgive me for troubling you.—

Yours affecti[y] | C. Darwin

The Correspondence of Charles Darwin, Volume 7: 1858–1859

Letter 2285—Darwin, C. R., to Lyell, Charles, 18 June 1858

Darwin indeed wrote the short sketch and a much longer one, but never published them. The longer one—1844—was read, however, by J. D. Hooker before the Linnaean Society in 1847. Darwin had his rude awakening in 1858 when A. R. Wallace sent him a preliminary paper of his own research in Indonesia. The startled Darwin sent the following letters to Lyell, thus setting off a chain of events that, at long last, prodded Darwin to complete and to publish his On the Origin of Species. *The poignancy of Darwin's distress is that he shows such remarkable honor.*

Down Bromley Kent
18th.

My dear Lyell
Some year or so ago, you recommended me to read a paper by Wallace in the Annals, which had interested you & as I was writing to him, I knew this would please him much, so I told him. He has to day sent me the enclosed & asked me to forward it to you. It seems to me well worth reading. Your words have come true with a vengeance that I sh[d]. be forestalled. You said this when I explained to you here very briefly my views of "Natural Selection" depending on the Struggle for existence.—I never saw a more striking coincidence. If Wallace had my M.S. sketch written out in 1842 he could not have made a better short abstract! Even his terms now stand as Heads of my Chapters.

Please return me the M.S. which he does not say he wishes me to publish; but I shall of course at once write & offer to send to any Journal. So all my originality, whatever it may amount to, will be smashed. Though my Book, if it will ever have any value, will not be deteriorated; as all the labour consists in the application of the theory.

I hope you will approve of Wallace's sketch, that I may tell him what you say.

My dear Lyell | Yours most truly | C. Darwin

Letter 2294—Darwin, C.R., to Lyell, Charles, 25 June 1858

Down Bromley Kent
Friday

My dear Lyell
I am very very sorry to trouble you, busy as you are, in so merely personal an affair. But if you will give me your deliberate opinion, you will do me as great a service, as ever man did, for I have entire confidence in your judgment & honour.—

I sh[d]. not have sent off your letter without further reflexion, for I am at present quite upset, but write now to get subject for time out of mind. But I confess it never did occur to me, as it ought, that Wallace could have made any use of your letter.

There is nothing in Wallace's sketch which is not written out much fuller in my sketch copied in 1844, & read by Hooker some dozen years ago. About a year ago I sent a short sketch of which I have copy of my views (owing to correspondence on several points) to Asa Gray, so that I could most truly say & prove that I take nothing from Wallace. I sh^{d}. be *extremely* glad **now** to publish a sketch of my general views in about a dozen pages or so. But I cannot persuade myself that I can do so honourably. Wallace says nothing about publication, & I enclose his letter.—But as I had not intended to publish any sketch, can I do so honourably because Wallace has sent me an outline of his doctrine?—I would far rather burn my whole book than that he or any man sh^{d}. think that I had behaved in a paltry spirit. Do you not think his having sent me this sketch ties my hands? I do not in least believe that he originated his views from anything which I wrote to him.

If I could honourably publish I would state that I was induced now to publish a sketch (& I sh^{d} be very glad to be permitted to say to follow your advice long ago given) from Wallace having sent me an outline of my general conclusions.—We differ only, that I was led to my views from what artificial selection has done for domestic animals. I could send Wallace a copy of my letter to Asa Gray to show him that I had not stolen his doctrine. But I cannot tell whether to publish now would not be base & paltry: this was my first impression, & I sh^{d}. have certainly acted on it, had it not been for your letter.—

This is a trumpery affair to trouble you with; but you cannot tell how much obliged I sh^{d}. be for your advice.—

By the way would you object to send this & your answer to Hooker to be forwarded to me, for then I shall have the opinion of my two best & kindest friends.—This letter is miserably written & I write it now, that I may for time banish whole subject. And I am worn out with musing.

I fear we have case of scarlet-fever in House with Baby.—Etty is weak but is recovering.—

My good dear friend forgive me.—This is a trumpery letter influenced by trumpery feelings.

Yours most truly | C. Darwin

I will never trouble you or Hooker on this subject again.—

Letter 2295—Darwin, C. R., to Lyell, Charles, 26 June 1858

Down.
26th.

My dear Lyell
Forgive me for adding P.S. to make the case as strong as possible against myself.

Wallace might say "you did not intend publishing an abstract of your views till you received my communication, is it fair to take advantage of my having freely, though unasked, communicated to you my ideas, & thus prevent me forestalling you?" The advantage which I should take being that I am induced to publish from privately knowing that Wallace is in the field. It seems hard on me that I should be thus compelled to lose my priority of many years standing, but I cannot feel at all sure that this alters the justice of the case. First impressions are generally right & I at first thought it w^{d}. be dishonourable in me now to publish.—

Yours most truly | C. Darwin

I have always thought you would have made a first-rate Lord Chancellor; & I now appeal to you as a Lord Chancellor

Emma desires her affectionate thanks, in which I heartily join, to Lady L. for her most kind note.—Etty is very weak but progressing well. The Baby has much fever but we hope not S. Fever.—What has frightened us so much is, that 3 children have died in village from Scarlet Fever, & others have been at death's door, with terrible suffering.

The baby, Charles Waring Darwin, died two days later, on Monday June 28. The youngest of their children, he had been born less than two years earlier, on December 6, 1856. Both Charles and Emma were disconsolate, and Charles had little energy or will to think about a manuscript for joint presentation with Wallace. The next day Darwin wrote to Hooker:

Letter 2297—Darwin, C. R., to Hooker, J. D., 29 June 1858

Down
Tuesday

My dearest Hooker
You will, & so will Mrs Hooker, be most sorry for us when you hear that poor Baby died yesterday evening. I hope to God he did not suffer so much as he appeared. He became quite suddenly worse. It was Scarlet-Fever. It was the most blessed relief to see his poor little innocent face resume its sweet expression in the sleep of death.—Thank God he will never suffer more in this world.

I have received your letters. I cannot think now on subject, but soon will. But I can see that you have acted with more kindness & so has Lyell even than I could have expected from you both most kind as you are.

I can easily get my letter to Asa Gray copied, but it is too short.—

Poor Emma behaved nobly & how she stood it all I cannot conceive. It was wonderful relief, when she could let her feelings break forth.—

God Bless you.—You shall hear soon as soon as I can think.

Yours affectionately | C. Darwin

During the day, Darwin hastily scrabbled together what documents he could find, including his 1857 letter to Asa Gray and his 1844 sketch, with Hooker's penciled notes. These established that Darwin's theory was thought out well before the correspondence from Wallace. Thus, Darwin sent the package to Hooker with the following letter, couriered that same night.

Letter 2298—Darwin, C. R., to Hooker, J. D., 29 June 1858

Tuesday Night

My dear Hooker
I have just read your letter, & see you want papers at once. I am quite prostrated & can do nothing but I send Wallace & my abstract of letter to Asa Gray, which gives most imperfectly **only** *the means of change & does not touch* on reasons for believing species do change. I daresay all is too late. I hardly care about it.—

But you are too generous to sacrifice so much time & kindness.—It is most generous, most kind. I send sketch of 1844 **solely** that you may see by your own handwriting that you did read it.—

I really cannot bear to look at it.—Do not waste much time. It is miserable in me to care at all about priority.—

The table of contents will show what it is. I would make a similar, but shorter & more accurate sketch for Linnean Journal.—I will do anything.

God Bless you my dear kind friend. I can write no more. I send this by servant to Kew.

Yours | C. Darwin

40. The Darwin-Wallace Papers

Communication from Lyell and Hooker

Three papers, known as the Darwin-Wallace papers, follow: the first is an extract of Darwin's 1844 summary of his theory; the second is the earlier-referenced letter to Asa Gray, American botanist, inserted to demonstrate that Darwin's ideas were earlier than Wallace's; the third is Wallace's paper summarizing his own theory. It will be evident that there are some important differences between natural selection as envisioned by each man, reflected in Darwin's subsequent letter to Wallace. Wallace believed that "natural" selection and stock breeding—"artificial" selection—were not only different, but would not produce the same results. He further viewed selection as acting upon populations rather than individuals, with obvious implications for evolutionary adaptations and diversification. To Wallace, for example, an adapted species would not submit to variations at all unless the environment changed. Hence, the two used the term variety *differently.*

London,
June 30th, 1858

My Dear Sir,—The accompanying papers, which we have the honour of communicating to the Linnean Society, and which all relate to the same subject, viz. the Laws which affect the Production of Varieties, Races, and Species, contain the results of the investigations of two indefatigable naturalists, Mr. Charles Darwin and Mr. Alfred Wallace.

These gentlemen having, independently and unknown to one another, conceived the very same very ingenious theory to account for the appearance and perpetuation of varieties and of specific forms on our planet, may both fairly claim the merit of being original thinkers in this important line of inquiry; but neither of them having published his views, though Mr. Darwin has for many years past been repeatedly urged by us to do so, and both authors having now unreservedly placed their papers in our hands, we think it would best promote the interests of science that a selection from them should be laid before the Linnean Society.

Taken in the order of their dates, they consist of:-

I. Extracts from a MS. work on Species*, by Mr. Darwin, which was sketched in 1839, and copied in 1844, when the copy was read by Dr. Hooker, and its contents afterwards communicated to Sir Charles Lyell. The first Part is devoted to "The Variation of Organic Beings under Domestication and in their Natural State;" and the second chapter of that Part, from which we propose to read to the Society the extracts referred to, is headed, "On the Variation of Organic Beings in a state of Nature; on the Natural Means of Selection; on the Comparison of Domestic Races and true Species."

II. An abstract of a private letter addressed to Professor Asa Gray, of Boston, U.S., in October 1857, by Mr. Darwin, in which he repeats his views, and which shows that these remained unaltered from 1839 to 1857.

III. An Essay by Mr. Wallace, entitled "On the Tendency of Varieties to depart indefinitely from the Original Type." This was written at Ternate in February 1858, for the perusal of his friend and correspondent Mr. Darwin, and sent to him with the expressed wish that it should be forwarded to Sir Charles Lyell, if Mr. Darwin thought it sufficiently novel and interesting. So highly did Mr. Darwin appreciate the value of the views therein set forth, that he proposed in a letter to Sir Charles Lyell, to obtain Mr. Wallace's consent to allow the Essay to be published as soon as possible. Of this step we highly approved, provided Mr. Darwin did not withhold from the public, as he was strongly inclined to do (in favour of Mr. Wallace), the memoir which he had himself written on the same subject, and which as before stated, one of us had perused in 1844, and the contents of which we had both of us been privy to for many years. On representing this to Mr. Darwin, he gave us permission to make what use thought proper of his memoir, &c.: and in adopting our present course, of presenting to the Linnean Society, we have explained to him that we are not solely considering the relative claims to priority of himself and his friend, but the interests of science generally; for we feel it to be desirable that views founded on a wide deduction from the facts, and matured by years of reflection, should constitute at once a goal from which others may start, and that, while the scientific world is waiting for the appearance of Mr. Darwin's complete work, some of the leading results of his

*This MS. work was never intended for publication, and therefore was not written with care.—C.D. 1858.

labours, as well as those of his able correspondent, should together be laid before the public.

We have the honour to be yours very obediently,

Charles Lyell.
Jos. D. Hooker.
J.J. Bennett, Esq.,
Secretary of the Linnean Society

I. Extract from an unpublished Work on Species, by C. Darwin, Esq., etc.

De Candolle, in an eloquent passage has declared that all nature is at war, one organism with another, or with external nature. Seeing the contented face of nature, this may at first be well doubted; but reflection will inevitably prove it to be true. The war, however, is not constant, but recurrent in a slight degree at short periods, and more severely at occasional more distant periods; and hence its effects are easily overlooked. It is the doctrine of Malthus applied in most cases with tenfold force. As in every climate there are seasons, for each of its inhabitants, of greater and less abundance, so all annually breed; and the moral restraint which in some small degree checks the increase of mankind is entirely lost. Even slow-breeding mankind has doubled in twenty-five years; and if he could increase his food with greater ease, he would double in less time. But for animals without artificial means, the amount of food for each species must, on the average, be constant, whereas the increase of all organisms tends to be geometrical, and in a vast majority of cases at an enormous ratio. Suppose in a certain spot there are eight pairs of birds, and that only four pairs of them annually (including double hatches) rear only four young, and that these go on rearing their young at the same rate, then at the end of seven years (a short life, excluding violent deaths, for any bird) there will be 2048 birds, instead of the original sixteen. As this increase is quite impossible, we must conclude either that birds do not nearly half their young, or that the average life of a bird is, from accident, not nearly seven years. Both checks probably concur. The same kind of calculation applied to all plants and animals affords results more or less striking, but in very few instances more striking than in man.

Many practical illustrations of this rapid tendency to increase are on record, among which, during peculiar seasons, are the extraordinary numbers of certain animals; for instance, during the years 1826 to 1828, in La Plata, when from drought some millions of cattle perished, the whole country actually swarmed with mice. Now I think it cannot be doubted that during the breeding-season all the mice (with the exception of a few males or females in excess) ordinarily pair, and therefore that this astounding increase during three years must be attributed to a greater number than usual surviving the first year, and then breeding, and so on till the third year, when their numbers were brought down to their usual limits on the return of wet weather. Where man has introduced plants and animals into a new and favourable country, there are many accounts in how surprisingly few years the whole country has become stocked with them. This increase would necessarily stop as soon as the country was fully stocked; and yet we have every reason to believe, from what is known of wild animals, that all would pair in the spring. In the majority of cases it is most difficult to imagine where the checks fall—though generally, no doubt, on the seeds, eggs, and young; but when we remember how impossible, even in mankind (so much better known than any other animal), it is to infer from repeated casual observations what the average duration of life is, or to discover the different percentages of deaths to births in different countries, we ought to feel no surprise at our being unable to discover where the check falls in any animal or plant. It should always be remembered, that in most cases the checks are recurrent yearly in a small, regular degree, and in an extreme degree during unusually cold, hot, dry, or wet years, according to the constitution of the being in question. Lighten any check in the least degree, and the geometrical powers of increase in every organism will almost instantly increase the average number of the favoured species. Nature may be compared to a surface on which rest ten thousand sharp wedges touching each other and driven inwards by incessant blows. Fully to realize those views much reflection is requisite. Malthus on man should be studied; and all such cases as those of the mice in La Plata, of the cattle and horses when first turned out in South America, of the birds by our calculation, &c., should be well considered. Reflect on the enormous multiplying power inherent and annually in action in all animals; reflect on the countless seeds scattered by a hundred ingenious contrivances, year after year, over the whole face of the land; and yet we have every reason to suppose that the average percentage of each of the inhabitants of a country usually remains constant. Finally, let it be borne in mind that this average number of individuals (the external conditions remaining the same) in each country is kept up by recurrent struggles against other species or against external nature (as on the borders of the Arctic regions, where the cold checks life), and that ordinarily each individual of every species holds its place, either by its own struggle and capacity of acquiring nourishment in some period of its life, from the egg upwards; or by the struggle of its parents (in short-lived organisms, when the main check occurs at longer intervals) with other individuals of the same or different species.

But let the external conditions of a country alter. If in a small degree, the relative proportions of the inhabitants will in most cases simply be slightly changed; but let the number of inhabitants be small, as on an island, and free access to it from other countries be circumscribed, and let the change of conditions continue progressing (forming new stations), in such a case the original inhabitants must cease to be as perfectly adapted to the changed conditions as they were originally. It has been shown in a former part of this work, that such changes of external conditions would from their acting on the reproductive system, probably cause the organization of those beings which were most affected to become, as under domestication, plastic. Now, can it be doubted, from the struggle each individual has to obtain subsistence, that any minute variation in structure, habits, or instincts, adapting that individual better to the new conditions, would tell upon its vigour and health? In the struggle it would have a better chance of surviving; and those of its offspring which inherited the variation, be it ever so slight, would also have a better chance. Yearly more are bred than can survive; the smallest grain in the balance, in the long run, must tell on which death shall fall, and which shall survive. Let this work of selection on the one hand, and death on the other, go on for a thousand generations, who will pretend to affirm that it would produce no effect, when we remember what, in a few years, Bakewell effected in cattle, and Western in sheep, by this identical principle of selection?

To give an imaginary example from changes in progress on an island:—let the organization of a canine animal which preyed chiefly on rabbits, but sometimes on hares, become slightly plastic; let these same changes cause the number of rabbits very slowly to decrease, and the number of hares to increase; the effect of this would be that the fox or dog would be driven to try to catch more hares: his organization, however, being slightly plastic, those individuals with the lightest forms, longest limbs, and best eyesight, let the differences be ever so small, would be slightly favoured, and would tend to live longer, and to survive during that time of the year when food was scarcest; they would also rear more young, which would tend to inherit these slight peculiarities. The less fleet ones would be rigidly destroyed. I can see no more reason to doubt that these causes in a thousand generations would produce a marked effect, and adapt the form of the fox or dog to the catching of hares instead of rabbits, than that greyhounds can be improved by selection and careful breeding. So would it be with plants under similar circumstances. If the number of individuals of a species with plumed seeds could be increased by greater powers of dissemination within its own area (that is if the check to increase fell chiefly on the seeds), those seeds which were provided with ever so little more down, would in the long run be most disseminated; hence a greater number of seeds thus formed would germinate, and would tend to produce plants inheriting the slightly better-adapted down*.

Besides this natural means of selection, by which those individuals are preserved, whether in their egg, or larval, or mature state, which are best adapted to the place they fill in nature, there is a second agency at work in most unisexual animals, tending to produce the same effect, namely, the struggle of the males for the females. These struggles are generally decided by the law of battle, but in the case of birds, apparently, by the charms of their song, by their beauty or their power of courtship, as in the dancing rock-thrush of Guiana. The most vigorous and healthy males, implying perfect adaptation, must generally gain the victory in their contests. This kind of selection, however, is less rigorous than the other; it does not require the death of the less successful, but gives to them fewer descendants. The struggle falls, moreover, at a time of year when food is generally abundant, and perhaps the effect chiefly produced would be the modification of the secondary sexual characters, which are not related to the power of obtaining food, or to defence from enemies, but to fighting with or rivalling other males. The result of this struggle amongst the males may be compared in some respects to that produced by those agriculturists who pay less attention to the careful selection of all their young animals, and more to the occasional use of a choice mate.

*I can see no more difficulty in this, than in the planter improving his varieties of the cotton plant.—C.D. 1858.

II. Abstract of a Letter from C. Darwin, Esq., to Prof. Asa Gray, etc.

1. It is wonderful what the principle of selection by man, that is the picking out of individuals with any desired quality, and breeding from them, and again picking out, can do. Even breeders have been astounded at their own results. They can act on differences inappreciable to an uneducated eye. Selection has been methodically followed in Europe for only the last half century; but it was occasionally, and even in some degree methodically, followed in the most ancient times. There must have been also a kind of unconscious selection from a remote period, namely in the preservation of the individual animals (without any thought of their offspring) most useful to each race of man in his particular circumstances. This "roguing," as nurserymen call the destroying of varieties which depart from their type, is a kind of selection. I am convinced that intentional and occasional selection has been the main agent in the production of our domestic races; but however this may be, its great power of modification has been indisputably shown in later times. Selection acts only by the accumulation of slight or greater variations, caused by external conditions, or by the mere fact that in generation the child is not absolutely similar to its parent. Man, by this power of accumulating variations, adapts living beings to his wants—may be said to make the wool of one sheep good for carpets, of another for cloth, &c.

2. Now suppose there were a being who did not judge by mere external appearances, but who could study the whole internal organization, who was never capricious, and should go on selecting for one object during millions of generations; who will say what he might not effect? In nature we have some slight variation occasionally in all parts; and I think it can be shown that changed conditions of existence is the main cause of the child not exactly resembling its parents; and in nature geology shows us what changes have taken place, and are taking place. We have almost unlimited time; no one but a practical geologist can fully appreciate this. Think of the Glacial period, during the whole of which the same species at least of shells have existed; there must have been during this period millions on millions of generations.

3. I think it can be shown that there is such an unerring power at work in Natural Selection (the title of my book),

which selects exclusively for the good of each organic being. The elder De Candolle, W. Herbert, and Lyell have written excellently on the struggle for life; but even they have not written strongly enough. Reflect that every being (even the elephant) breeds at such a rate, that in a few years, or at most a few centuries, the surface of the earth would not hold the progeny of one pair. I have found it hard constantly to bear in mind that the increase of every single species is checked during some part of its life, or during some shortly recurrent generation. Only a few of those annually born can live to propagate their kind. What a trifling difference must often determine which shall survive, and which perish!

4. Now take the case of a country undergoing some change. This will tend to cause some of its inhabitants to vary slightly—not but that I believe most beings vary at all times enough for selection to act on them. Some of its inhabitants will be exterminated; and the remainder will be exposed to the mutual action of a different set of inhabitants, which I believe to be far more important to the life of each being than mere climate. Considering the infinitely various methods which living beings follow to obtain food by struggling with other organisms, to escape danger at various times of life, to have their eggs or seeds disseminated, &c. &c., I cannot doubt that during millions of generations individuals of a species will be occasionally born with some slight variation, profitable to some part of their economy. Such individuals will have a better chance of surviving, and of propagating their new and slightly different structures; and the modification may be slowly increased by the accumulative action of natural selection to any profitable extent. The variety thus formed will either coexist with, or, more commonly, will exterminate its parent form. An organic being, like the woodpecker or misseltoe, may thus come to be adapted to a score of contingences—natural selection accumulating those slight variations in all parts of its structure, which are in any way useful to it during any part of its life.

5. Multiform difficulties will occur to every one, with respect to this theory. Many can, I think, be satisfactorily answered. Natura non facit saltum answers some of the most obvious. The slowness of the changes, and only a very few individuals undergoing change at any one time, answers others. The extreme imperfection of our geological record answers others.

6. Another principle, which may be called the principle of divergence, plays, I believe, an important part in the origin of species. The same spot will support more life if occupied by very diverse forms. We see this in the many generic forms in a square yard of turf, and in the plants or insects on any little uniform islet, belonging almost invariably to as many genera and families as species. We can understand the meaning of this fact amongst the higher animals, whose habits we understand. We know that it has been experimentally shown that a plot of land will yield a greater weight if sown with several species and genera of grasses, than if sown with only two or three species. Now, every organic being, by propagating so rapidly, may be said to be striving its utmost to increase in numbers. So it will be with the offspring of any species after it has become diversified into varieties, or subspecies, or true species. And it follows, I think, from the foregoing facts, that the varying offspring of each species will try (only few will succeed) to seize on as many and as diverse places in the economy of nature as possible. Each new variety or species, when formed, will generally take the place of, and thus exterminate its less well-fitted parent. This I believe to be the origin of the classification and affinities of organic beings at all times; for organic beings always seem to branch and sub-branch like the limbs of a tree from a common trunk, the flourishing and diverging twigs destroying the less vigorous—the dead and lost branches rudely representing extinct genera and families.

This sketch is most imperfect; but in so short a space I cannot make it better. Your imagination must fill up very wide blanks.

C. Darwin.

III. On the Tendency of Varieties to depart indefinitely from the Original Type. By Alfred Russel Wallace.

One of the strongest arguments which have been adduced to prove the original and permanent distinctness of species is, that varieties produced in a state of domesticity are more or less unstable, and often have a tendency, if left to themselves, to return to the normal form of the parent species; and this instability is considered to be a distinctive peculiarity of all varieties, even of those occurring among wild animals in a state of nature, and to constitute a provision for preserving unchanged the originally created distinct species.

In the absence or scarcity of facts and observations as to varieties occurring among wild animals, this argument has had great weight with naturalists, and has led to a very general and somewhat prejudiced belief in the stability of species. Equally general, however, is the belief in what are called "permanent or true varieties,"—races of animals which continually propagate their like, but which differ so slightly (although constantly) from some other race, that the one is considered to be a variety of the other. Which is the variety and which the original species, there is generally no means of determining, except in those rare cases in which one race has been known to produce an offspring unlike itself and resembling the other. This, however, would seem quite incompatible with the "permanent invariability of species," but the difficulty is overcome by assuming that such varieties have strict limits, and can never again vary further from the original type, although they may return to it, which, from the analogy of the domesticated animals, is considered to be highly probable, if not certainly proved.

It will be observed that this argument rests entirely on the assumption, that varieties occurring in a state of nature are in all respects analogous to or even identical with those of domestic animals, and are governed by the same laws as regards their permanence of further variation. But this is the object of the present paper to show that this assumption is altogether false, that there is a general principle in nature which will cause many varieties to survive the parent species, and to give rise to successive variations departing further and further from the original type, and which also produces, in domesticated animals, the tendency of varieties to return to the parent form.

The life of wild animals is a struggle for existence. The full exertion of all their faculties and all their energies is required to preserve their own existence and provide for that of their infant offspring. The possibility of procuring food during the least favourable seasons, and of escaping the attacks of their most dangerous enemies, are the primary conditions which determine the existence both of individuals and of entire species. These conditions will also determine the population of a species; and by a careful consideration of all the circumstances we may be enabled to comprehend, and in some degree to explain, what at first sight appears so inexplicable—the excessive abundance of some species, while others closely allied to them are very rare.

The general proportion that must obtain between certain groups of animals is readily seen. Large animals cannot be so abundant as small ones; the carnivora must be less numerous than the herbivora; eagles and lions can never be so plentiful as pigeons and antelopes; the wild asses of the Tartarian deserts cannot equal in numbers the horses of the more luxuriant prairies and pampas of America. The greater or less fecundity of an animal is often considered to be one of the chief causes of its abundance or scarcity; but a consideration of these facts will show us that it really has little or nothing to do with the matter. Even the least prolific of animals would increase rapidly if unchecked, whereas it is evident that the animal population of the globe must be stationary, or perhaps, through the influence of man, decreasing. Fluctuations there may be; but permanent increase, except in restricted localities, is almost impossible. For example, our own observation must convince us that birds do not go on increasing every year in a geometrical ratio, as they would do, were there not some powerful check to their natural increase. Very few birds produce less than two young ones each year, while many have six, eight, or ten; four will certainly be below the average; and if we suppose that each pair produce young only four times in their life, that will also be below the average, supposing them not to die either by violence or want of food. Yet at this rate how tremendous would be the increase in a few years from a single pair! A simple calculation will show that in fifteen years each pair of birds would have increased to nearly ten millions! whereas we have no reason to believe that the number of birds of any country increases at all in fifteen or in one hundred and fifty years. With such powers of increase the population must have reached its limits, and have become stationary, in a very few years after the origin of each species. It is evident, therefore, that each year an immense number of birds must perish—as many in fact as are born; and as in the lowest calculation the progeny are each year twice as numerous as their parents, it follows that, whatever be the average number of individuals existing in any given country, twice that number must perish annually,—a striking result, but one which seems at least highly probable, and is perhaps under rather than over the truth. It would therefore appear that, as far as the continuance of the species and the keeping up the average number of individuals are concerned, large broods are superfluous. On the average all above one become food for hawks and kites, wild cats and weasels, or perish of cold and hunger as winter comes on. This is strikingly proved by the case of particular species; for we find that their abundance in individuals bears no relation whatever to their fertility in producing offspring. Perhaps the most remarkable instance of an immense bird population is that of the passenger pigeon of the United States, which lays only one, or at most two eggs, and is said to rear generally but one young one. Why is this bird so extraordinarily abundant, while others producing two or three times as many young are much less plentiful? The explanation is not difficult. The food most congenial to this species, and on which it thrives best, is abundantly distributed over a very extensive region, offering such differences of soil and climate, that in one part or another of the area the supply never fails. The bird is capable of a very rapid and long-continued flight, so that it can pass without fatigue over the whole of the district it inhabits, and as soon as the supply of food begins to fail in one place is able to discover a fresh feeding-ground. This example strikingly shows us that the procuring a constant supply of wholesome food is almost the sole condition requisite for ensuring the rapid increase of a given species, since neither the limited fecundity, nor the unrestricted attacks of birds of prey and of man are here sufficient to check it. In no other birds are these peculiar circumstances so strikingly combined. Either their food is more liable to failure, or they have not sufficient power of wing to search for it over an extensive area, or during some season of the year it becomes very scarce, and less wholesome substitutes have to be found; and thus, though more fertile in offspring, they can never increase beyond the supply of food in the least favourable seasons. Many birds can only exist by migrating, when their food becomes scarce, to regions possessing a milder, or at least a different climate, though, as these migrating birds are seldom excessively abundant, it is evident that the countries they visit are still deficient in a constant and abundant supply of wholesome food. Those whose organization does not permit them to migrate when their food becomes periodically scarce, can never attain a large population. This is probably the reason why woodpeckers are scarce with us, while in the tropics they are among the most abundant of solitary birds. Thus the house sparrow is more abundant than the redbreast, because its food is more constant and plentiful,—seeds of grasses being preserved during the winter, and our farm-yards and stubble-fields furnishing an almost inexhaustible supply. Why, as a general rule, are aquatic, and especially sea birds, very numerous in individuals? Not because they are more prolific than the others, generally the contrary; but because their food never fails, the sea-shores and river-banks daily swarming with a fresh supply of small mollusca and crustacea. Exactly the same law applies to mammals. Wild cats are prolific and have few enemies; why then are they never as abundant as rabbits? The only intelligible answer is, that their supply of food is more precarious. It appears evident, therefore, that so long as a country remains physically unchanged, the numbers of its animal population cannot materially increase. If one species does so, some others requiring the same kind of food must diminish in proportion. The numbers that die annually must be immense; and as the individual existence of each animal depends upon itself, those that must die must be the weakest—the very young, the aged, and the diseased,—while those that prolong their existence can only be the most perfect in health and vigour—those who are best able to obtain food regularly, and avoid their numerous enemies. It is, as we commenced by remarking, "a struggle for existence," in which the weakest and least perfectly organized must always succumb.

Now it is clear that what takes place among the individuals of a species must also occur among the several allied species of a group,—viz. that those which are best adapted to obtain a regular supply of food, and to defend themselves against the attacks of their enemies and the vicissitudes of the seasons must necessarily obtain and preserve a superiority in population; while those species which from some defect of power or organization are the least capable of counteracting the vicissitudes of food supply, &c., must diminish in numbers, and, in extreme cases, become altogether extinct. Between these extremes the species will present various degrees of capacity for ensuring the means of preserving life; and it is thus we account for the abundance or rarity of species. Our ignorance will generally prevent us from accurately tracing the effects to their causes; but could we become perfectly acquainted with the organization and habits of the various species of animals, and could we measure the capacity of each for performing the different acts necessary to its safety and existence under all the varying circumstances by which it is surrounded, we might be able even to calculate the proportionate abundance of individuals which is the necessary result.

If now we have succeeded in establishing these two points—1st , that the animal population of a country is generally stationary, being kept down by a periodical deficiency of food, and other checks; and, 2nd , that the comparative abundance or scarcity of the individuals of the several species is entirely due to their organization and resulting habits, which, rendering it more difficult to procure a regular supply of food and to provide for their personal safety in some cases than in others, can only be balanced by a difference in the population which have to exist in a given area—we shall be in a condition to proceed to the consideration of varieties, to which the preceding remarks have a direct and very important application.

Most or perhaps all the variations from the typical form of a species must have some definable effect, however slight, on the habits or capacities of the individuals. Even a change of colour might, by rendering them more or less distinguishable, affect their safety; a greater or less development of hair might modify their habits. More important changes, such as an increase in the power or dimensions of the limbs or any of the external organs, would more or less affect their mode of procuring food or the range of country which they inhabit. It is also evident that most changes would affect, either favourably or adversely, the powers of prolonging existence. An antelope with shorter or weaker legs must necessarily suffer more from the attacks of the feline carnivora; the passenger pigeon with less powerful wings would sooner or later be affected in its powers of procuring a regular supply of food; and in both cases the result must necessarily be a diminution of the population of the modified species. If, on the other hand, any species should produce a variety having slightly increased powers of preserving existence, that variety must inevitably in time acquire a superiority in numbers. These results must follow as surely as old age, intemperance, or scarcity of food produces an increased mortality. In both cases there may be many individual exceptions; but on the average the rule will invariably be found to hold good. All varieties will therefore fall into two classes—those which under the same conditions would never reach the population of the parent species, and those which would in time obtain and keep a numerical superiority. Now let some alteration of physical conditions occur in the district—a long period of drought, a destruction of vegetation by locusts, the irruption of some new carnivorous animal seeking "pastures new"—any change in fact tending to render existence more difficult to the species in question, and taking its utmost powers to avoid complete extermination; it is evident that, of all the individuals composing the species, those forming the least numerous and most feebly organized variety would suffer first, and, were the pressure severe, must soon become extinct. The same causes continuing in action, the parent species would next suffer, would gradually diminish in numbers, and with a recurrence of similar unfavourable conditions might also become extinct. The superior variety would then alone remain, and on a return to favourable circumstances would rapidly increase in numbers and occupy the place of the extinct species and variety.

The variety would now have replaced the species, of which it would be a more perfectly developed and more highly organized form. It would be in all respects better adapted to secure its safety, ad to prolong its individual existence and that of the race. Such a variety could not return to the original form; for that form is an inferior one, and could never compete with it for existence. Granted, therefore, a "tendency" to reproduce the original type of species, still the variety must ever remain preponderant in numbers, and under adverse physical conditions again alone survive. But this new, improved, and populous race might itself, in course of time, give rise to new varieties, exhibiting several diverging modifications of form, any of which, tending to increase the facilities for preserving existence, must, by the same general law, in their turn become predominant. Here, then, we have progression and continued divergence deduced from the general laws which regulate the existence of animals in a state of nature, and from the undisputed fact that varieties do frequently occur. It is not, however, contended that this result would be invariable; a change of physical conditions in the district might at times materially modify it, rendering the race which had been the most capable of supporting existence under the former conditions now the least so, and even causing the extinction of the newer and, for a time, superior race, while the old or parent species and its first inferior varieties continued to flourish. Variations in unimportant parts might also occur, having no perceptible effect of the life-preserving powers; and the varieties so furnished might run a course parallel with the parent species, either giving rise to further variations or returning to the former type. All we argue for is, that certain varieties have a tendency to maintain their existence longer than the original species, and this tendency must make itself felt; for though the doctrine of chances or averages can never be trusted to on a limited scale, yet, if applied to high numbers, the results come nearer to what theory demands, and, as we approach to an infinity of examples, becomes strictly accurate. Now the scale on which nature works is so vast—the numbers of individuals and periods of time with which she deals approach so near to infinity, that any cause, however slight, and however liable to be veiled and counteracted by accidental circumstances, must in the end produce its full legitimate results.

Let us now turn to domesticated animals, and inquire how varieties produced among them are affected by the principles here enunciated. The essential difference in the condition of wild and domestic animals is this—that among the former, their well-being and very existence depend upon the full exercise and healthy condition of all their senses and physical powers, whereas, among the latter, these are only partially exercised, and in some cases are absolutely unused. A wild animal has to search, and often to labour, for every mouthful of food—to exercise sight, hearing, and smell in seeking it, and in avoiding dangers, in procuring shelter from the inclemency of the seasons, and in providing for the subsistence and safety of its offspring. There is no muscle of its body that is not called into daily and hourly activity; there is no sense or faculty that is not strengthened by continual exercise. The domestic animal, on the other hand, has food provided for it, is sheltered, and often confined, to guard against the vicissitudes of the seasons, is carefully secured from the attacks of its natural enemies, and seldom even rears its young without human assistance. Half of its senses and faculties are quite useless; and the other half are but occasionally called into feeble exercise, while even its muscular system is only irregularly called into action.

Now when a variety of such an animal occurs, having increased power or capacity in any organ or sense, such increase is totally useless, is never called into action, and may even exist without the animal ever becoming aware of it. In the wild animal, on the contrary, all its faculties and powers being brought into full action for the necessities of existence, any increase becomes immediately available, is strengthened by exercise, and must even slightly modify the food, the habits, and the whole economy of the race. It creates as it were a new animal, one of superior powers, and which will necessarily increase in numbers and outlive those inferior to it.

Again, in the domesticated animal all variations have an equal chance of continuance; and those which would decidedly render a wild animal unable to compete with its fellows and continue its existence are not disadvantaged whatever in a state of domesticity. Our quickly fattening pigs, short-legged sheep, pouter pigeons, and poodle dogs could never have come into existence in a state of nature, because the very first step towards such inferior forms would have led to rapid extinction of the race; still less could they now exist in competition with their wild allies. The great speed but slight endurance of the race horse, the unwieldy strength of the ploughman's team, would both be useless in a state of nature. If turned wild on the pampas, such animals would probably soon become extinct, or under favourable circumstances might each lose those extreme qualities which would never be called into action, and in a few generations would revert to a common type, which must be that in which the various powers and faculties are so proportioned to each other as to be best adapted to procure food and secure safety,—that in which by the full exercise of every part of his organization the animal can alone continue to live. Domestic varieties, when turned wild, must return to something near the type of the original wild stock, or become altogether extinct.

We see, then, that no inferences as to varieties in a state of nature can be deduced from the observation of those occurring among domestic animals. The two are so much opposed to each other in every circumstance of their existence, that what applies to the one is almost sure not to apply to the other. Domestic animals are abnormal, irregular, artificial; they are subject to varieties which never occur and never can occur in a state of nature; their very existence depends altogether on human care; so far are many of them removed from that just proportion of faculties, that true balance of organization, by means of which alone an animal left to its own resources can preserve its existence and continue its race.

The hypothesis of Lamarck—that progressive changes in species have been produced by the attempts of animals to increase the development of their own organs, and thus modify their structure and habits—has been repeatedly and easily refuted by all writers on the subject of varieties and species, and it seems to have been considered that when this was done the whole question has been finally settled; but the view here developed renders such an hypothesis quite unneccessary, by showing that similar results must be produced by the action of principles constantly at work in nature. The powerful retractile talons of the falcon—and the cat—tribes have not been produced or increased by the volition of those animals; but among different varieties which occurred in the earlier and less highly organized forms of these groups, those always survived longest which had the greatest facilities for seizing their prey. Neither did the giraffe acquire its long neck by desiring to reach the foliage of the more lofty shrubs, and constantly stretching its neck for the purpose, but because any varieties which occurred among its antitypes with a longer neck than usual at once secured a fresh range of pasture over the same ground as their shorter-necked companions, and on the first scarcity of food were thereby enabled to outlive them. Even the peculiar colours of many animals, especially insects, so closely resembling the soil or the leaves or the trunks on which they habitually reside, are explained on the same principle; for though in the course of ages varieties of many tints may have occurred, yet those races having colours best adapted to concealment from their enemies would inevitably survive the longest. We have also here an acting cause to account for that balance so often observed in nature,—a deficiency in one set of organs always being compensated by an increased development of some others—powerful wings accompanying weak feet, or great velocity making up for the absence of defensive weapons; for it has been shown that all varieties in which an unbalanced deficiency occurred could not long continue their existence. The action of this principle is exactly like that of the centrifugal governor of the steam engine, which checks and corrects any irregularities almost before they become evident; and in like manner no unbalanced deficiency in the animal kingdom can ever reach any conspicuous magnitude, because it would make itself felt at the very first step, by rendering existence difficult and extinction almost sure soon to follow. An origin such as is here advocated will also agree with the peculiar character of the modifications of form and structure which obtain in organized beings—the many lines of divergence from a central type, the increasing efficiency and power of a particular organ through a succession of allied species, and the remarkable persistence of unimportant parts such as colour, texture of plumage and hair, form of horns or crests, through a series of species differing considerably in more essential characters. It also furnishes us with a reason for that "more specialized structure" which Professor Owen states to be a characteristic of recent compared with extinct forms, and which would evidently be the result of the progressive modification of any organ applied to a special purpose in the animal economy.

We believe we have now shown that there is a tendency in nature to the continued progression of certain classes of varieties further and further from the original type—a progression to which there appears no reason to assign any definite limits—and that the same principle which produces this result is a state of nature will also explain why domestic varieties have a tendency to revert to the original type. This progression, by minute steps, in various directions, but always checked and balanced by the necessary conditions, subject to which alone existence can be preserved, may, it is believed, be followed out so as to agree with all the phenomena presented by organized beings, their extinction and succession in past ages, and all the extraordinary modifications of form, instinct, and habits which they exhibit.

Ternate, February, 1858.

---

41. CHARLES DARWIN

The Correspondence of Charles Darwin, Volume 7: 1858–1859

Letter 2337—Wallace, A. R., to Hooker, J. D., 6 Oct 1858

Ternate, Moluccas,
Oct. 6. 1858.

My dear Sir
I beg leave to acknowledge the receipt of your letter of July last, sent me by Mr. Darwin, & informing me of the steps you had taken with reference to a paper I had communicated to that gentleman. Allow me in the first place sincerely to thank yourself & Sir Charles Lyell for your kind offices on this occasion, & to assure you of the gratification afforded me both by the course you have pursued, & the favourable opinions of my essay which you have so kindly expressed. I cannot but consider myself a favoured party in this matter, because it has hitherto been too much the practice in cases of this sort to impute *all* the merit to the first discoverer of a new fact or a new theory, & little or none to any other party who may, quite independently, have arrived at the same result a few years or a few hours later.

I also look upon it as a most fortunate circumstance that I had a short time ago commenced a correspondence with Mr. Darwin on the subject of "Varieties", since it has led to the earlier publication of a portion of his researches & has secured to him a claim to priority which an independent publication either by myself or some other party might have injuriously effected;—for it is evident that the time has now arrived when these & similar views *will* be promulgated & *must* be fairly discussed.

It would have caused me much pain & regret had Mr. Darwin's excess of generosity led him to make public my paper unaccompanied by his own much earlier & I doubt not much more complete views on the same subject, & I must again thank you for the course you have adopted, which while strictly just to both parties, is so favourable to myself.

Being on the eve of a fresh journey I can now add no more than to thank you for your kind advice as to a speedy return to England;—but I dare say you well know & feel, that to induce a Naturalist to quit his researches at their most interesting point requires some more cogent argument than the prospective loss of health.

I remain | My dear Sir | Yours very sincerely | Alfred R. Wallace J. D. Hooker, M.D.

Letter 2405—Darwin, C. R., to Wallace, A. R., 25 Jan 1859

Down Bromley Kent
Jan. 25th

My dear Sir
I was extremely much pleased at receiving three days ago your letter to me & that to Dr. Hooker. Permit me to say how heartily I admire the spirit in which they are written. Though I had absolutely nothing whatever to do in leading Lyell & Hooker to what they thought a fair course of action, yet I naturally could not but feel anxious to hear what your impression would be. I owe indirectly much to you & them; for I almost think that Lyell would have proved right & I shd. never have completed my larger work, for I have found my abstract hard enough with my poor health, but now thank God I am in my last chapter, but one. My abstract will make a small vol. of 400 or 500 pages.—Whenever published, I will of course send you a copy, & then you will see what I mean about the part which I believe Selection has played with domestic productions. It is a very different part, as you suppose, from that played by "Natural Selection".—

Most cordially do I wish you health & entire success in all your pursuits & God knows if admirable zeal & energy deserve success, most amply do you deserve it.

I look at my own career as nearly run out: if I can publish my abstract & perhaps my greater work on same subject, I shall look at my course as done.

Believe me, my dear Sir | Yours very sincerely | C. Darwin

Ideas to Think About

1. What observations and assumptions led Darwin to believe that—if species did transmute—they did so in discontinuous, sudden leaps ("per saltum")?
2. Describe the observations that led Darwin, finally, to the conclusion that "descent with modification" had actually occurred. How was his acknowledgement of extinction tied to this belief?
3. Be prepared to discuss the source of Darwin's contention that extinction is internally programmed rather than externally fostered.
4. Can you find in Darwin's observations in South America the definitive change in his thinking from the Lamarckian model to the adaptation model for evolution?
5. In Darwin's argument for selection, why was the example of domestication inadequate? What were its shortcomings, which prevented a simple change from breeding practices to nature?
6. Why did the unsolved problem of diversification upon speciation cause Darwin problems, and how was this problem reconciled?
7. Darwin was obsessively concerned with the detailed evidence for his theory, and for this reason resisted premature or summary publication. In what way does this reflect his Baconian approach to doing science?

For Further Reading

The books on or about Darwin—particularly at the 2009 anniversary—are numerous and excellent. I hesitate to choose only three or four, but have done so nevertheless!

Charles Darwin: The Power of Place, Janet Browne. Princeton, NJ: Princeton University Press (2002). With its biographical companion, volume I, subtitled "Voyaging," Browne firmly established her status as a peerless authority on Darwin. This book is incomparable in both detail and scope, and the world of Darwiniana is fortunate indeed that Browne's impeccable scholarship is readily matched by her narrative ability.

Darwin: The Life of a Tormented Evolutionist, Adrian Desmond and James Moore. New York: Warner Books (1991). This is an equally acclaimed biography. Exclaimed the *London Sunday Times*, "Over the past 20 years, a veritable Darwin industry has mushroomed. Desmond and Moore have crowned all this new scholarship."

Darwin, Marx, Wagner: Critique of a Heritage, Jacques Barzun. New York: Doubleday Anchor Books (1958). It was prophetic that *Origin*, Marx's *Critique of Political Economy*, and Wagner's *Tristan and Isolde* all appeared in the same year. Barzun sees in the three disparate works—the scientific, the political, and the artistic—a singular reflection of mechanistic and materialistic views of reality common to the era, and he does an intriguing job of weaving them into the fabric of history. All three were, in their time and in their own ways, revolutionary. A tour de force, typical of Barzun.

The Origin, Irving Stone. New York: Doubleday (1980). Stone is the master of the biographical novel, and this bold and massive (over seven hundred pages) effort does not lessen that accolade. This is a typical Stone integration of factual detail (including correspondence) and dramatic license.

UNIT 9

The Age of Darwin, III: The Origin

> If I lived twenty more years and was able to work, how I should have to modify the *Origin*, and how much the views on all points will have to be modified! Well it is a beginning, and that is something....
>
> —Charles Darwin to J. D. Hooker, *1869*

That *is* something, indeed! Darwin was ever diffident. Never insecure in his assertions and beliefs, he was forever insecure in responding effectively to criticism or doubt. He knew, despite his obsession to amass still more data to make his case, it would never be enough to erode deeply etched convictions. So he delayed. He wrote letters to scientific colleagues, many not personally known, asking for any information on natural hybrids, on the geographic distribution of varieties, on the tendency to generate "sports," on the stability of species when isolated. He advertised for information in botanical journals and magazines (see Barrett 1977). He tested ideas cautiously through correspondence. He carefully and strategically laid the foundations for what he knew was to be a radical departure in scientific understanding. In his extreme caution, and the fear that fellow scientists would not accept his revolutionary views, Darwin reminds us of Copernicus, writing three hundred years earlier in his dedication, admitting "the scorn which I had reason to fear on account of the novelty and unconventionality of my opinion almost induced me to abandon completely the work which I had undertaken" (see Unit 1 of this volume).

He understood what he was about to produce for all to see and read was a virtual bombshell, and he feared he might become one of its victims. He sought not only information to support his ideas, but as a true scientist, information that challenged them. He well anticipated all of the objections that were to bombard him. He continued this search even after the first edition of *The Origin*, making additional points or deleting objectionable passages. He was never satisfied.

The Linnaean Society publication of the Darwin-Wallace papers changed all of that. No longer was Darwin hiding a ferocious scientific beast that threatened its master. It was now set loose upon the world, and regardless of its reception, an oppressive weight had been lifted from his shoulders. His exhaustive search for a flawless argument had been preempted by Wallace, and Darwin was visibly relieved. Following the deaths of Charles's youngest, Charles Waring, on June 28, 1858, and his sister Marianne (who died shortly afterward at age sixty), the entire family escaped Downe and the pestilence to spend a few weeks on the Isle of Wight. Darwin's confidence returned, and, with it, his consuming correspondence and writing. Back at Down House, he worked feverishly on his larger "abstract," what was to become *The Origin*.

The production and exchange of knowledge through correspondence—a hallmark of Victorian Britain—was new to intellectual life, and Darwin took advantage of it from the comfort of rural life. This time in his life corresponded with the time of postal advance: Mail trains, introduced in 1830, began to replace the slower mail coaches, and by mid-century, London had up to six deliveries per day, doubling by the turn of the century. The Uniform Penny Post was established in 1840, wherein senders paid a single rate for delivery throughout the Isles, affixing the world's first postage stamp, the "penny black," to confirm that postage was paid. Darwin's postal expenses in 1851, we are told by Janet Browne, were the equivalent of £1,000 (about US$1,500) at modern rates. His currently identified total of fourteen thousand letters was most likely twice that (Browne 2002, 11–13).

His raised spirits were helped immensely by favorable comments in journals and magazines over the Darwin-Wallace papers. Even more conservative Bible believers gave him fair marks, for the most part. Natural selection became a favorite topic of conversation, and Darwin was anxious to set all misunderstandings right and to expand on and clarify the concept of species evolution. Public and professional response convinced Darwin that a publication through the Linnaean Society was too limiting; furthermore, the burgeoning size of his manuscript exceeded what the society could produce. His theory needed to stand alone in book form. He was hopeful that the interest would provide a good return on the publishing investment. His hopes were realized.

His first edition of *On the Origin of Species* came out on November 24, 1859. It sold out on its first day, according to his publisher, all 1,170 copies. The second edition of 3,000 copies, released on January 7, 1860, reached an even broader audience. The third edition, in 1861, was followed in five years by the fourth. Each edition added or subtracted, depending upon reviews and critiques. Natural Selection had suddenly become a cause célèbre and John Murray Publishers had found an author to market and take to the bank. Recognition abounded in all of the journals, newspapers, and critical reviews, some carefully orchestrated by friends of Darwin to counter the more radical opposition. Greatly enhancing the spread of readership to the Victorian book-loving public were orders of hundreds of copies from lending libraries, which arose at midcentury—along with newspapers, magazines, and journals—in response to the voracious reading appetites of the educated classes.

Evolution was here to stay, even though Darwin would first use the term in his *Descent of Man* in 1871, and in the sixth edition of *Origin*, in 1872. In Darwin's lifetime, *Origin* appeared in the United States (thanks to Asa Gray) and eleven foreign translations. Some of these, notably the German and French, were liberally scattered with their translators' philosophical or political biases.

Each of the following highly selected chapter segments from *Origin* comes from the first edition. Chapter I, omitted from these selections, is on "Variation Under Domestication" and provided Darwin's first line of evidence supporting natural selection—to which Wallace seriously objected, as we shall see in unit 11 of this book. A selection from *Origin*'s chapter 13 on embryology is reproduced in Unit 15 of the present work, as it is historically relevant to the topic of that unit.

---

42. CHARLES DARWIN

On the Origin of Species by Means of Natural Selection

Chapter II: Variation Under Nature

[50] When a young naturalist commences the study of a group of organisms quite unknown to him, he is at first much perplexed to determine what differences to consider as specific, and what as varieties; for he knows nothing of the amount and kind of variation to which the group is subject; and this shows, at least, how very generally there is some variation. But if he confine his attention to one class within one country, he will soon make up his mind how to rank most of the doubtful forms. His [51] general tendency will be to make many species, for he will become impressed, just like the pigeon or poultry-fancier before alluded to, with the amount of difference in the forms which he is continually studying; and he has little general knowledge of analogical variation in other groups and in other countries, by which to correct his first impressions. As he extends the range of his observations, he will meet with more cases of difficulty; for he will encounter a greater number of closely-allied forms. But if his observations be widely extended, he will in the end generally be enabled to make up his own mind which to call varieties and which species; but he will succeed in this at the expense of admitting much variation,—and the truth of this admission will often be disputed by other naturalists. When, moreover, he comes to study allied forms brought from countries not now continuous, in which case he can hardly hope to find the intermediate links between his doubtful forms, he will have to trust almost entirely to analogy, and his difficulties will rise to a climax.

Certainly no clear line of demarcation has as yet been drawn between species and sub-species—that is, the forms which in the opinion of some naturalists come very near to, but do not quite arrive at the rank of species; or, again, between sub-species and well-marked varieties, or between lesser varieties and individual differences. These differences blend into each other in an insensible series; and a series impresses the mind with the idea of an actual passage....

[52] It need not be supposed that all varieties or incipient species necessarily attain the rank of species. They may whilst in this incipient state become extinct, or they may endure as varieties for very long periods, as has been shown to be the case by Mr. Wollaston with the varieties of certain fossil land-shells in Madeira. If a variety were to flourish so as to exceed in numbers the parent species, it would then rank as the species, and the species as the variety; or it might come to supplant and exterminate the

parent species; or both might co-exist, and both rank as independent species. But we shall hereafter have to return to this subject.

From these remarks it will be seen that I look at the term species, as one arbitrarily given for the sake of convenience to a set of individuals closely resembling each other, and that it does not essentially differ from the term variety, which is given to less distinct and more fluctuating forms. The term variety, again, in comparison with mere individual differences, is also applied arbitrarily, and for mere convenience sake....

[58] Finally, then, varieties have the same general characters as species, for they cannot be distinguished from species,—except, firstly, by the discovery of intermediate linking forms, and the occurrence of such links cannot affect the actual characters of the forms which they connect; and except, secondly, by a certain amount of [59] difference, for two forms, if differing very little, are generally ranked as varieties, notwithstanding that intermediate linking forms have not been discovered; but the amount of difference considered necessary to give to two forms the rank of species is quite indefinite. In genera having more than the average number of species in any country, the species of these genera have more than the average number of varieties. In large genera the species are apt to be closely, but unequally, allied together, forming little clusters round certain species. Species very closely allied to other species apparently have restricted ranges. In all these several respects the species of large genera present a strong analogy with varieties. And we can clearly understand these analogies, if species have once existed as varieties, and have thus originated: whereas, these analogies are utterly inexplicable if each species has been independently created.

Chapter III: Struggle for Existence

[60] Before entering on the subject of this chapter, I must make a few preliminary remarks, to show how the struggle for existence bears on Natural Selection. It has been seen in the last chapter that amongst organic beings in a state of nature there is some individual variability; indeed I am not aware that this has ever been disputed. It is immaterial for us whether a multitude of doubtful forms be called species or sub-species or varieties; what rank, for instance, the two or three hundred doubtful forms of British plants are entitled to hold, if the existence of any well-marked varieties be admitted. But the mere existence of individual variability and of some few well-marked varieties, though necessary as the foundation for the work, helps us but little in understanding how species arise in nature. How have all those exquisite adaptations of one part of the organisation to another part, and to the conditions of life, and of one distinct organic being to another being, been perfected? We see these beautiful co-adaptations most plainly in the woodpecker and missletoe; and only a little less plainly in the humblest parasite which clings [61] to the hairs of a quadruped or feathers of a bird; in the structure of the beetle which dives through the water; in the plumed seed which is wafted by the gentlest breeze; in short, we see beautiful adaptations everywhere and in every part of the organic world.

Again, it may be asked, how is it that varieties, which I have called incipient species, become ultimately converted into good and distinct species, which in most cases obviously differ from each other far more than do the varieties of the same species? How do those groups of species, which constitute what are called distinct genera, and which differ from each other more than do the species of the same genus, arise? All these results, as we shall more fully see in the next chapter, follow inevitably from the struggle for life. Owing to this struggle for life, any variation, however slight and from whatever cause proceeding, if it be in any degree profitable to an individual of any species, in its infinitely complex relations to other organic beings and to external nature, will tend to the preservation of that individual, and will generally be inherited by its offspring. The offspring, also, will thus have a better chance of surviving, for, of the many individuals of any species which are periodically born, but a small number can survive. I have called this principle, by which each slight variation, if useful, is preserved, by the term of Natural Selection, in order to mark its relation to man's power of selection. We have seen that man by selection can certainly produce great results, and can adapt organic beings to his own uses, through the accumulation of slight but useful variations, given to him by the hand of Nature. But Natural Selection, as we shall hereafter see, is a power incessantly ready for action, and is as immeasurably superior to man's feeble efforts, as the works of Nature are to those of Art.

[62] We will now discuss in a little more detail the struggle for existence....

I should premise that I use the term Struggle for Existence in a large and metaphorical sense, including dependence of one being on another, and including (which is more important) not only the life of the individual, but success in leaving progeny. Two canine animals in a time of dearth, may be truly said to struggle with each other which shall get food and live. But a plant on the edge of a desert is said to struggle for life against the drought, though more properly it should be said to be dependent on the moisture. A [63] plant which annually produces a thousand seeds, of which on an average only one comes to maturity, may be more truly said to struggle with the plants of the same and other kinds which already clothe the ground. The missletoe is dependent on the apple and a few other trees, but can only in a far-fetched sense be said to struggle with these trees, for if too many of these parasites grow on the same tree, it will languish and die. But several seedling missletoes, growing close together on the same branch, may more truly be said to struggle with each other. As the missletoe is disseminated by birds, its existence depends on birds; and it may metaphorically be said to struggle with other fruit-bearing plants, in order to tempt birds to devour and thus disseminate its seeds rather than those of other plants. In these several senses, which pass into each other,

I use for convenience sake the general term of struggle for existence.

A struggle for existence inevitably follows from the high rate at which all organic beings tend to increase. Every being, which during its natural lifetime produces several eggs or seeds, must suffer destruction during some period of its life, and during some season or occasional year, otherwise, on the principle of geometrical increase, its numbers would quickly become so inordinately great that no country could support the product. Hence, as more individuals are produced than can possibly survive, there must in every case be a struggle for existence, either one individual with another of the same species, or with the individuals of distinct species, or with the physical conditions of life. It is the doctrine of Malthus applied with manifold force to the whole animal and vegetable kingdoms; for in this case there can be no artificial increase of food, and no prudential restraint from marriage. Although some species may [64] be now increasing, more or less rapidly, in numbers, all cannot do so, for the world would not hold them.

There is no exception to the rule that every organic being naturally increases at so high a rate, that if not destroyed, the earth would soon be covered by the progeny of a single pair. Even slow-breeding man has doubled in twenty-five years, and at this rate, in a few thousand years, there would literally not be standing room for his progeny. Linnæus has calculated that if an annual plant produced only two seeds—and there is no plant so unproductive as this—and their seedlings next year produced two, and so on, then in twenty years there would be a million plants. . . .

[76] As species of the same genus have usually, though by no means invariably, some similarity in habits and constitution, and always in structure, the struggle will generally be more severe between species of the same genus, when they come into competition with each other, than between species of distinct genera. We see this in the recent extension over parts of the United States of one species of swallow having caused the decrease of another species. The recent increase of the missel-thrush in parts of Scotland has caused the decrease of the song-thrush. How frequently we hear of one species of rat taking the place of another species under the most different climates! In Russia the small Asiatic cockroach has everywhere driven before it its great congener. One species of charlock will supplant another, and so in other cases. We can dimly see why the competition should be most severe between allied forms, which fill nearly the same place in the economy of nature; but probably in no one case could we precisely say why one species has been victorious over another in the great battle of life.

[78] It is good thus to try in our imagination to give any form some advantage over another. Probably in no single instance should we know what to do, so as to succeed. It will convince us of our ignorance on the mutual relations of all organic beings; a conviction as necessary, as it seems to be difficult to acquire. All that we can do, is to keep steadily in mind that each organic being is striving to increase at a geometrical [79] ratio; that each at some period of its life, during some season of the year, during each generation or at intervals, has to struggle for life, and to suffer great destruction. When we reflect on this struggle, we may console ourselves with the full belief, that the war of nature is not incessant, that no fear is felt, that death is generally prompt, and that the vigorous, the healthy, and the happy survive and multiply.

Chapter IV: Natural Selection

The following was Darwin's most important chapter, for it contained the seat of his argument. It was also the most contentious. Wallace, who objected to the phrase "natural selection" because it personified a natural force, strongly urged Darwin to use Herbert Spencer's "survival of the fittest" instead. In his fifth edition (with a change in the book's title to The Origin of Species, *Darwin compromised by changing the chapter heading to "Chapter IV: Natural Selection, or the Survival of the Fittest."*

[80] How will the struggle for existence, discussed too briefly in the last chapter, act in regard to variation? Can the principle of selection, which we have seen is so potent in the hands of man, apply in nature? I think we shall see that it can act most effectually. Let it be borne in mind in what an endless number of strange peculiarities our domestic productions, and, in a lesser degree, those under nature, vary; and how strong the hereditary tendency is. Under domestication, it may be truly said that the whole organisation becomes in some degree plastic. Let it be borne in mind how infinitely complex and close-fitting are the mutual relations of all organic beings to each other and to their physical conditions of life. Can it, then, be thought improbable, seeing that variations useful to man have undoubtedly occurred, that other variations useful in some way to each being in the great and complex battle of life, should sometimes occur in the course of thousands of generations? If such do occur, can we doubt (remem- [81] bering that many more individuals are born than can possibly survive) that individuals having any advantage, however slight, over others, would have the best chance of surviving and of procreating their kind? On the other hand, we may feel sure that any variation in the least degree injurious would be rigidly destroyed. This preservation of favourable variations and the rejection of injurious variations, I call Natural Selection. Variations neither useful nor injurious would not be affected by natural selection, and would be left a fluctuating element, as perhaps we see in the species called polymorphic. . . .

[107] To sum up the circumstances favourable and unfavourable to natural selection, as far as the extreme intricacy of the subject permits. I conclude, looking to the future, that for terrestrial productions a large continental area, which will probably undergo many oscillations of level, and which consequently will exist for long periods in a broken condition, will be the most favourable for the production of many new forms of life, likely to endure long and to spread widely. For the area will first have existed

as a continent, and the inhabitants, at this period numerous in individuals and kinds, will have been subjected to very severe competition. When converted by subsidence into large separate islands, there will still exist many individuals of the same species on each island: intercrossing on the confines of the range of each species will thus be checked: after physical changes of any kind, immigration will be pre- [108] vented, so that new places in the polity of each island will have to be filled up by modifications of the old inhabitants; and time will be allowed for the varieties in each to become well modified and perfected. When, by renewed elevation, the islands shall be re-converted into a continental area, there will again be severe competition: the most favoured or improved varieties will be enabled to spread: there will be much extinction of the less improved forms, and the relative proportional numbers of the various inhabitants of the renewed continent will again be changed; and again there will be a fair field for natural selection to improve still further the inhabitants, and thus produce new species.

That natural selection will always act with extreme slowness, I fully admit. Its action depends on there being places in the polity of nature, which can be better occupied by some of the inhabitants of the country undergoing modification of some kind. The existence of such places will often depend on physical changes, which are generally very slow, and on the immigration of better adapted forms having been checked. But the action of natural selection will probably still oftener depend on some of the inhabitants becoming slowly modified; the mutual relations of many of the other inhabitants being thus disturbed. Nothing can be effected, unless favourable variations occur, and variation itself is apparently always a very slow process. The process will often be greatly retarded by free intercrossing. Many will exclaim that these several causes are amply sufficient wholly to stop the action of natural selection. I do not believe so. On the other hand, I do believe that natural selection will always act very slowly, often only at long intervals of time, and generally on only a very few of the inhabitants of the same region at the same time. I further believe, that this very slow, intermit- [109] tent action of natural selection accords perfectly well with what geology tells us of the rate and manner at which the inhabitants of this world have changed....

[116] After the foregoing discussion, which ought to have been much amplified, we may, I think, assume that the modified descendants of any one species will succeed by so much the better as they become more diversified in structure, and are thus enabled to encroach on places occupied by other beings. Now let us see how this principle of great benefit being derived from divergence of character, combined with the principles of natural selection and of extinction, will tend to act.

The accompanying diagram will aid us in understanding this rather perplexing subject. Let A to L represent the species of a genus large in its own country; these species are supposed to resemble each other in unequal degrees, as is so generally the case in nature, and as is represented in the diagram by the letters standing at unequal distances. I have said a large genus, because we have seen in the second chapter, [117] that on an average more of the species of large genera vary than of small genera; and the varying species of the large genera present a greater number of varieties. We have, also, seen that the species, which are the commonest and the most widely-diffused, vary more than rare species with restricted ranges. Let (A) be a common, widely-diffused, and varying species, belonging to a genus large in its own country. The little fan of diverging dotted lines of unequal lengths proceeding from (A), may represent its varying offspring. The variations are supposed to be

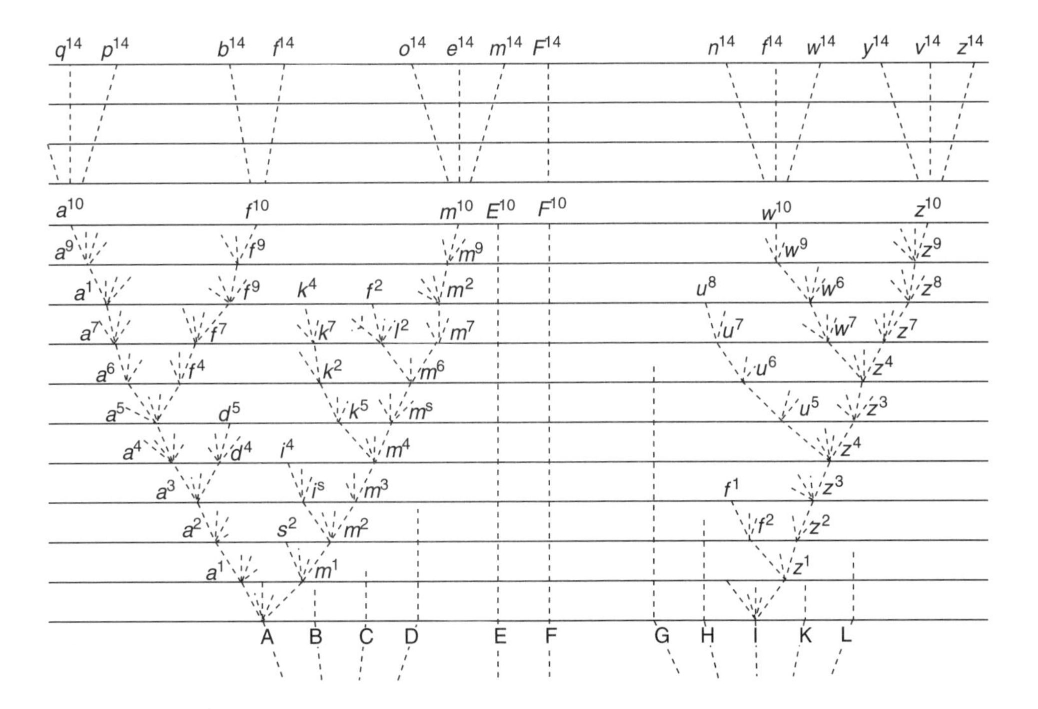

extremely slight, but of the most diversified nature; they are not supposed all to appear simultaneously, but often after long intervals of time; nor are they all supposed to endure for equal periods. Only those variations which are in some way profitable will be preserved or naturally selected. And here the importance of the principle of benefit being derived from divergence of character comes in; for this will generally lead to the most different or divergent variations (represented by the outer dotted lines) being preserved and accumulated by natural selection. When a dotted line reaches one of the horizontal lines, and is there marked by a small numbered letter, a sufficient amount of variation is supposed to have been accumulated to have formed a fairly well-marked variety, such as would be thought worthy of record in a systematic work.

The intervals between the horizontal lines in the diagram, may represent each a thousand generations; but it would have been better if each had represented ten thousand generations. After a thousand generations, species (A) is supposed to have produced two fairly well-marked varieties, namely a1 and m1. These two varieties will generally continue to be exposed to the same conditions which made their parents variable, [118] and the tendency to variability is in itself hereditary, consequently they will tend to vary, and generally to vary in nearly the same manner as their parents varied. Moreover, these two varieties, being only slightly modified forms, will tend to inherit those advantages which made their common parent (A) more numerous than most of the other inhabitants of the same country; they will likewise partake of those more general advantages which made the genus to which the parent-species belonged, a large genus in its own country. And these circumstances we know to be favourable to the production of new varieties.

If, then, these two varieties be variable, the most divergent of their variations will generally be preserved during the next thousand generations. And after this interval, variety a^1 is supposed in the diagram to have produced variety a^2, which will, owing to the principle of divergence, differ more from (A) than did variety a^1. Variety m^1 is supposed to have produced two varieties, namely m^2 and s^2, differing from each other, and more considerably from their common parent (A). We may continue the process by similar steps for any length of time; some of the varieties, after each thousand generations, producing only a single variety, but in a more and more modified condition, some producing two or three varieties, and some failing to produce any. Thus the varieties or modified descendants, proceeding from the common parent (A), will generally go on increasing in number and diverging in character. In the diagram the process is represented up to the ten-thousandth generation, and under a condensed and simplified form up to the fourteen-thousandth generation....

[119] After ten thousand generations, species (A) is supposed to have produced three forms, a^{10}, f^{10}, and m^{10}, which, from having diverged in character during the successive generations, will have come to differ largely, but perhaps unequally, from each other and from their common parent. If we suppose the amount of change between each horizontal line in our diagram to be excessively small, these three forms may still be only well-marked varieties; or they may have arrived at the doubtful category of sub-species; but we have only to suppose the steps in the process of modification to be more numerous or greater in amount, to convert these three forms into well-defined species: thus the diagram illustrates the steps by which the small differences distinguishing varieties are increased into the larger differences distinguishing species....

[122] If then our diagram be assumed to represent a considerable amount of modification, species (A) and all the earlier varieties will have become extinct, having been replaced by eight new species (a^{14} to m^{14}); and (I) will have been replaced by six (n^{14} to z^{14}) new species....

[126] *Summary of Chapter.*—If during the long course of ages and under varying conditions of life, organic beings [127] vary at all in the several parts of their organisation, and I think this cannot be disputed; if there be, owing to the high geometrical powers of increase of each species, at some age, season, or year, a severe struggle for life, and this certainly cannot be disputed; then, considering the infinite complexity of the relations of all organic beings to each other and to their conditions of existence, causing an infinite diversity in structure, constitution, and habits, to be advantageous to them, I think it would be a most extraordinary fact if no variation ever had occurred useful to each being's own welfare, in the same way as so many variations have occurred useful to man. But if variations useful to any organic being do occur, assuredly individuals thus characterised will have the best chance of being preserved in the struggle for life; and from the strong principle of inheritance they will tend to produce offspring similarly characterised. This principle of preservation, I have called, for the sake of brevity, Natural Selection. Natural selection, on the principle of qualities being inherited at corresponding ages, can modify the egg, seed, or young, as easily as the adult. Amongst many animals, sexual selection will give its aid to ordinary selection, by assuring to the most vigorous and best adapted males the greatest number of offspring. Sexual selection will also give characters useful to the males alone, in their struggles with other males.

Whether natural selection has really thus acted in nature, in modifying and adapting the various forms of life to their several conditions and stations, must be judged of by the general tenour and balance of evidence given in the following chapters. But we already see how it entails extinction; and how largely extinction has acted in the world's history, geology plainly declares. Natural selection, also, leads to divergence of [128] character; for more living beings can be supported on the same area the more they diverge in structure, habits, and constitution, of which we see proof by looking at the inhabitants of any small spot or at naturalised productions. Therefore during the modification of the descendants of any one species, and during the incessant struggle of all species to increase in numbers, the more diversified these descendants become, the better will be their chance of succeeding in the battle of life. Thus the small differences distinguishing varieties of the same species, will steadily tend to increase till they come to equal the greater differences between species of the same genus, or even of distinct genera.

Chapter VI: Difficulties on Theory

A significant strategy that Darwin followed was to preempt objections by anticipating and addressing them, as he does in chapter VI. In subsequent editions, Darwin addressed additional objections, as he was made aware of them. In chapter VI, Darwin identifies four classifications of objections. The third, on inheritance of instincts, and fourth, on sterility of hybrids, he treats separately in chapters VII and VIII. I omit discussion of them here.

[171] Long before having arrived at this part of my work, a crowd of difficulties will have occurred to the reader. Some of them are so grave that to this day I can never reflect on them without being staggered; but, to the best of my judgment, the greater number are only apparent, and those that are real are not, I think, fatal to my theory.

These difficulties and objections may be classed under the following heads:—Firstly, why, if species have descended from other species by insensibly fine gradations, do we not everywhere see innumerable transitional forms? Why is not all nature in confusion instead of the species being, as we see them, well defined?

Secondly, is it possible that an animal having, for instance, the structure and habits of a bat, could have been formed by the modification of some animal with wholly different habits? Can we believe that natural selection could produce, on the one hand, organs of trifling importance, such as the tail of a giraffe, which serves as a fly-flapper, and, on the other hand, organs of [172] such wonderful structure, as the eye, of which we hardly as yet fully understand the inimitable perfection?...

On the absence or rarity of transitional varieties.—As natural selection acts solely by the preservation of profitable modifications, each new form will tend in a fully-stocked country to take the place of, and finally to exterminate, its own less improved parent or other less-favoured forms with which it comes into competition. Thus extinction and natural selection will, as we have seen, go hand in hand. Hence, if we look at each species as descended from some other unknown form, both the parent and all the transitional varieties will generally have been exterminated by the very process of formation and perfection of the new form.

But, as by this theory innumerable transitional forms must have existed, why do we not find them embedded in countless numbers in the crust of the earth? It will be much more convenient to discuss this question in the chapter on the Imperfection of the geological record; and I will here only state that I believe the answer mainly lies in the record being incomparably less perfect than is generally supposed; the imperfection of the record being chiefly due to organic beings not inhabiting [173] profound depths of the sea, and to their remains being embedded and preserved to a future age only in masses of sediment sufficiently thick and extensive to withstand an enormous amount of future degradation; and such fossiliferous masses can be accumulated only where much sediment is deposited on the shallow bed of the sea, whilst it slowly subsides. These contingencies will concur only rarely, and after enormously long intervals....

But it may be urged that when several closely-allied species inhabit the same territory we surely ought to find at the present time many transitional forms. Let us take a simple case: in travelling from north to south over a continent, we generally meet at successive intervals with closely allied or representative species, evidently filling nearly the same place in the natural economy of the land. These representative species often meet and interlock; and as the one becomes rarer and rarer, the other becomes more and more frequent, till the one replaces the other. But if we compare these species where they intermingle, they are generally as absolutely distinct from each other in every detail of structure as are specimens taken from the metropolis inhabited by each. By my theory these allied species have descended from a common parent; and during the process of modification, each has become adapted to the conditions of life of its own region, and has supplanted and exterminated its original parent and all the transitional varieties between its past and present states. Hence we ought not to expect at the [174] present time to meet with numerous transitional varieties in each region, though they must have existed there, and may be embedded there in a fossil condition....

[179] *On the origin and transitions of organic beings with peculiar habits and structure.*—It has been asked by the opponents of such views as I hold, how, for instance, a land carnivorous animal could have been converted into one with aquatic habits; for how could the animal in its transitional state have subsisted? It would be easy to show that within the same group carnivorous animals exist having every intermediate grade between truly aquatic and strictly terrestrial habits; and as each exists by a struggle for life, it is clear that each is well adapted in its habits to its place in nature. Look at the Mustela vison [*the mink.—RKW*] of North America, which has webbed feet and which resembles an otter in its fur, short legs, and form of tail; during summer this animal dives for and preys on fish, but during the long winter [180] it leaves the frozen waters, and preys like other polecats on mice and land animals. If a different case had been taken, and it had been asked how an insectivorous quadruped could possibly have been converted into a flying bat, the question would have been far more difficult, and I could have given no answer. Yet I think such difficulties have very little weight.

Here, as on other occasions, I lie under a heavy disadvantage, for out of the many striking cases which I have collected, I can give only one or two instances of transitional habits and structures in closely allied species of the same genus; and of diversified habits, either constant or occasional, in the same species. And it seems to me that nothing less than a long list of such cases is sufficient to lessen the difficulty in any particular case like that of the bat....

[185] He who believes that each being has been created as we now see it, must occasionally have felt surprise when he has met with an animal having habits and structure not at all in agreement. What can be plainer than that the webbed feet of ducks and geese are formed for swimming; yet there are upland geese with webbed feet which rarely or never go near the water; and no one except Audubon has seen the frigate-bird, which has all its four toes webbed, alight on the surface of the sea. On the other hand, grebes and coots are eminently aquatic, although their toes are only bordered by membrane. What seems plainer than that the long toes of grallatores are formed for walking over swamps and floating plants, yet the water-hen is nearly as aquatic as the coot; and the landrail nearly as terrestrial as the quail or partridge. In such cases, and many others could be given, habits have changed without a corresponding change of structure. The webbed feet of the upland goose may be said to have become rudimentary in function, though not in structure. In the frigate-bird, the deeply-scooped membrane between the toes shows that structure has begun to change.

He who believes in separate and innumerable acts of creation will say, that in these cases it has pleased the [186] Creator to cause a being of one type to take the place of one of another type; but this seems to me only restating the fact in dignified language. He who believes in the struggle for existence and in the principle of natural selection, will acknowledge that every organic being is constantly endeavouring to increase in numbers; and that if any one being vary ever so little, either in habits or structure, and thus gain an advantage over some other inhabitant of the country, it will seize on the place of that inhabitant, however different it may be from its own place. Hence it will cause him no surprise that there should be geese and frigate-birds with webbed feet, either living on the dry land or most rarely alighting on the water; that there should be long-toed corncrakes living in meadows instead of in swamps; that there should be woodpeckers where not a tree grows; that there should be diving thrushes, and petrels with the habits of auks.

Organs of extreme perfection and complication.—To suppose that the eye, with all its inimitable contrivances for adjusting the focus to different distances, for admitting different amounts of light, and for the correction of spherical and chromatic aberration, could have been formed by natural selection, seems, I freely confess, absurd in the highest possible degree. Yet reason tells me, that if numerous gradations from a perfect and complex eye to one very imperfect and simple, each grade being useful to its possessor, can be shown to exist; if further, the eye does vary ever so slightly, and the variations be inherited, which is certainly the case; and if any variation or modification in the organ be ever useful to an animal under changing conditions of life, then the difficulty of believing that a perfect and complex eye could be formed by natural [187] selection, though insuperable by our imagination, can hardly be considered real. How a nerve comes to be sensitive to light, hardly concerns us more than how life itself first originated; but I may remark that several facts make me suspect that any sensitive nerve may be rendered sensitive to light, and likewise to those coarser vibrations of the air which produce sound.

In looking for the gradations by which an organ in any species has been perfected, we ought to look exclusively to its lineal ancestors; but this is scarcely ever possible, and we are forced in each case to look to species of the same group, that is to the collateral descendants from the same original parent-form, in order to see what gradations are possible, and for the chance of some gradations having been transmitted from the earlier stages of descent, in an unaltered or little altered condition....

[188] He who will go thus far, if he find on finishing this treatise that large bodies of facts, otherwise inexplicable, can be explained by the theory of descent, ought not to hesitate to go further, and to admit that a structure even as perfect as the eye of an eagle might be formed by natural selection, although in this case he does not know any of the transitional grades. His reason ought to conquer his imagination; though I have felt the difficulty far too keenly to be surprised at any degree of hesitation in extending the principle of natural selection to such startling lengths.

Chapter XIV: Recapitulation and Conclusion

This final chapter I include largely intact. It provides an excellent summary of the book.

[459] As this whole volume is one long argument, it may be convenient to the reader to have the leading facts and inferences briefly recapitulated.

That many and grave objections may be advanced against the theory of descent with modification through natural selection, I do not deny. I have endeavoured to give to them their full force. Nothing at first can appear more difficult to believe than that the more complex organs and instincts should have been perfected, not by means superior to, though analogous with, human reason, but by the accumulation of innumerable slight variations, each good for the individual possessor. Nevertheless, this difficulty, though appearing to our imagination insuperably great, cannot be considered real if we admit the following propositions, namely,—that gradations in the perfection of any organ or instinct, which we may consider, either do now exist or could have existed, each good of its kind,—that all organs and instincts are, in ever so slight a degree, variable,—and, lastly, that there is a struggle for existence leading to the preservation of each profitable deviation of structure or instinct. The truth of these propositions cannot, I think, be disputed.

[460] It is, no doubt, extremely difficult even to conjecture by what gradations many structures have been perfected, more especially amongst broken and failing groups of organic beings; but we see so many strange gradations in nature, as is proclaimed by the canon, "Natura non facit saltum," that we ought to be extremely cautious in saying that any organ or instinct, or any whole being, could not have arrived at its present state by many graduated steps. There are, it must be admitted, cases of special difficulty on the theory of natural selection; and one of the most curious of these is the existence of two or three defined castes of workers or sterile females in the same community of ants; but I have attempted to show how this difficulty can be mastered....

The fertility of varieties when intercrossed and of their mongrel offspring cannot be considered as universal; nor is their very general fertility surprising when we remember that it is not likely that either their constitutions or their reproductive systems should have been profoundly modified. Moreover, most of the [461] varieties which have been experimentised on have been produced under domestication; and as domestication apparently tends to eliminate sterility, we ought not to expect it also to produce sterility.

The sterility of hybrids is a very different case from that of first crosses, for their reproductive organs are more or less functionally impotent; whereas in first crosses the organs on both sides are in a perfect condition. As we continually see that organisms of all kinds are rendered in some degree sterile from their constitutions having been disturbed by slightly different and new conditions of life, we need not feel surprise at hybrids being in some degree sterile, for their constitutions can hardly fail to have been disturbed from being compounded of two distinct organisations. This parallelism is supported by another parallel, but directly opposite, class of facts; namely, that the vigour and fertility of all organic beings are increased by slight changes in their conditions of life, and that the offspring of slightly modified forms or varieties acquire from being crossed increased vigour and fertility. So that, on the one hand, considerable changes in the conditions of life and crosses between greatly modified forms, lessen fertility; and on the other hand, lesser changes in the conditions of life and crosses between less modified forms, increase fertility.

Turning to geographical distribution, the difficulties encountered on the theory of descent with modification are grave enough. All the individuals of the same species, and all the species of the same genus, or even higher group, must have descended from common parents; and therefore, in however distant and isolated parts of the world they are now found, they must in the course of successive generations have passed from some one part to the others. We are often wholly unable [462] even to conjecture how this could have been effected. Yet, as we have reason to believe that some species have retained the same specific form for very long periods, enormously long as measured by years, too much stress ought not to be laid on the occasional wide diffusion of the same species; for during very long periods of time there will always be a good chance for wide migration by many means. A broken or interrupted range may often be accounted for by the extinction of the species in the intermediate regions. It cannot be denied that we are as yet very ignorant of the full extent of the various climatal and geographical changes which have affected the earth during modern periods; and such changes will obviously have greatly facilitated migration. As an example, I have attempted to show how potent has been the influence of the Glacial period on the distribution both of the same and of representative species throughout the world. We are as yet profoundly ignorant of the many occasional means of transport. With respect to distinct species of the same genus inhabiting very distant and isolated regions, as the process of modification has necessarily been slow, all the means of migration will have been possible during a very long period; and consequently the difficulty of the wide diffusion of species of the same genus is in some degree lessened.

As on the theory of natural selection an interminable number of intermediate forms must have existed, linking together all the species in each group by gradations as fine as our present varieties, it may be asked, Why do we not see these linking forms all around us? Why are not all organic beings blended together in an inextricable chaos? With respect to existing forms, we should remember that we have no right to expect (excepting in rare cases) to discover directly connecting [463] links between them, but only between each and some extinct and supplanted form. Even on a wide area, which has during a long period remained continuous, and of which the climate and other conditions of life change insensibly in going from a district occupied by one species into another district occupied by a closely allied species, we have no just right to expect often to find intermediate varieties in the intermediate zone. For we have reason to believe that only a few species are undergoing change at any one period; and all changes are slowly effected. I have also shown that the intermediate varieties which will at first probably exist in the intermediate zones, will be liable to be supplanted by the allied forms on either hand; and the latter, from existing in greater numbers, will generally be modified and improved at a quicker rate than the intermediate varieties, which exist in lesser numbers; so that the intermediate varieties will, in the long run, be supplanted and exterminated.

On this doctrine of the extermination of an infinitude of connecting links, between the living and extinct inhabitants of the world, and at each successive period between the extinct and still older species, why is not every geological formation charged with such links? Why does not every collection of fossil remains afford plain evidence of the gradation and mutation of the forms of life? We meet with no such evidence, and this is the most obvious and forcible of the many objections which may be urged against my theory. Why, again, do whole groups of allied species appear, though certainly they often falsely appear, to have come in suddenly on the several geological stages? Why do we not find great piles of strata beneath the Silurian system, stored with the remains of the progenitors of the Silurian groups of fossils? For certainly on my theory such [464] strata must somewhere have been deposited at these ancient and utterly unknown epochs in the world's history.

I can answer these questions and grave objections only on the supposition that the geological record is far more imperfect than most geologists believe. It cannot be objected that there has not been time sufficient for any amount of organic change; for the lapse of time has been so great as to be utterly inappreciable by the human intellect. The number of specimens in all our museums is absolutely as nothing compared with the countless generations of countless species which certainly have existed. We should not be able to recognise a species as the parent of any one or more species if we were to examine them ever so closely, unless we likewise possessed many of the intermediate links between their past or parent and present states; and these many links we could hardly ever expect to discover, owing to the imperfection of the geological record. Numerous existing doubtful forms could be named which are probably varieties; but who will pretend that in future ages so many fossil links will be discovered, that naturalists will be able to decide, on the common view, whether or not these doubtful forms are varieties? As long as most of the links between any two species are unknown, if any one link or intermediate variety be discovered, it will simply be classed as another and distinct species. Only a small portion of the world has been geologically explored. Only organic beings of certain classes can be preserved in a fossil condition, at least in any great number. Widely ranging species vary most, and varieties are often at first local,—both causes rendering the discovery of intermediate links less likely. Local varieties will not spread into other and distant regions until they are considerably modified and im- [465] proved; and when they do spread, if discovered in a geological formation, they will appear as if suddenly created there, and will be simply classed as new species. Most formations have been intermittent in their accumulation; and their duration, I am inclined to believe, has been shorter than the average duration of specific forms. Successive formations are separated from each other by enormous blank intervals of time; for fossiliferous formations, thick enough to resist future degradation, can be accumulated only where much sediment is deposited on the subsiding bed of the sea. During the alternate periods of elevation and of stationary level the record will be blank. During these latter periods there will probably be more variability in the forms of life; during periods of subsidence, more extinction.

With respect to the absence of fossiliferous formations beneath the lowest Silurian strata, I can only recur to the hypothesis given in the ninth chapter. That the geological record is imperfect all will admit; but that it is imperfect to the degree which I require, few will be inclined to admit. If we look to long enough intervals of time, geology plainly declares that all species have changed; and they have changed in the manner which my theory requires, for they have changed slowly and in a graduated manner. We clearly see this in the fossil remains from consecutive formations invariably being much more closely related to each other, than are the fossils from formations distant from each other in time.

Such is the sum of the several chief objections and difficulties which may justly be urged against my theory; and I have now briefly recapitulated the answers and explanations which can be given to them. I have felt these difficulties far too heavily during many years to [466] doubt their weight. But it deserves especial notice that the more important objections relate to questions on which we are confessedly ignorant; nor do we know how ignorant we are. We do not know all the possible transitional gradations between the simplest and the most perfect organs; it cannot be pretended that we know all the varied means of Distribution during the long lapse of years, or that we know how imperfect the Geological Record is. Grave as these several difficulties are, in my judgment they do not overthrow the theory of descent with modification.

Now let us turn to the other side of the argument. Under domestication we see much variability. This seems to be mainly due to the reproductive system being eminently susceptible to changes in the conditions of life; so that this system, when not rendered impotent, fails to reproduce offspring exactly like the parent-form. Variability is governed by many complex laws,—by correlation of growth, by use and disuse, and by the direct action of the physical conditions of life. There is much difficulty in ascertaining how much modification our domestic productions have undergone; but we may safely infer that the amount has been large, and that modifications can be inherited for long periods. As long as the conditions of life remain the same, we have reason to believe that a modification, which has already been inherited for many generations, may continue to be inherited for an almost infinite number of generations. On the other hand we have evidence that variability, when it has once come into play, does not wholly cease; for new varieties are still occasionally produced by our most anciently domesticated productions.

Man does not actually produce variability; he only [467] unintentionally exposes organic beings to new conditions of life, and then nature acts on the organisation, and causes variability. But man can and does select the variations given to him by nature, and thus accumulate them in any desired manner. He thus adapts animals and plants for his own benefit or pleasure. He may do this methodically, or he may do it unconsciously by preserving the individuals most useful to him at the time, without any thought of altering the breed. It is certain that he can largely influence the character of a breed by selecting, in each successive generation, individual differences so slight as to be quite inappreciable by an uneducated eye. This process of selection has been the great agency in the production of the most distinct and useful domestic breeds. That many of the breeds produced by man have to a large extent the character of natural species, is shown by the inextricable doubts whether very many of them are varieties or aboriginal species.

There is no obvious reason why the principles which have acted so efficiently under domestication should not have acted under nature. In the preservation of favoured individuals and races, during the constantly-recurrent Struggle for Existence, we see the most powerful and ever-acting means of selection. The struggle for existence inevitably follows from the high geometrical ratio of increase which is common to all organic beings. This high rate of increase is proved by calculation, by the effects of a succession of peculiar seasons, and by the results of naturalisation, as explained in the third chapter. More individuals are born than can possibly survive. A grain in the balance will determine which individual shall live and which shall die,—which variety or species shall increase in number, and which shall decrease, or finally become extinct. As the individuals [468] of the same species come in all respects into the closest competition with each other, the struggle will generally be most severe between them; it will be almost equally severe between the varieties of the same species, and next in severity between the species of the same genus. But the struggle will often be very severe between beings most remote in the scale of nature. The slightest advantage in one being, at any age or during any season, over those with which it comes into competition, or better adaptation in however slight a degree to the surrounding physical conditions, will turn the balance.

With animals having separated sexes there will in most cases be a struggle between the males for possession of the females. The most vigorous individuals, or those which have most successfully struggled with their conditions of life, will generally leave most progeny. But success will often depend on having special weapons or means of defence, or on the charms of the males; and the slightest advantage will lead to victory.

As geology plainly proclaims that each land has undergone great physical changes, we might have expected that organic beings would have varied under nature, in the same way as they generally have varied under the changed conditions of domestication. And if there be any variability under nature, it would be an unaccountable fact if natural selection had not come into play. It has often been asserted, but the assertion is quite incapable of proof, that the amount of variation under nature is a strictly limited quantity. Man, though acting on external characters alone and often capriciously, can produce within a short period a great result by adding up mere individual differences in his domestic productions; and every one admits that there are at least individual differences in species under nature. But, besides such differences, all naturalists [469] have admitted the existence of varieties, which they think sufficiently distinct to be worthy of record in systematic works. No one can draw any clear distinction between individual differences and slight varieties; or between more plainly marked varieties and sub-species, and species. Let it be observed how naturalists differ in the rank which they assign to the many representative forms in Europe and North America.

If then we have under nature variability and a powerful agent always ready to act and select, why should we doubt that variations in any way useful to beings, under their excessively complex relations of life, would be preserved, accumulated, and inherited? Why, if man can by patience select variations most useful to himself, should nature fail in selecting variations useful, under changing conditions of life, to her living products? What limit can be put to this power, acting during long ages and rigidly scrutinising the whole constitution, structure, and habits of each creature,—favouring the good and rejecting the bad? I can see no limit to this power, in slowly and beautifully adapting each form to the most complex relations of life. The theory of natural selection, even if we looked no further than this, seems to me to be in itself probable. I have already recapitulated, as fairly as I could, the opposed difficulties and objections: now let us turn to the special facts and arguments in favour of the theory.

On the view that species are only strongly marked and permanent varieties, and that each species first existed as a variety, we can see why it is that no line of demarcation can be drawn between species, commonly supposed to have been produced by special acts of creation, and varieties which are acknowledged to have been produced by secondary laws. On this same view we can understand how it is that in each region [470] where many species of a genus have been produced, and where they now flourish, these same species should present many varieties; for where the manufactory of species has been active, we might expect, as a general rule, to find it still in action; and this is the case if varieties be incipient species. Moreover, the species of the larger genera, which afford the greater number of varieties or incipient species, retain to a certain degree the character of varieties; for they differ from each other by a less amount of difference than do the species of smaller genera. The closely allied species also of the larger genera apparently have restricted ranges, and they are clustered in little groups round other species—in which respects they resemble varieties. These are strange relations on the view of each species having been independently created, but are intelligible if all species first existed as varieties.

As each species tends by its geometrical ratio of reproduction to increase inordinately in number; and as the modified descendants of each species will be enabled to increase by so much the more as they become more diversified in habits and structure, so as to be enabled to seize on many and widely different places in the economy of nature, there will be a constant tendency in natural selection to preserve the most divergent offspring of any one species. Hence during a long-continued course of modification, the slight differences, characteristic of varieties of the same species, tend to be augmented into the greater differences characteristic of species of the same genus. New and improved varieties will inevitably supplant and exterminate the older, less improved and intermediate varieties; and thus species are rendered to a large extent defined and distinct objects. Dominant species belonging to the larger groups tend to give birth to new and dominant [471] forms; so that each large group tends to become still larger, and at the same time more divergent in character. But as all groups cannot thus succeed in increasing in size, for the world would not hold them, the more dominant groups beat the less dominant. This tendency in the large groups to go on increasing in size and diverging in character, together with the almost inevitable contingency of much extinction, explains the arrangement of all the forms of life, in groups subordinate to groups, all within a few great classes, which we now see everywhere around us, and which has prevailed throughout all time. This grand fact of the grouping of all organic beings seems to me utterly inexplicable on the theory of creation.

As natural selection acts solely by accumulating slight, successive, favourable variations, it can produce no great or sudden modification; it can act only by very short and slow steps. Hence the canon of "Natura non facit saltum," which every fresh addition to our knowledge tends to make more strictly correct, is on this theory simply intelligible. We can plainly see why nature is prodigal in variety, though niggard in innovation. But why this should be a law of nature if each species has been independently created, no man can explain....

[473] On the ordinary view of each species having been independently created, why should the specific characters, or those by which the species of the same genus differ from each other, be more variable than the generic characters in which they all agree? Why, for instance, should the colour of a flower be more likely to vary in any one species of a genus, if the other species, supposed to have been created independently, have differently coloured flowers, than if all the species of the genus have the same coloured flowers? If species are only well-marked varieties, of which the characters have become in a high degree permanent, we can understand this fact; for they have already varied since they branched off from a common progenitor in certain characters, by which they have come to be specifically distinct from each other; [474] and therefore these same characters would be more likely still to be variable than the generic characters which have been inherited without change for an enormous period. It is inexplicable on the theory of creation why a part developed in a very unusual manner in any one species of a genus, and therefore, as we may naturally infer, of great importance to the species, should be eminently liable to variation; but, on my view, this part has undergone, since the several species branched off from a common progenitor, an unusual amount of variability and modification, and therefore we might expect this part generally to be still variable. But a part may be developed in the most unusual manner, like the wing of a bat, and yet not be more variable than any other structure, if the part be common to many subordinate forms, that is, if it has been inherited for a very long period; for in this case it will have been rendered constant by long-continued natural selection....

[475] If species be only well-marked and permanent varieties, we can at once see why their crossed offspring should follow the same complex laws in their degrees and kinds of resemblance to their parents,—in being absorbed into each other by successive crosses, and in other such points,—as do the crossed offspring of acknowledged varieties. On the other hand, these would be strange facts if species have been independently created, and varieties have been produced by secondary laws.

If we admit that the geological record is imperfect in an extreme degree, then such facts as the record gives, support the theory of descent with modification. New species have come on the stage slowly and at successive intervals; and the amount of change, after equal intervals of time, is widely different in different groups. The extinction of species and of whole groups of species, which has played so conspicuous a part in the history of the organic world, almost inevitably follows on the principle of natural selection; for old forms will be supplanted by new and improved forms. Neither single species nor groups of species reappear when the chain of ordinary generation has once been broken. The gradual diffusion of dominant forms, with the slow modification of their descendants, causes the forms of life, after long intervals of time, to appear as if they had changed simultaneously throughout the world. The fact of the fossil remains of each formation being in some degree intermediate in character between the [476] fossils in the formations above and below, is simply explained by their intermediate position in the chain of descent. The grand fact that all extinct organic beings belong to the same system with recent beings, falling either into the same or into intermediate groups, follows from the living and the extinct being the offspring of common parents. As the groups which have descended from an ancient progenitor have generally diverged in character, the progenitor with its early descendants will often be intermediate in character in comparison with its later descendants; and thus we can see why the more ancient a fossil is, the oftener it stands in some degree intermediate between existing and allied groups. Recent forms are generally looked at as being, in some vague sense, higher than ancient and extinct forms; and they are in so far higher as the later and more improved forms have conquered the older and less improved organic beings in the struggle for life. Lastly, the law of the long endurance of allied forms on the same continent,—of marsupials in Australia, of edentata in America, and other such cases,—is intelligible, for within a confined country, the recent and the extinct will naturally be allied by descent....

[478] The fact, as we have seen, that all past and present organic beings constitute one grand natural system, with group subordinate to group, and with extinct groups often falling in between recent groups, is intelligible on the theory of natural selection with its contingencies of extinction and divergence of character. On these same principles we see how it is, that the mutual affinities of the species and genera within each class are so complex and circuitous. We

see why certain characters are far more serviceable than others for classification;—why adaptive characters, though of paramount importance to the being, are of hardly any [479] importance in classification; why characters derived from rudimentary parts, though of no service to the being, are often of high classificatory value; and why embryological characters are the most valuable of all. The real affinities of all organic beings are due to inheritance or community of descent. The natural system is a genealogical arrangement, in which we have to discover the lines of descent by the most permanent characters, however slight their vital importance may be.

The framework of bones being the same in the hand of a man, wing of a bat, fin of the porpoise, and leg of the horse,—the same number of vertebræ forming the neck of the giraffe and of the elephant,—and innumerable other such facts, at once explain themselves on the theory of descent with slow and slight successive modifications. The similarity of pattern in the wing and leg of a bat, though used for such different purpose,—in the jaws and legs of a crab,—in the petals, stamens, and pistils of a flower, is likewise intelligible on the view of the gradual modification of parts or organs, which were alike in the early progenitor of each class. On the principle of successive variations not always supervening at an early age, and being inherited at a corresponding not early period of life, we can clearly see why the embryos of mammals, birds, reptiles, and fishes should be so closely alike, and should be so unlike the adult forms. We may cease marvelling at the embryo of an air-breathing mammal or bird having branchial slits and arteries running in loops, like those in a fish which has to breathe the air dissolved in water, by the aid of well-developed branchiæ.

Disuse, aided sometimes by natural selection, will often tend to reduce an organ, when it has become useless by changed habits or under changed conditions [480] of life; and we can clearly understand on this view the meaning of rudimentary organs. But disuse and selection will generally act on each creature, when it has come to maturity and has to play its full part in the struggle for existence, and will thus have little power of acting on an organ during early life; hence the organ will not be much reduced or rendered rudimentary at this early age. The calf, for instance, has inherited teeth, which never cut through the gums of the upper jaw, from an early progenitor having well-developed teeth;...

I have now recapitulated the chief facts and considerations which have thoroughly convinced me that species have changed, and are still slowly changing by the preservation and accumulation of successive slight favourable variations. Why, it may be asked, have all the most eminent living naturalists and geologists rejected this view of the mutability of species? It cannot be [481] asserted that organic beings in a state of nature are subject to no variation; it cannot be proved that the amount of variation in the course of long ages is a limited quantity; no clear distinction has been, or can be, drawn between species and well-marked varieties. It cannot be maintained that species when intercrossed are invariably sterile, and varieties invariably fertile; or that sterility is a special endowment and sign of creation. The belief that species were immutable productions was almost unavoidable as long as the history of the world was thought to be of short duration; and now that we have acquired some idea of the lapse of time, we are too apt to assume, without proof, that the geological record is so perfect that it would have afforded us plain evidence of the mutation of species, if they had undergone mutation.

But the chief cause of our natural unwillingness to admit that one species has given birth to other and distinct species, is that we are always slow in admitting any great change of which we do not see the intermediate steps. The difficulty is the same as that felt by so many geologists, when Lyell first insisted that long lines of inland cliffs had been formed, and great valleys excavated, by the slow action of the coast-waves. The mind cannot possibly grasp the full meaning of the term of a hundred million years; it cannot add up and perceive the full effects of many slight variations, accumulated during an almost infinite number of generations.

Although I am fully convinced of the truth of the views given in this volume under the form of an abstract, I by no means expect to convince experienced naturalists whose minds are stocked with a multitude of facts all viewed, during a long course of years, from a point of view directly opposite to mine. It is so easy [482] to hide our ignorance under such expressions as the "plan of creation," "unity of design," &c., and to think that we give an explanation when we only restate a fact. Any one whose disposition leads him to attach more weight to unexplained difficulties than to the explanation of a certain number of facts will certainly reject my theory. A few naturalists, endowed with much flexibility of mind, and who have already begun to doubt on the immutability of species, may be influenced by this volume; but I look with confidence to the future, to young and rising naturalists, who will be able to view both sides of the question with impartiality. Whoever is led to believe that species are mutable will do good service by conscientiously expressing his conviction; for only thus can the load of prejudice by which this subject is overwhelmed be removed.

Several eminent naturalists have of late published their belief that a multitude of reputed species in each genus are not real species; but that other species are real, that is, have been independently created. This seems to me a strange conclusion to arrive at. They admit that a multitude of forms, which till lately they themselves thought were special creations, and which are still thus looked at by the majority of naturalists, and which consequently have every external characteristic feature of true species,—they admit that these have been produced by variation, but they refuse to extend the same view to other and very slightly different forms. Nevertheless they do not pretend that they can define, or even conjecture, which are the created forms of life, and which are those produced by secondary laws.

They admit variation as a vera causa in one case, they arbitrarily reject it in another, without assigning any distinction in the two cases. The day will come when this will be given as a curious illustration of [483] the blindness of preconceived opinion. These authors seem no more startled at a miraculous act of creation than at an ordinary birth. But do they really believe that at innumerable periods in the earth's history certain elemental atoms have been commanded suddenly to flash into living tissues? Do they believe that at each supposed act of creation one individual or many were produced? Were all the infinitely numerous kinds of animals and plants created as eggs or seed, or as full grown? and in the case of mammals, were they created bearing the false marks of nourishment from the mother's womb? Although naturalists very properly demand a full explanation of every difficulty from those who believe in the mutability of species, on their own side they ignore the whole subject of the first appearance of species in what they consider reverent silence....

[484] Analogy would lead me one step further, namely, to the belief that all animals and plants have descended from some one prototype. But analogy may be a deceitful guide. Nevertheless all living things have much in common, in their chemical composition, their germinal vesicles, their cellular structure, and their laws of growth and reproduction. We see this even in so trifling a circumstance as that the same poison often similarly affects plants and animals; or that the poison secreted by the gall-fly produces monstrous growths on the wild rose or oak-tree. Therefore I should infer from analogy that probably all the organic beings which have ever lived on this earth have descended from some one primordial form, into which life was first breathed.

When the views entertained in this volume on the origin of species, or when analogous views are generally admitted, we can dimly foresee that there will be a considerable revolution in natural history....

[486] A grand and almost untrodden field of inquiry will be opened, on the causes and laws of variation, on correlation of growth, on the effects of use and disuse, on the direct action of external conditions, and so forth. The study of domestic productions will rise immensely in value. A new variety raised by man will be a far more important and interesting subject for study than one more species added to the infinitude of already recorded species. Our classifications will come to be, as far as they can be so made, genealogies; and will then truly give what may be called the plan of creation. The rules for classifying will no doubt become simpler when we have a definite object in view. We possess no pedigrees or armorial bearings; and we have to discover and trace the many diverging lines of descent in our natural genealogies, by characters of any kind which have long been inherited. Rudimentary organs will speak infallibly with respect to the nature of long-lost structures. Species and groups of species, which are called aberrant, and which may fancifully be called living fossils, will aid us in forming a picture of the ancient forms of life. Embryology will reveal to us the structure, in some degree obscured, of the prototypes of each great class....

[488] In the distant future I see open fields for far more important researches. Psychology will be based on a new foundation, that of the necessary acquirement of each mental power and capacity by gradation. Light will be thrown on the origin of man and his history.

Authors of the highest eminence seem to be fully satisfied with the view that each species has been independently created. To my mind it accords better with what we know of the laws impressed on matter by the Creator, that the production and extinction of the past and present inhabitants of the world should have been due to secondary causes, like those determining the birth and death of the individual. When I view all beings not as special creations, but as the lineal descendants of some few beings which lived long before the [489] first bed of the Silurian system was deposited, they seem to me to become ennobled. Judging from the past, we may safely infer that not one living species will transmit its unaltered likeness to a distant futurity. And of the species now living very few will transmit progeny of any kind to a far distant futurity; for the manner in which all organic beings are grouped, shows that the greater number of species of each genus, and all the species of many genera, have left no descendants, but have become utterly extinct. We can so far take a prophetic glance into futurity as to foretel [*sic*] that it will be the common and widely-spread species, belonging to the larger and dominant groups, which will ultimately prevail and procreate new and dominant species. As all the living forms of life are the lineal descendants of those which lived long before the Silurian epoch, we may feel certain that the ordinary succession by generation has never once been broken, and that no cataclysm has desolated the whole world. Hence we may look with some confidence to a secure future of equally inappreciable length. And as natural selection works solely by and for the good of each being, all corporeal and mental endowments will tend to progress towards perfection.

It is interesting to contemplate an entangled bank, clothed with many plants of many kinds, with birds singing on the bushes, with various insects flitting about, and with worms crawling through the damp earth, and to reflect that these elaborately constructed forms, so different from each other, and dependent on each other in so complex a manner, have all been produced by laws acting around us. These laws, taken in the largest sense, being Growth with Reproduction; Inheritance which is almost implied by reproduction; Variability from the indirect and direct action of the external con- [490] ditions of life, and from use and disuse; a Ratio of Increase so high as to lead to a Struggle for Life, and as a consequence to Natural Selection, entailing Divergence of Character and the Extinction of less-improved forms. Thus, from the war of nature, from famine and death, the most exalted object which we are capable of conceiving, namely, the production of the higher animals, directly follows. There is grandeur in this view of life, with its several powers, having been originally breathed into a few forms or into one; and that, whilst this planet has gone cycling on according to the fixed law of gravity, from so simple a beginning endless forms most beautiful and most wonderful have been, and are being, evolved.

This last sentence is probably the most often quoted passage in the entire book. Indeed, the entire last paragraph is a heightened example of the persuasive and passionate style often encountered in Darwin's writing. In response to criticisms that Darwin's views were atheistic, he changed the final sentence in the second edition of 1860, as follows: "There is grandeur in this view of life, with its several powers, having been originally breathed by the Creator into a few forms or into one;..." Darwin came to regret this. In a letter to Hooker in 1863, Darwin wrote "I have long regretted that I truckled to public opinion & used Pentateuchal term of creation, by which I really meant 'appeared' by some wholly unknown process."

Ideas to Think About

1. What influence do you think Darwin's Galapagos experience had on his views of the frequent intergradation between varieties and species and their "restricted ranges"?
2. Discuss both the value and the danger of using domestication as an "analogy" with nature. What does Darwin say about this?
3. One can readily see how Darwin's 1838 sketch grew, with his thoughts, into the branching diagram in his chapter IV. Describe how his arguments for more intense selection among closely related species and the diversification of varieties in space are represented visually here.
4. How does Darwin make use of the distinction between extinction from competition and extinction through replacement (succession)?
5. What distinction does Darwin make, in describing natural selection, between mortality and fertility as selective mechanisms? Why do you think his initial and predominant explanations were based on the former?
6. In his recapitulation chapter, Darwin answers most of the criticisms his critics had presented. Which criticisms were based on faulty information and which on missing information? Does Darwin do an adequate job in answering these? What evidence does he bring to bear against creationist claims, and how effective is his argument?

For Further Reading

Darwin's Century, Loren Eiseley. New York: Doubleday Anchor (1958). This remarkable book, even after more than fifty years have passed, remains a hallmark of erudition and scholarship on the impact of Darwin on science. Clear and even elegant prose combine with original insights to reveal some of the philosophical obstacles Darwin faced, and the ambiguities of his closest friends, including Lyell, in fully accepting Darwin's remarkable ideas.

The Living Thoughts of Darwin, Julian Huxley. Greenwich, CT: Fawcett Publications (1963). This small book (176 pages) is notable for its author. Julian Huxley, grandson of Thomas H. Huxley ("Darwin's Bulldog"), and part of the famous Huxley dynasty, was a notable evolutionary biologist and one of the framers of the great evolutionary synthesis of the early twentieth century (see Unit 11). This astute interpretation of Darwin's work was originally published in 1939.

UNIT 10

The Age of Darwin, IV: The Aftermath

> This review, however, & Harveys [*sic*] letter have convinced me that I must be a very bad explainer. Neither really understand what I mean by natural selection.—I am inclined to give up attempt as hopeless.—Those who do not understand, it seems, cannot be made to understand.
>
> —Darwin to Hooker, *5 June 1860*

Darwin was referring to a letter from William Henry Harvey criticizing *The Origin*, and to a review published by Samuel Haughton claiming that natural selection was inadequate as the explanation that it claimed to be. Charles was more frustrated by the misunderstandings associated with natural selection than he was by the opposition to his theory on religious grounds. He feared that broad ignorance of what he was attempting to explain could doom his ideas. Such opposition would multiply, of course. What was surprising was the initial mildness of criticism. The book had been a bombshell, and until people had taken the time to discuss and understand its meaning an d its implications, the negative response was rather subdued.

There quickly gathered an inner circle in strong support of Darwin, however, and he was comforted and humbled by this. Charles Lyell and J. D. Hooker, Asa Gray and Thomas Henry Huxley—these were his champions, and they rallied in strong defense against any negative comment.

Asa Gray was born in New York in 1810—the year after Darwin's birth—and was instrumental in fostering American support for natural selection. He became the foremost American botanist of the century, publishing a manual of botany for the northern United States that is still used. Gray and his wife, Jane Loring Gray, met Charles Darwin when they visited Kew Gardens in London in 1851. They were introduced by Joseph Hooker, who was then director of Kew Gardens. In 1855, Charles established correspondence with Gray and discovered a like-minded naturalist. Gray was the prime mover in getting the first American edition of *Origin* and defended Darwin against American geologist Louis Agassiz, who was the first to promote the existence of the ice ages. Agassiz, a devout Christian, vehemently opposed Darwin's theory as atheistic. Gray argued that theism and natural law were not incompatible, although he tried to encourage Darwin to return to his past religious faith.

Thomas Henry Huxley, born in Middlesex in 1825, became a self-taught biologist and comparative anatomist. He was a voracious reader, including the classics, and learned to read both German and Greek. In 1854, Huxley wrote a scathing review of Chambers's *Vestiges*, and the following year gave a talk at his appointment as Fullerian Professor at the Royal Institution, advancing a skeptical attitude toward transmutation. The erudite and articulate style of Huxley encouraged Charles Darwin to befriend and convert him.

Huxley, in turn, was personally affected by Darwin's brilliance and persuasive attitude—eventually becoming widely known as "Darwin's bulldog." Although he remained skeptical of natural selection privately, he publically defended the theory with both skill and enthusiasm. They had a lifelong relationship, as Huxley became a member of the intimate circle of Darwin's friends. He went on to serve as president of the British Association for the Advancement of Science (1869–70) and president of the Royal Society (1883–85). He died in 1895, three years after Darwin.

Despite the sense of confidence these close friends gave him, Darwin continued to be troubled by an opposition he could not quite fathom; hence his doubts about his effectiveness as a communicator. This gnawed deeply into his intellect. It also affected him physically. He unfortunately was too ill to attend the infamous BAAS meetings in Oxford in

June 1860. "My stomach has utterly failed," he wrote Hooker, "I cannot think of Oxford." Although still apprehensive of Wallace's reaction to the joint Linnean papers of 1858, Darwin wrote early in 1859 to Syms Covington, his assistant during the *Beagle* voyage and thereafter for a while, "My health keeps very poor & I never know 24 hours comfort. I force myself to try & bear this as incurable misfortune. We all have our unhappiness, only some are worse than others."

The most telling admission, however, comes from a letter to Hooker on March 26, 1861: "It is strange how immediately any mental excitement upsets & utterly prostrate; seeing [son] George chloroformed for his teeth brought on my eternal sickness for 24 hours." Despite any organic etiology, it is obvious that Darwin's chronic illness responded dramatically to emotional stress.

Even evolutionists today appear baffled by the reluctance of so many to accept natural selection's role in species transformation, and, at least in America, this resistance currently appears to be no less common than in England during the 1860s. Excluding the religious element in this rejection does not diminish the numbers in any extraordinary way, nor did it in Darwin's time.

A good part of the resistance had to do with the strict materialism and the coldness it brought to the heart and spirit of so many. Quite aside from the religious objection in the strictest sense, it was a deeply humanistic objection: Many people thought Darwin's new system was overreaching by attempting to explain such uniquely human conditions as emotional expression, moral sentiment, and the power of abstract thought. Such reductionism, after all, expelled us from an exclusive club membership. This is reflected in Darwin's subsequent work on the *Descent of Man* (in two volumes, 1871, which included treatment of sexual selection), and *The Expression of Emotion in Man and Animals* (1872) Here is how George Bernard Shaw expressed it in *Back to Methuselah,* in 1921:

> [Darwinism] seems simple, because you do not at first realize all that it involves. But when its whole significance dawns on you, your heart sinks into a heap of sand within you. There is a hideous fatalism about it, a ghastly and damnable reduction of beauty and intelligence, of strength and purpose, of honor and aspiration.

Shaw was not known for his discretion—nor for his good sense—in castigating scientists. He frequently expressed disbelief in the germ theory, labeling Pasteur and Lister as frauds.

Not only did Huxley harbor doubts about the new Darwinism, but so did Hooker, Gray, and Lyell. Faithful to their old friend, they couched their doubts in patience and slowly came around to acceptance. "I fully believe that I owe the comfort of the next few years of my life to your generous support," Darwin wrote to Lyell in early December 1859, "& that of a very few others: I do not think I am brave enough to have stood being odious without support. Now I feel as bold as a Lion." He added his admission that he was perhaps too sensitive to criticism: "But there is one thing I can see I must learn, viz. to think less of myself & my book."

The resistance to natural selection led to alternative theories that sought to retain some inherent purpose in evolution—particularly human evolution—or at least some process more sympathetic to our moral nature and less mechanistic in its action. One reaction was a renewed interest in Lamarckism. Another was a host of teleological theories that accepted evolution but challenged Darwinism. Driesch's application of the old Aristotelian "entelechy," an inward drive for realization of potential, and Bergson's "creative evolution" are two examples. But these reactions emerged much later. In the immediate aftermath, there were all of these responses with which Darwin was confronted.

What follows, then, are some of the more prominent critiques and reviews, positive and negative. They have been selected, from among many, in proportion to their affect on Darwin and his friends.

43. Charles Darwin

The Correspondence of Charles Darwin Volume 7: 1858–1859

Sedgwick's was the first formal response Darwin received after publishing The Origin. *Although Darwin knew that Sedgwick would oppose his conclusions, he did not expect such passionate condemnation. Sedgwick, after all, had been his original field mentor and had initiated him in the science of geology. He had even stayed in Darwin's home as an honored guest.*

Letter 2548—Sedgwick, Adam, to Darwin, C. R., 24 Nov 1859

Cambridge
Novr 24 1859

My dear Darwin
I write to thank you for your work on the origin of Species. It came, I think, in the latter part of last week; but it *may* have come a few days sooner, & been overlooked among my bookparcels, which often remain unopened when I am lazy, or busy with any work before me. So soon as I opened it I began to read it, & I finished it, after many interruptions, on tuesday. Yesterday I was employed 1st. in preparing for my lecture—2dly. In attending a meeting of my brother Fellows to discuss the final propositions of the Parliamentary Commissions. 3rd. In lecturing 4^{thly}. In hearing the conclusion of the discussion & the College reply whereby in conformity with my own wishes we accepted the scheme of the Commission 5th. in Dining with an old friend at Clare College—6^{thly} In adjourning to the weekly meeting of the Ray Club, from which I returned at 10. P.M.—dog-tired & hardly able to climb my staircase—Lastly in looking thro' the Times to see what was going on in the busy world—

I do not state this to fill space (tho' I believe that Nature does abhor a vacuum); but to prove that my reply & my thanks are sent to you by the earliest leisure I have; tho' this is but a very contracted opportunity.— If I did not think you a good tempered & truth loving man I should not tell you that, (spite of the great knowledge; store of facts; capital views of the corelations of the various parts of organic nature; admirable hints about the diffusions, thro' wide regions, of nearly related organic beings; &c &c) I have read your book with more pain than pleasure. Parts of it I admired greatly; parts I laughed at till my sides were almost sore; other parts I read with absolute sorrow; because I think them utterly false & grievously mischievous—You have *deserted*—after a start in that tram-road of all solid physical truth—the the true method of induction—& started up a machinery as wild I think as Bishop Wilkin's locomotive that was to sail with us to the Moon. Many of your wide conclusions are based upon assumptions which can neither be proved nor disproved. Why then express them in the language & arrangements of philosophical induction?.—

As to your grand principle—*natural selection*—what is it but a secondary consequence of supposed, or known, primary facts. Development is a better word because more close to the cause of the fact. For you do not deny causation. I call (in the abstract) causation the will of God: & I can prove that He acts for the good of His creatures. He also acts by laws which we can study & comprehend—Acting by law, & under what is called final cause, comprehends, I think, your whole principle. You write of "natural selection" as if it were done consciously by the selecting agent. 'Tis but a consequence of the pre*supposed* development, & the subsequent battle for life.—

This view of nature you have stated admirably; tho' admitted by all naturalists & denied by no one of common sense. We all admit development as a fact of history; but how came it about? Here, in language, & still more in logic, we are point blank at issue—There is a moral or metaphysical part of nature as well as a physical A man who denies this is deep in the mire of folly. Tis the crown & glory of organic science that it *does* thro' *final cause*, link material to moral; & yet *does not* allow us to mingle them in our first conception of laws, & our classification of such laws whether we consider one side of nature or the other—You have ignored this link; &, if I do not mistake your meaning, you have done your best in one or two pregnant cases to break it. Were it possible (which thank God it is not) to break it, humanity in my mind, would suffer a damage that might brutalize it—& sink the human race into a lower grade of degradation than any into which it has fallen since its written records tell us of its history. Take the case of the bee cells. If your development produced the successive modification of the bee & its cells (which no mortal can prove) final cause would stand good as the directing cause under which the successive generations acted & gradually improved—Passages in your book, like that to which I have alluded (& there are others almost as bad) greatly shocked my moral taste. I think in speculating upon organic descent, you *over* state the evidence of geology; & that you *under* state it while you are talking of the broken links of your natural pedigree: but my paper is nearly done, & I must go to my lecture room—

Lastly then, I greatly dislike the concluding chapter—not as a summary—for in that light it appears good—but I dislike it from the tone of triumphant confidence in which you appeal to the rising generation (in a tone I condemned in the author of the Vestiges), & prophesy of things not yet in the womb of time; nor, (if we are to trust the accumulated experience of human sense & the inferences of its logic) ever likely to be found any where but in the fertile womb of man's imagination.—

And now to say a word about a son of a monkey & an old friend of yours. I am better, far better than I was last year. I have been lecturing three days a week (formerly I gave six a week) without much fatigue but I find, by the loss of activity & memory, & of all productive powers, that my bodily frame is sinking slowly towards the earth. But I have visions of the future. They are as much a part of myself as my stomach & my heart; & tho visions are to have their antitype in solid fruition of what is best & greatest But on one condition only—that I humbly accept God's revelation of himself both in His works & in His word; & do my best to act in conformity with that knowledge which He only can give me, & He only can sustain me in doing If you & I do all this we shall meet in heaven

I have written in a hurry & in a spirit of brotherly love. Therefore forgive any sentence you happen to dislike; & believe me, spite of our disagreement in some points of the deepest moral interest, your true-hearted old friend

A. Sedgwick

Darwin's response was respectful, as befits correspondence with an old friend, but he was also obviously wounded.

Letter 2555—Darwin, C. R., to Sedgwick, Adam, 26 Nov 1859

Ilkley Wells House | Otley, Yorkshire
Nov. 26

My dear Prof. Sedgwick
I did not at all expect that you would have written to me.—You could not possibly have paid me a more honourable compliment than in expressing freely your strong disapprobation of my Book.—I fully expected it. I can only say that I have worked like a slave on the subject for above 20 years & am not conscious that bad motives have influenced the conclusions at which I have arrived. I grieve to have shocked a man whom I sincerely honour. But I do not think you would wish anyone to conceal the results at which he has arrived after he has worked, according to the best ability which may be in him. I do not think my book will be mischievous; for there are so many workers that, if I be wrong I shall soon be annihilated; & surely you will agree that truth can be known only by rising victorious from every attack.

I daresay I may have written too confidently from feeling so confident of the truth of my main doctrine. I have made already a few converts of good & tried naturalists & oddly enough two of them compliment me on my cautious mode of expression! This will make you laugh. My notion of young men being best judges of new doctrines was not invented for occasion; for however erroneous, I remember nearly twenty years ago laughing with Lyell over the idea.—I have tried to be honest in giving all the many & grave difficulties which occurred to me, or I met in published works. I cannot think a false theory would explain so many classes of facts, as the theory seems to me to do. But magna est veritas & thank God, prevalebit.

Forgive me for scribbling at such length, & let me say again how grieved I am to have encountered your severe disapprobation & ridicule. Your kind & noble heart shows itself throughout your letter. I thank you for writing, & remain with sincere respect

Yours truly obliged
Charles Darwin

---

44. RICHARD OWEN

Darwin on the Origin of Species

Richard Owen's review was another matter altogether. It was maliciously negative, a mixture of ridicule and reason. Writing as an anonymous third person, which allowed him to quote himself in opportunistically favorable ways, his authorship was hardly mysterious to Darwin and others who well knew Owen's writing style. It cut Darwin deeply. In December 1859, Charles had visited Owen in London, genuinely wanting to get his opinion of Origin. *The visit barely shielded Owen's displeasure. According to Janet Browne (2002, 98), the two never spoke again after that meeting. They had once been friends and, beyond this, shared mutual respect. Now, both sentiments had vanished. "It is painful," wrote Charles, "to be hated in the intense degree with which Owen hates me."*

...The octavo volume, of upwards of 500 pages, which made its appearance towards the end of last year, has been received and perused with avidity, not only by the professed naturalist, but by that far wider intellectual class which now takes interest in the higher generalisations of all the sciences. The same pleasing style which marked Mr. Darwin's earliest work, and a certain artistic disposition and sequence of his principal arguments, have more closely recalled the attention of thinking men to the hypothesis of the inconstancy and transmutation of species, than had been done by the writings of previous advocates of similar views. Thus several, and perhaps the majority, of our younger naturalists have been seduced into the acceptance of the homoeopathic form of the transmutative hypothesis now presented to them by Mr. Darwin, under the phrase of 'natural selection.'

...The scientific world has looked forward with great interest to the facts which Mr. Darwin might finally deem adequate to the support of his theory on this supreme question in biology, and to the course of inductive original research which might issue in throwing light on 'that mystery of mysteries.' But having now cited the chief, if not the whole, of the original observations adduced by its author in the volume now before us, our disappointment may be conceived. Failing the adequacy of such observations, not merely to carry conviction, but to give a colour to the hypothesis, we were then left to confide in the superior grasp of mind, strength of intellect, clearness and precision of thought and expression, which raise one man so far above his contemporaries, as to enable him to discern in the common stock of facts, of coincidences, correlations and analogies in Natural History, deeper and truer conclusions than his fellow-labourers had been able to reach....

The origin of species is the question of questions in Zoology; the supreme problem which the most striking of our original labourers, the clearest zoological thinkers, and the most successful generalisers, have never lost sight of, whilst they have approached it with due reverence. We have a right to expect that the mind proposing to treat of, and assuming to have solved, the problem, should show its equality to the task. The signs of such intellectual power we look for in clearness of expression, and in the absence of all ambiguous or unmeaning terms. Now, the present work is occupied by arguments, beliefs, and speculations on the origin of species, in which, as it seems to us, the fundamental mistake is committed, of confounding the questions, of species being the result of a secondary cause or law, and of the nature of that creative law. Various have been the ideas promulgated respecting its mode of operation; such as the reciprocal action of an impulse from within, and an influence from without, upon the organization (Demaillet, Lamarck); premature birth of an embryo at a phase of development, so distinct from that of the parents, as, with the power of life and growth, under that abortive phase, to manifest differences equivalent to specific (*Vestiges of Creation*); the hereditary transmission of what are called 'accidental monstrosities;' the principle of gradual transmutation by 'degeneration' (Buffon) as contrasted with the 'progressional' view....

Lasting and fruitful conclusions have, indeed, hitherto been based only on the possession of knowledge; now we are called upon to accept an hypothesis on the plea of want of knowledge. The geological record, it is averred, is so imperfect! But what human record is not? Especially must the record of past organisms be much less perfect than of present ones. We freely admit it. But when Mr. Darwin, in reference to the absence of the intermediate fossil forms required by his hypothesis—and only the zootomical zoologist can approximatively appreciate their immense numbers—the countless hosts of transitional links which, on 'natural selection,' must certainly have existed at one period or another of the world's history—when Mr. Darwin exclaims what may be, or what may not be, the forms yet forthcoming out of the graveyards of strata, we would reply, that our only ground for prophesying of what may come, is by the analogy of what has come to light. We may expect, e.g., a chambered-shell from a secondary rock; but not the evidence of a creature linking on the cuttle-fish to the lump-fish.

Mr. Darwin asks, 'How is it that varieties, which I have called incipient species, become ultimately good and distinct species?' To which we rejoin with the question:—Do they become good and distinct species? Is there any one instance proved by observed facts of such transmutation? We have searched the volume in vain for such. When we see the intervals that divide most species from their nearest congeners, in the recent and especially the fossil series, we either doubt the fact of progressive conversion, or, as Mr. Darwin remarks in his letter to Dr. Asa Gray, one's 'imagination must fill up very wide blanks.'

...To this, of course, the transmutationists reply that a still longer period of time might do what thirty thousand years have not done.

Professor Baden Powell, for example, affirms;—'Though each species may have possessed its peculiarities unchanged for a lapse of time, the fact that when long periods are considered, all those of our earlier period are replaced by new ones at a later period, proves that species change in the end, provided a sufficiently long time is granted.' But here lies the fallacy: it merely proves that species are changed, it gives us no evidence as to the mode of change; transmutation, gradual or abrupt, is in this case mere assumption....

The essential element in the complex idea of species, as it has been variously framed and defined by naturalists, viz., the blood-relationship between all the individuals of such species, is annihilated on the hypothesis of 'natural selection.' According to this view a genus, a family, an order, a class, a sub-kingdom,—the individuals severally representing these grades of difference or relationship,—now differ from individuals of the same species only in degree: the species, like every other group, is a mere creature of the brain; it is no longer from nature. With the present evidence from form, structure, and procreative phenomena, of the truth of the opposite proposition, that 'classification is the task of science, but species the work of nature,' we believe that this aphorism will endure; we are certain that it has not yet been refuted; and we repeat in the words of Linnæus, 'Classis et Ordo est sapientiæ, Species naturæ opus'.

---

45. Charles Darwin

The Correspondence of Charles Darwin Volume 8: 1860

Letter 2751—Darwin, C. R., to Huxley, T. H., 9 Apr. 1860

Darwin was exasperated with this review, because he feared that Owen's reputation would garner sympathetic support from those who had otherwise no special expertise with which to judge the work. He wrote the following to Huxley (extracts from a longer letter).

I never saw such an amount of misrepresentation. At p 530 he says we are called on to accept the hypothesis on the plea of ignorance, whereas I think I could not have made it clearer that I admit the imperfection of geological record as a great difficulty.—

The quotation at p. 512 of Review about "young & rising naturalists with plastic minds", attributed to "nature of limbs" is a false quotation, as I do not use words "plastic minds" At p. 501 the quotation is garbled, for I only ask whether naturalists believe about elemental atoms flashing &c, & he changes it into that I state that they do believe.

At p. 500 It is very false to say that I imply by "blindness of preconceived opinion" the simple belief of creation.—And so on in other cases.—But I beg pardon for troubling you.—I am heartily sorry that in your unselfish endeavours to spread what you believe to be truth, you sh^{d}. have incurred so brutal an attack. And now I will not think any more of this false & malignant attack

Ever yours
C. Darwin

The following excerpts of the review by Bishop of Oxford, Samuel Wilberforce, is—for a man so passionately fighting against the heresy of Darwin's theory—rather restrained. Darwin himself respected this restraint and the courtesy with which Wilberforce treated him in the review. He also saw the unmistakable hand of Richard Owen in the more specific details of particular scientific arguments: Samuel Wilberforce had no training in science. The following excerpt omits these detailed treatments, largely between pages 226 and 256.

The substance of Wilberforce's address at Oxford, from all accounts, was the text of this review. That Oxford confrontation follows this.

---

46. Samuel Wilberforce

On the Origin of Species, by means of Natural Selection, ...

Any contribution to our Natural History literature from the pen of Mr. C. Darwin is certain to command attention. His scientific attainments, his insight and carefulness as an observer, blended with no scanty measure of imaginative sagacity, and his clear and lively style, make all his writings unusually attractive. His present volume on the 'Origin of Species' is the [226] result of many years of observation, thought, and speculation; and is manifestly regarded by him as the 'opus' upon which his future fame is to rest. It is true that he announces it modestly enough as the mere precursor of a mightier volume. But that volume is only intended to supply the facts which are to support the completed argument of the present essay. In this we have a specimen-collection of the vast accumulation; and, working from these as the high analytical mathematician may work from the admitted results of his conic sections, he proceeds to deduce all the conclusions to which he wishes to conduct his readers.

The essay is full of Mr. Darwin's characteristic excellences. It is a most readable book; full of facts in natural history, old and new, of his collecting and of his observing; and all of these are told in his own perspicuous language, and all thrown into picturesque combinations, and all sparkle with the colours of fancy and the lights of imagination. It assumes, too, the grave proportions of a sustained argument upon a matter of the deepest interest, not to naturalists only, or even to men of science exclusively, but to every one who is interested in the history of man and of the relations of nature around him to the history and plan of creation.

With Mr. Darwin's 'argument' we may say in the outset that we shall have much and grave fault to find. But this does not make us the less disposed to admire the singular excellences of his work; and we will seek *in limine* to give our readers a few examples of these. Here, for instance, is a beautiful illustration of the wonderful interdependence of nature—of the golden chain of unsuspected relations which bind together all the mighty web which stretches from end to end of this full and most diversified earth. Who, as he listened to the musical hum of the great bumble-bees, or marked their ponderous flight from flower to flower, and watched the unpacking of their trunks for their work of suction, would have supposed that the multiplication or diminution of their race, or the fruitfulness and sterility of the red clover, depend as directly on the vigilance of our cats as do those of our well-guarded game-preserves on the watching of our keepers? Yet this Mr. Darwin has discovered to be literally the case:...

[256] He who is as sure as he is of his own existence that the God of Truth is at once the God of Nature and the God of Revelation, cannot believe it to be possible that His voice in either, rightly understood, can differ, or deceive His creatures. To oppose facts in the natural world because they seem to oppose Revelation, or to humour them so as to compel them to speak its voice, is, he knows, but another form of the ever-ready feebleminded dishonesty of lying for God, and trying by fraud or falsehood to do the work of the God of truth. It is with another and a nobler spirit that the true believer walks amongst the works of nature. The words graven on the everlasting rocks are the words of God, and they are graven by His hand. No more can they contradict His Word written in His book, than could the words of the old [257] covenant graven by His hand on the stony tables contradict the writings of His hand in the volume of the new dispensation. There may be to man difficulty in reconciling all the utterances of the two voices. But what of that? He has learned already that here he knows only in part, and that the day of reconciling all apparent contradictions between what must agree is nigh at hand. He rests his mind in perfect quietness on this assurance, and rejoices in the gift of light without a misgiving as to what it may discover....

[257] Mr. Darwin writes as a Christian, and we doubt not that he is one. We do not for a moment believe him to be one of those who retain in some corner of their hearts a secret unbelief which they dare not vent; and we therefore pray him to consider well the grounds on which we brand his speculations with the charge of such a tendency. First, then, he not obscurely declares that he applies his scheme of the action of the principle of natural selection to MAN himself, as well as to the animals around him. [258] Now, we must say at once, and openly, that such a notion is absolutely incompatible not only with single expressions in the word of God on that subject of natural science with which it is not immediately concerned, but, which in our judgment is of far more importance, with the whole representation of that moral and spiritual condition of man which is its proper subject-matter. Man's derived supremacy over the earth; man's power of articulate speech; man's gift of reason; man's free-will and responsibility; man's fall and man's redemption; the incarnation of the Eternal Son; the indwelling of the Eternal Spirit,—all are equally and utterly irreconcilable with the degrading notion of the brute origin of him who was created in the image of God, and redeemed by the Eternal Son assuming to himself his nature. Equally inconsistent, too, not with any passing expressions, but with the whole scheme of God's dealings with man as recorded in His word, is Mr. Darwin's daring

notion of man's further development into some unknown extent of powers, and shape, and size, through natural selection acting through that long vista of ages which he casts mistily over the earth upon the most favoured individuals of his species....

Nor can we doubt, secondly, that this view, which thus contradicts the revealed relation of creation to its Creator, is equally inconsistent with the fulness of His glory. It is, in truth, an ingenious theory for diffusing throughout creation the working and so the personality of the Creator. And thus, however unconsciously to him who holds them, such views really tend inevitably to banish from the mind most of the peculiar attributes of the Almighty.

How, asks Mr. Darwin, can we possibly account for the manifest plan, order, and arrangement which pervade creation, except we allow to it this self-developing power through modified descent?...

[259] How can we account for all this? By the simplest and yet the most comprehensive answer. By declaring the stupendous fact that all creation is the transcript in matter of ideas eternally existing in the mind of the Most High—that order in the utmost perfectness of its relation pervades His works, because it exists as in its centre and highest fountainhead in Him the Lord of all. Here is the true account of the fact which has so utterly misled shallow observers, that Man himself, the Prince and Head of this creation, passes in the earlier stages of his being through phases of existence closely analogous, so far as his earthly tabernacle is concerned, to those in which the lower animals ever remain. At that point of being the development of the protozoa is arrested. Through it the embryo of their chief passes to the perfection of his earthly frame. But the types of those lower forms of being must be found in the animals which never advance beyond them —not in man for whom they are but the foundation for an after-development; whilst he too, Creation's crown and perfection, thus bears witness in his own frame to the law of order which pervades the universe....

---

47. The Oxford Confrontation

The Correspondence of Charles Darwin Volume 8: 1860

Letter 2852—Hooker, J. D. to Darwin, C. R., 2 July 1860

Events often converge serendipitously to create legends. They did so in Oxford in June 1860. Ancient Oxford University, having fallen behind Cambridge in science, had entered the contest in earnest by constructing an imposing monument to the sciences: the Oxford Museum of Natural History. Standing in grandeur on Parks Road, set back some one hundred meters, created in fine gothic style with pillars whose capitals feature carved plants, insects, and other animals, it had just been completed—awaiting a proper investiture.

The British Association for the Advancement of Science annual meeting was scheduled for Oxford in 1860. These were week-long meetings, held in a different location each year (the first was in York in 1831), in which scientific reports were given for both public and professional consumption—as it still is today. It was designed to celebrate science. Most papers and colloquia were summaries of recent research, and these would have various venues among the proud colleges of Oxford.

Barely six months after his Origin *began igniting conversations across the country this was Darwin's year. Furthermore, this was Samuel Wilberforce's diocese. And Wilberforce fervently desired more exposure before the societal matrons, political wags, and government dignitaries who would attend. Besides, he was a vice president of the association.*

A session on Darwin's new theory was scheduled. It was in Session D, which was devoted to the life sciences, and held in the museum in a grand hall (that has since been partitioned and now serves as the department of entomology). Its collections include original insect specimens collected by Darwin and Wallace. This hall was the original site of the Radcliffe Science Library, and the library sign remains in position above the entry door on the museum's second floor.

The legendary event occurred on Saturday, June 30, 1860. The previous day's papers focused on Darwinian ideas, both pro and con, promising heated debate on Saturday. Accounts vary and have been embellished, as legends always are. Darwin was too ill to attend ("My stomach has utterly failed; & I cannot think of Oxford," he wrote Hooker on the 26th).

Hooker was quick to recount the meeting in a letter to Darwin (grammatical errors and misspellings not tagged).

Botanic Gardens Oxford
July 2/60

Dear Darwin
I have just come in from my last moonlight saunter at Oxford & been soliloquizing over the Ratcliffe & our old rooms at the corner & cannot go to bed without inditing a few lines to you my dear old Darwin. I came here on Thursday afternoon & immediately fell into a lengthened revirie: without you & my wife I was as dull as ditch water & crept about the once familiar streets feeling like a fish out of water—I swore I would not go near a Section & did not for two days—but amused myself with the Colleges buildings & alternate sleeps in the sleepy gardens & rejoiced in my indolence. Huxley & Owen had had a furious battle over Darwins absent body at Section D., before my arrival,—of which more anon. H. was triumphant—You & your book forthwith became the topics of the day, & I d—d the days & double d—d the topics too, & like a craven felt bored out of my life by being woke out of my reveries to become referee on Natural Selection &c &c &c—On Saturday I walked with my old friend of the Erebus Capt Dayman to the Sections & swore as usual I would not go in; but getting equally bored of doing nothing I did. A paper of a yankee donkey called Draper on "civilization

according to the Darwinian hypothesis" or some such title was being read, & it did not mend my temper; for of all the flatulent stuff and all the self sufficient stuffers—these were the greatest, it was all a pie of Herb[t] Spenser & Buckle without the seasoning of either—however hearing that Soapy Sam was to answer I waited to hear the end. The meeting was so large that they had adjourned to the Library which was crammed with between 700 & 1000 people, for all the world was there to hear Sam Oxon—Well Sam Oxon got up & spouted for half an hour with inimitable spirit uglyness & emptyness & unfairness, I saw he was coached up by Owen & knew nothing & he said not a syllable but what was in the Reviews—he ridiculed you badly & Huxley savagely—Huxley answered admirably & turned the tables, but he could not throw his voice over so large an assembly, nor command the audience; & he did not allude to *Sam's* weak points nor put the matter in a form or way that carried the audience. The battle waxed hot. Lady Brewster fainted, the excitement increased as others spoke—my blood boiled, I felt myself a dastard; now I saw my advantage—I swore to myself I would smite that Amalekite Sam hip & thigh if my heart jumped out of my mouth & I handed my name up to the President (Henslow) as ready to throw down the gauntlet—must tell you that Henslow as president would have none speak but those who had *arguments* to use, & 4 persons had been burked [*quietly suppressed.—RKW*] by the audience & President for mere declamation: it moreover became necessary for each speaker to mount the platform & so there I was cocked up with Sam at my right elbow, & there & then I smashed him amid rounds of aplause—I hit him in the wind at the first shot in 10 words taken from his own ugly mouth—& then proceeded to demonstrate in as few more 1 that he could never have read your book & 2 that he was absolutely ignorant of the rudiments of Bot. Science—I said a few more on the subject of my own experience, & conversion & wound up with a very few observations on the relative position of the old & new hypotheses, & with some words of caution to the audience—Sam was shut up—had not one word to say in reply & the meeting *was dissolved forthwith* leaving you master of the field after 4 hours battle. Huxley who had borne all the previous brunt of the battle & who never before (thank God) praised me to my face, told me it was splendid, & that he did not know before what stuff I was made of—I have been congratulated & thanked by the blackest coats & whitest stocks in Oxford....

48. Thomas Huxley

An alternative—and the most popular—form of this legend has Huxley, who had become known as "Darwin's bulldog," responding to Wilberforce, who asked whether Huxley's descent from the ape was on his father's or his mother's side. This was Huxley's account in a letter shortly after the event:

> If then, said I, the question is put to me would I rather have a miserable ape for a grandfather or a man highly endowed by nature and possessing great means and influence and yet who employs those faculties and that influence for the mere purpose of introducing ridicule into a grave scientific discussion—I unhesitatingly affirm my preference for the ape. Whereupon there was unextinguishable laughter among the people, and they listened to the rest of my argument with the greatest attention (Reproduced in Gardiner 1994, 130).

Among the more lyrical accounts, also in Gardiner (1994), is that of poet Alfred Noyes, from his The Book of Earth *(1925):*

> *The lean tall figure of Huxley quietly rose.*
> *He looked, for a moment, thoughtfully, at the crowd;*
> *Saw rows of hostile faces, caught the grin*
> *Of ignorant curiosity; here and there,*
> *A hopeful gleam of friendship; and, far back,*
> *The young, swift-footed, waiting for the fire.*
> *He fixed his eyes on these—then, in low tones,*
> *Clear, cool, incisive,* "I have come here," *he said,*
> "In the cause of Science only."

It may be fairly claimed that the Oxford meeting was a watershed for Victorian science, and one of epic proportions, at that. It quickly galvanized proponents and opponents and crystallized the salient issue for years—indeed, centuries—to come: The issue was science versus religion as the two claimed authority to speak for the natural order. The issue was where the proper boundaries were between the two realms.

Almost all of the protagonists were present except the author himself. Sir John Lubbock, Robert Chambers, John S. Henslow, and numerous other esteemed scientists spoke out in support of Huxley and Darwin. Adam Sedgwick, Richard Owen, and other defenders of the faith rose against them. Robert FitzRoy, now an admiral, shouted from the audience against the heresy of natural selection, lamenting his role in fostering it by having invited Darwin on the journey. The crowd quickly polarized, and chairman Henslow declared the meeting adjourned. But the attendees carried the topic with them into Oxford's cobbled streets and venerable pubs. The association could justly claim the annual meeting a ripping success in its ultimate goal of popularizing science.

The meeting was also a turning point in popularizing the arcane points of the theory, in sharpening the edges of argument, in broadening the fires of debate. The public became involved. Educated readers throughout England and the Western world quickly made Darwin a household name, and Darwinism a respectable, if contentious, topic of conversation. In an era dominated by the church, a new challenge had forcefully emerged. As Janet Browne (2002, 124) writes, "In this dispute, the challenge was clear. Any success for the Darwinian scheme would require renegotiating—often with bitter controversy—the lines to be drawn between cultural domains."

Darwin was well aware of the enormous portent of the Oxford meeting for a continuing dialogue about evolution. He was immensely pleased. Shortly afterward, in a letter to Asa Gray (Letter 2855, 3 July 1860) he wrote:

> Owen will not prove right, when he said that the whole subject would be forgotten in 10 years. My book has stirred up the mud with a vengeance; & it will be a blessing to me if all my friends do not get to hate me. But I look at it as certain, if I had not stirred up the mud some one else would very soon; so that the sooner the battle is fought the sooner it will be settled,—not that the subject will be settled in our lives' times. It will be an immense gain, if the question becomes a fairly open one; so that each man may try his new facts on it pro & contra.

Darwin was prophetic in writing this letter. He had indeed stirred up the mud. He had systematically applied an elegant reductionism, which had served the physical sciences so well since far more ancient times, to an understanding of living nature. In so doing, he had relieved the historical sciences of their methodological dependence on deductive reasoning. Part of this reduction was to reposition the place of humans as equal—not superior—partners in the narrative, as we see in the following selection from the Descent of Man. Indeed, it anthropomorphized nature. But in so doing, it subtly created a new biocentric view of that same natural world, eliminating the previous anthropocentric consensus (as reflected, for example, in the quote from Notebook B at the beginning of Unit 8). Darwin was also correct in suggesting that the battle would not end in his lifetime. Nor has it yet ended in ours.

Huxley, in the following review, expresses his doubts about the sufficiency of natural selection as the primary force in species evolution and disagrees with the exclusive gradualism of Darwin's theory. Many others shared these same concerns, as we shall see.

Darwin on the Origin of Species

[541] Mr. Darwin's long-standing and well-earned scientific eminence probably renders him indifferent to that social notoriety which passes by the name of success; but if the calm spirit of the philosopher have not yet wholly superseded the ambition and the vanity of the carnal man within him, he must be well satisfied with the results of his venture in publishing the "Origin of Species." Overflowing the narrow bounds of purely scientific circles, the "species question" divides with Italy and the Volunteers the attention of general society. Everybody has read Mr. Darwin's book, or, at least, has given an opinion upon its merits or demerits; pietists, whether lay or ecclesiastic, decry it with the mild railing which sounds so charitable; bigots denounce it with ignorant invective; old ladies of both sexes consider it a decidedly dangerous book, and even savants, who have no better mud to throw, quote antiquated writers to show that its author is no better than an ape himself; while every philosophical thinker hails it as a veritable Whitworth gun in the armoury of liberalism; and all competent naturalists and physiologists, whatever their opinions as to the ultimate fate of the doctrines put forth, acknowledge that the work in which they are embodied is a solid contribution to knowledge and inaugurates a new epoch in natural history....

[542] But it may be doubted if the knowledge and acumen of prejudged scientific opponents, and the subtlety of orthodox special pleaders, have yet exerted their full force in mystifying the real issues of the great controversy which has been set afoot, and whose end is hardly likely to be seen by this generation; so that, at this eleventh hour, and even failing anything new, it may be useful to state afresh that which is true, and to put the fundamental positions advocated by Mr. Darwin in such a form that they may be grasped by those whose special studies lie in other directions. And the adoption of this course may be the more advisable, because, notwithstanding its great deserts, and indeed partly on account of them, the "Origin of Species" is by no means an easy book to read if by reading is implied the full comprehension of an author's meaning.

We do not speak jestingly in saying that it is Mr. Darwin's misfortune to know more about the question he has taken up than any man living. Personally and practically exercised in zoology, in minute anatomy, in geology; a student of geographical distribution, not on maps and in museums only, but by long voyages and laborious collection; having largely advanced each of these branches of science, and having spent many years in gathering and sifting materials for his present work, the store of accurately registered facts upon which the author of the "Origin of Species" is able to draw at will is prodigious.

But this very superabundance of matter must have been embarrassing to a writer who, for the present, can only put forward an abstract of his views; and thence it arises, perhaps, that notwithstanding the clearness of the style, those who attempt fairly to digest the book find much of it a sort of intellectual pemmican a mass of facts crushed and pounded into shape, rather than held together by the ordinary medium of an obvious logical bond; due attention will, without doubt, discover this bond, but it is often hard to find....

[555] (I)t must not be forgotten that the really important fact, so far as the inquiry into the origin of species goes, is, that there are such things in Nature as groups of animals and of plants, the members of which are incapable of fertile union with those of other groups; and that there are such things as hybrids, which are absolutely sterile when crossed with other hybrids. For, if such phænomena as these were exhibited by only two of those assemblages of living objects, to which the name of species (whether it be used in its physiological or in its morphological sense) is given, it would have to be accounted for by any theory of the origin of species, and every theory which could not account for it would be, so far, imperfect....

[557] The hypotheses respecting the origin of species which profess to stand upon a scientific basis, and, as such, alone demand serious attention, are of two kinds. The one, the "special creation" hypothesis, presumes every species to have originated from one or more stocks, these not being the result of the modification of any other form of living

matter or arising by natural agencies but being produced, as such, by a supernatural creative act.

The other, the so-called "transmutation" hypothesis, considers that all existing species are the result of the modification of pre-existing species, and those of their predecessors, by agencies similar to those which at the present day produce varieties and races, and therefore in an altogether natural way; and it is a probable, though not a necessary consequence of this hypothesis, that all living beings have arisen from a single stock....

[566] The Darwinian hypothesis has the merit of being eminently simple and comprehensible in principle, and its essential positions may be stated in a very few words: all species have been produced by the development of varieties from common stocks; by the conversion of these, first into permanent races and then into new species, by the process of *natural selection,* which process is essentially identical with that artificial selection by which man has originated the races of domestic animals the *struggle for existence* taking the place of man, and exerting, in the case of natural selection, that selective action which he performs in artificial selection.

The evidence brought forward by Mr. Darwin in support of his hypothesis is of three kinds. First, he endeavours to prove that species may be originated by selection; secondly, he attempts to show that natural causes are competent to exert selection; and thirdly, he tries to prove that the most remarkable and apparently anomalous phænomena exhibited by the distribution, development, and mutual relations of species, can be shown to be deducible from the general doctrine of their origin, which he propounds, combined with the known facts of geological change; and that, even if all these phænomena are not at present explicable by it, none are necessarily inconsistent with it.

There cannot be a doubt that the method of inquiry which Mr. Darwin has adopted is not only rigorously in accordance with the canons of scientific logic, but that it is the only adequate method. Critics exclusively trained in classics or in mathematics, who have never determined a scientific fact in their lives by induction from experiment or observation, prate learnedly about Mr. Darwin's method, which is not inductive enough, not Baconian enough, forsooth, for them. But even if practical acquaintance with the process of scientific investigation is denied them, they may learn, by the perusal of Mr. Mill's admirable chapter "On the Deductive Method," that there are multitudes of scientific inquiries in which the method of pure induction helps the investigator but a very little way.

[567] Now, the conditions which have determined the existence of species are not only exceedingly complex, but, so far as the great majority of them are concerned, are necessarily beyond our cognisance. But what Mr. Darwin has attempted to do is in exact accordance with the rule laid down by Mr. Mill; he has endeavoured to determine certain great facts inductively, by observation and experiment; he has then reasoned from the data thus furnished; and lastly, he has tested the validity of his ratiocination by comparing his deductions with the observed facts of Nature. Inductively, Mr. Darwin endeavours to prove that species arise in a given way. Deductively, he desires to show that, if they arise in that way, the facts of distribution, development, classification, &c., may be accounted for, *i.e.* may be deduced from their mode of origin, combined with admitted changes in physical geography and climate, during an indefinite period. And this explanation, or coincidence of observed with deduced facts, is, so far as it extends, a verification of the Darwinian view.

There is no fault to be found with Mr. Darwin's method, then; but it is another question whether he has fulfilled all the conditions imposed by that method. Is it satisfactorily proved, in fact, that species may be originated by selection? that there is such a thing as natural selection? that none of the phænomena exhibited by species are inconsistent with the origin of species in this way? If these questions can be answered in the affirmative, Mr. Darwin's view steps out of the rank of hypotheses into those of proved theories; but, so long as the evidence at present adduced falls short of enforcing that affirmation, so long, to our minds, must the new doctrine be content to remain among the former an extremely valuable, and in the highest degree probable, doctrine, indeed the only extant hypothesis which is worth anything in a scientific point of view; but still a hypothesis, and not yet the theory of species.

After much consideration, and with assuredly no bias against Mr. Darwin's views, it is our clear conviction that, as the evidence stands, it is not absolutely proven that a group of animals, having all the characters exhibited by species in Nature, has ever been originated by selection, whether artificial or natural. Groups having the morphological character of species distinct and permanent races in fact have been so produced over and over again; but there is no positive evidence, at present, that any group of animals has, by variation and selective breeding, given rise to [568] another group which was, even in the least degree, infertile with the first. Mr. Darwin is perfectly aware of this weak point, and brings forward a multitude of ingenious and important arguments to diminish the force of the objection. We admit the value of these arguments to their fullest extent; nay, we will go so far as to express our belief that experiments, conducted by a skilful physiologist, would very probably obtain the desired production of mutually more or less infertile breeds from a common stock, in a comparatively few years; but still, as the case stands at present, this "little rift within the lute" is not to be disguised nor overlooked.

In the remainder of Mr. Darwin's argument our own private ingenuity has not hitherto enabled us to pick holes of any great importance; and judging by what we hear and read, other adventurers in the same field do not seem to have been much more fortunate. It has been urged, for instance, that in his chapters on the struggle for existence and on natural selection, Mr. Darwin does not so much prove that natural selection does occur, as that it must occur; but, in fact, no other sort of demonstration is attainable. A race does not attract our attention in Nature until it has, in all probability, existed for a considerable time, and then it is too late to inquire into the conditions of its origin. Again, it is said that there is no real analogy between the selection which takes place under domestication, by human influence, and any operation which can be effected by Nature, for man interferes intelligently. Reduced to its elements, this argument implies that an effect produced with trouble by an intelligent agent must, *a fortiori*, be more troublesome, if not impossible, to an unintelligent agent. Even putting aside the question whether Nature, acting as she does according to definite and invariable laws, can be rightly called an unintelligent agent, such a position as this is wholly untenable. Mix salt and sand, and it shall puzzle the wisest of men, with his mere natural appliances, to separate all the grains of sand from all the grains of salt; but a shower of rain will effect the same object in ten minutes. And so, while man may find it tax all his intelligence to separate any variety which arises, and to breed selectively from it, the destructive agencies incessantly at work in Nature, if they find one variety to be more soluble in circumstances than the other, will inevitably, in the long run, eliminate it.

[569]...Mr. Darwin's position might, we think, have been even stronger than it is if he had not embarrassed himself with the aphorism, *"Natura non facit saltum,"* which turns up so often in his pages. We believe, as we have said above, that Nature does make jumps now and then, and a recognition of the fact is of no small importance in disposing of many minor objections to the doctrine of transmutation....

Our object has been attained if we have given an intelligible, however brief, account of the established facts connected with species, and of the relation of the explanation of those facts offered by Mr. Darwin to the theoretical views held by his predecessors and his contemporaries, and, above all, to the requirements of scientific logic. We have ventured to point out that it does not, as yet, satisfy all those requirements; but we do not hesitate to assert that it is as superior to any preceding or contemporary hypothesis, in the extent of observational and experimental basis on which it rests, in its rigorously scientific method, and in its power of explaining biological phænomena, as was the hypothesis of Copernicus to the speculations of Ptolemy. But the planetary orbits turned out to be not quite circular after all, and, grand as was the service Copernicus rendered to science, Kepler and Newton had to come after him. What if the orbit of Darwinism should be a little too circular? What if species should offer residual phænomena, here and there, not explicable by natural selection? Twenty years hence naturalists may be in a position to say whether this is, or is not, the case; but in either event they will owe the author of "The Origin of Species" an immense debt of gratitude. We should leave a very wrong impression on the reader's mind if we permitted him to suppose that the value of that work depends wholly on the ultimate justification of the theoretical views which it contains. On the contrary, if they were disproved to-morrow, the book would still be the best of its kind; the most compendious statement of well-sifted facts bearing on the doctrine of species that has ever appeared. The chapters on Variation, on the Struggle for Existence, on Instinct, on Hybridism, on the Imperfection of the Geological Record, on [570] Geographical Distribution, have not only no equals, but, so far as our knowledge goes, no competitors, within the range of biological literature. And viewed as a whole, we do not believe that, since the publication of Von Baer's "Researches on Development," thirty years ago, any work has appeared calculated to exert so large an influence, not only on the future of Biology, but in extending the domination of Science over regions of thought into which she has, as yet, hardly penetrated.

49. Fleeming Jenkin

The Origin of Species

Fleeming Jenkin (1833–85) was the first professor of engineering at the University of Edinburgh, but he was intensely interested in a great variety of subjects, including evolution. His review of Darwin criticized natural selection as not enabling the spread of individual variations in a population due to the swamping effect of blending inheritance: The blending of maternal and paternal traits among offspring would quickly overwhelm a unique "sport," preventing its spread. Darwin, who also accepted blending inheritance as the way traits were passed on, had been aware of this difficulty and was struck by the intelligence of Jenkin's critique. Jenkin also claimed that geological age of the earth did not allow sufficient time for all species to have evolved.

[277] The theory proposed by Mr. Darwin as sufficient to account for the origin of species has been received as probably, and even as certainly true, by many who from their knowledge of physiology, natural history, and geology, are competent to form an intelligent opinion. The facts, they think, are consistent with the theory. Small differences are observed between animals and their offspring. Greater differences are observed between varieties known to be sprung from a common stock. The differences between what have been termed species are sometimes hardly greater in appearance than those between varieties owning a common origin. Even when species differ more widely, the difference they say, is one of degree only, not of kind. They can see no clear, definite distinction by which to decide in all cases, whether two animals have sprung from a common ancestor or not. They feel warranted in concluding, that for aught the structure of animals shows to the contrary, they may be descended from a few ancestors only,—nay, even from a single pair.

The most marked differences between varieties known to have sprung from one source have been obtained by artificial breeding. Men have selected, during many generations, those individuals possessing the desired attributes in the highest degree. They have thus been able to add, as it were, small successive differences, till they have at last produced marked varieties. Darwin shows that by a process, which he calls natural selection, animals more favourably constituted than their fellows will survive in the struggle for life, will produce descendants resembling themselves, of which the strong will [278] live, the weak will die; and so, generation after generation, nature, by a metaphor, may be said to choose certain animals, even as man does when he desires to raise a special breed. The device of nature is based on the attributes most useful to the animal; the device of man on the attributes useful to man, or admired by him. All must agree that the process termed natural selection is in universal operation. The followers of Darwin believe that by that process differences might be added even as they are added by man's selection, though more slowly, and that this addition might in time be carried to so great an extent as to produce every known species of animal from one or two pairs, perhaps from organisms of the lowest type.

A very long time would be required to produce in this way the great differences observed between existing beings. Geologists say their science shows no ground for doubting that the habitable world has existed for countless ages. Drift and inundation, proceeding at the rate we now observe, would require cycles of ages to distribute the materials of the surface of the globe in their present form and order; and they add, for aught we know, countless ages of rest may at many places have intervened between the ages of action.

But if all beings are thus descended from a common ancestry, a complete historical record would show an unbroken chain of creatures, reaching from each one now known back to the first type, with each link differing from its neighbour by no more than the several offspring of a single pair of animals now differ. We have no such record; but geology can produce vestiges which may be looked upon as a few out of the innumerable links of the whole conceivable chain, and what, say the followers of Darwin, is more certain than that the record of geology must necessarily be imperfect? The records we have show a certain family likeness between the beings living at each epoch, and this is at least consistent with our views....

[279] *Variability.*—Darwin's theory requires that there shall be no limit to the possible differences between descendants and their progenitors, or, at least, that if there be limits, they shall be at [280] so great a distance as to comprehend the utmost differences between any known forms of life. The variability required, if not infinite, is indefinite. Experience with domestic animals and cultivated plants shows that great variability exists. Darwin calls special attention to the differences between the various fancy pigeons, which, he says, are descended from one stock; between various breeds of cattle and horses, and some other domestic animals. He states that these differences are greater than those which induce some naturalists to class many specimens as distinct species. These differences are infinitely small as compared with the range required by his theory, but he assumes that by accumulation of successive difference any degree of variation may be produced; he says little in proof of the possibility of such an accumulation, seeming rather to take for granted that if Sir John Sebright could with pigeons produce in six years a certain head and beak of say half the bulk possessed by the original stock, then in twelve years this bulk could be reduced to a quarter, in twenty-four to an eighth, and so farther. Darwin probably never believed or intended to teach so extravagant a proposition, yet by substituting a few myriads of years for that poor period of six years, we obtain a proposition fundamental in his theory. That theory rests on the assumption that natural selection can do slowly what man's selection does quickly; it is by showing how much man can do, that Darwin hopes to prove how much can be done without him. But if man's selection cannot double, treble, quadruple, centuple, any special divergence from a parent stock, why should we imagine that natural selection should have that power?...

[294] *Lapse of Time.*—Darwin says with candour that he 'who does not admit how incomprehensibly vast have been the past periods of time,' may at once close his volume, admitting thereby that an indefinite, if not infinite time is required by his theory. Few will on this point be inclined to differ from the ingenious author. We are fairly certain that a thousand years has made no very great change in plants or animals living in a state of nature. The mind cannot conceive a multiplier vast enough to convert this trifling change by accumulation into differences commensurate with those between a butterfly and an elephant, or even between a horse and a hippopotamus. A believer in Darwin can only say to himself, Some little change does take place every thousand years; these changes accumulate, and if there be no limit to the continuance of the process, I must admit that in course of time any conceivable differences may be produced. He cannot think that a thousand-fold the difference produced in a thousand years would suffice, according to our present observation, to breed even a dog from a cat. He may perhaps think that by careful selection, continued for this million years, man might do quite as much as this; but he will readily admit that natural selection does take a much longer time, and that a million years must by the true believer [295] be looked upon as a minute. Geology lends her aid to convince him that countless ages have elapsed, each bearing countless generations of beings, and each differing in its physical conditions very little from the age we are personally acquainted with. This view of past time is, we believe, wholly erroneous. So far as this world is concerned, past ages are far from countless; the ages to come are numbered; no one age has resembled its predecessor, nor will any future time repeat the past. The estimates of geologists must yield before more accurate methods of computation, and these show that our world cannot have been habitable for more than an infinitely insufficient period for the execution of the Darwinian transmutation.

50. St. George Mivart

On the Genesis of Species

George Jackson Mivart (1827–1900) was a reputable biologist who, after initially fully espousing Darwinism, turned against it with fervor. Earlier a convert from the Anglican Communion to Catholicism, he claimed that evolution was too regular, and the parallel evolution of complex organs such as the eye too noncoincidental, to be the result of random forces. Instead, life must be under supernatural guidance. Mivart's objections, however, went beyond this to include some that actually concerned Darwin.

[60]...It has been here contended that a certain few facts, out of many which might have been brought forward, are inconsistent with the origination of species by "Natural Selection" only or mainly.

Mr. Darwin's theory requires minute, indefinite, fortuitous variations of all parts in all directions, and he insists that the sole operation of "Natural Selection" upon such is sufficient to account for the great majority of organic forms, with their most complicated structures, intricate mutual adaptations and delicate adjustments.

To this conception has been opposed the difficulties presented by such a structure as the form of the giraffe, which ought not to have been the solitary structure it is; also the minute beginnings and the last refinements of protective mimicry equally difficult or rather impossible to account for by "Natural Selection." Again the difficulty as to the heads of flat-fishes has been insisted on, as also the origin, and at the same time the constancy, of the limbs of the highest animals. Reference has also been made to the whalebone of whales, and to the [61] impossibility of understanding its origin through "Natural Selection" only; the same as regards the infant kangaroo, with its singular deficiency of power compensated for by maternal structures on the one hand, to which its own breathing organs bear direct relation on the other. Again, the delicate and complex pedicellariæ of Echinoderms, with a certain process of development (through a secondary larva) found in that class, together with certain other exceptional modes of development, have been brought forward. The development of colour in certain apes, the hood of the cobra, and the rattle of the rattlesnake have also been cited. Again, difficulties as to the process of formation of the eye and ear, and as to the fully developed condition of those complex organs, as well as of the voice, have been considered. The beauty of certain shell-fish; the wonderful adaptations of structure, and variety of form and resemblance, found in orchids; together with the complex habits and social conditions of certain ants, have been hastily passed in review. When all these complications are duly weighed and considered, and when it is borne in mind how necessary it is for the permanence of a new variety that many individuals in each case should be simultaneously modified, the cumulative argument seems irresistible.

The Author of this book can say that though by no means disposed originally to dissent from the theory of "Natural Selection," if only its difficulties could be solved, he has found each successive year that deeper consideration and more careful examination have more and more brought home to him the inadequacy of Mr. Darwin's theory to account for the preservation and intensification of incipient, specific, and generic characters. That minute, fortuitous, and indefinite variations could have brought about such special forms and modifications as have been enumerated in this chapter, seems to contradict not imagination, but reason. [62]

51. Charles Darwin

The Descent of Man, and Selection in Relation to Sex

In Darwin's time and, indeed, much before this, the term evolution *had a distinctly teleological meaning—the life was progressing resolutely upward through progressive states. Darwin generally rejected this, which is one reason he neglected to use the term until later editions of* Origin. *Though this belief dovetailed with the concept of orthogenesis—that an inner force drove evolutionary change in a uniform direction—the two beliefs did not necessarily conflate. A form of orthogenesis that Darwin accepted was that internal development guided evolutionary change in addition to external adaptation. We see this in Darwin's treatment of embryological parallels in mammalian and nonmammalian development in the following passage.*

The recognition of such parallels, which Darwin got from Karl Ernst von Baer, never suggested the evolutionary recapitulation that Ernst Haeckel saw in these comparisons and that led to the "ontogeny recapitulates phylogeny" hypothesis. The reader is encouraged to briefly examine the works of these two nineteenth-century scientists.

[14] *Embryonic Development.*—Man is developed from an ovule, about the 125th of an inch in diameter, which differs in no respect from the ovules of other animals. The embryo itself at a very early period can hardly be distinguished from that of other members of the vertebrate kingdom. At this period the arteries run in arch-like branches, as if to carry the blood to branchiæ which are not present in the higher vertebrata, though the slits on the sides of the neck still remain [*illustration not included*], marking their former position. At a somewhat later period, when the extremities are developed, "the feet of lizards and mammals," as the illustrious Von Baer remarks, "the wings and feet of birds, no less than the hands and feet of man, all arise from the same fundamental form." It is, says Prof. Huxley,[10] "quite in the later stages of development that the young human being presents marked differences from the young ape, while the latter departs as much from the dog in its developments, as

[10] 'Man's Place in Nature,' 1863, p. 67.

the man does. Startling as this last assertion may appear to be, it is demonstrably true."

[32] Thus we can understand how it has come to pass that man and all other vertebrate animals have been constructed on the same general model, why they pass through the same early stages of development, and why they retain certain rudiments in common. Consequently we ought frankly to admit their community of descent: to take any other view, is to admit that our own structure and that of all the animals around us, is a mere snare laid to entrap our judgment. This conclusion is greatly strengthened, if we look to the members of the whole animal series, and consider the evidence derived from their affinities or classification, their geographical distribution and geological succession. It is only our natural prejudice, and that arrogance which made our forefathers declare that they were descended from demi-gods, which leads us to demur to [33] this conclusion. But the time will before long come when it will be thought wonderful, that naturalists, who were well acquainted with the comparative structure and development of man and other mammals, should have believed that each was the work of a separate act of creation.

Ideas to Think About

1. Why were Darwin's close friends and fellow evolutionists initially so reluctant to accept natural selection as the mechanism for evolution?
2. Sedgwick first claims that Darwin's chief fault is to have misused the scientific method in inducing natural selection, but follows this with the charge that Darwin has neglected the "moral or metaphysical" link to the physical. In what way is this a consistent argument, and in what way is it not?
3. Examine critically the final paragraph of Owen's review. It identifies a problem that Darwin himself tangled with and that has plagued biologists off and on since then.
4. Examine Wilberforce's argument that the facts of nature cannot contradict scripture with his case against Darwin. To what extent is his argument not mainly against natural selection itself but against its application to humankind?
5. The various nonselection theorists that reacted to Darwin's publication contain some substantive criticisms mixed with some that Darwin considered either misrepresentations or misunderstandings. How are these reactions related?

For Further Reading

Darwin's Plots: Evolutionary Narrative in Darwin, George Eliot and Nineteenth-Century Fiction, Gillian Beer. Cambridge: Cambridge University Press (2009). As Darwin redefined evolution of the natural order, he likewise redefined man's place in that order and thus stimulated the public and literary imagination. Beer interprets Darwin's uses of analogy (domestication and natural selection), metaphor (the ever-branching tree of life, with its dying limbs), and narrative in storytelling to examine his influence on literature. A singularly provocative read!

Narratives of Human Evolution, Misia Landau. New Haven, CT: Yale University Press (1993). Darwin's narratives, along with those of other early evolutionists, Landau claims, represent a form based on the universal hero tale. Understanding this can provide insights into scientific theories.

Beasts of the Modern Imagination: Darwin, Nietzsche, Kafka, Ernst, & Lawrence, Margot Norris. Baltimore, MD: The Johns Hopkins University Press (1985). This is a linguistically dense text—not one I would normally include as further reading. However, it produces strong insights on the changing perspective immediately following Darwin and, in a way, evidences the counterreaction to follow upon Darwin's death.

UNIT 11

Early Twentieth Century: The Rise of Genetics and the Evolutionary Synthesis

> That we are in the presence of a new principle of the highest importance is, I think, manifest. To what further conclusions it may lead us cannot yet be foretold.
>
> —William Bateson, *1900*

Gregor Mendel truly did introduce a new principle, but neither he nor the rest of the scientific world knew it back in 1865. When, in 1900, Hugo de Vries, Carl Correns, and Erich von Tschermak independently rediscovered particulate inheritance, Mendel was belatedly recognized as a pioneer. His work almost immediately laid the basis for the new science of genetics, fostered a rapid flurry of new discoveries, and stimulated what would—some thirty years later—become an evolutionary synthesis that reinforced and expanded Darwin's original discovery.

But the fledgling science of genetics was almost immediately at odds with evolutionary theory, particularly Darwinian theory, and the story of this tension and the final resolution of the conflict is the subject of this unit. That period, from 1900 to the early1940s, has fortunately been chronicled and critiqued by numerous scholars here and abroad—such as Mayr, *The Growth of Biological Thought* (1982); Sober, *Conceptual Issues in Evolutionary Biology* (1984, 1993); Mayr and Provine, *The Evolutionary Synthesis* (1980)—so we can take advantage of a wealth of serious contemplation. This is important, because the road to the synthesis was not a straight and logical one and even today not all scholars agree on its pertinent components.

But the story has an important introduction, which is reflected in some earlier readings—most notably in Unit 6. It will serve us to make that body of thought briefly more explicit here.

Darwin made three major contributions to the future of biology: First, he provided overwhelming evidence for evolution in an astounding variety of different forms, both present and past. Second, he laid at the center of evolutionary change (transformism) the critical importance of variation. Third, he provided a mechanism—natural selection—to explain how evolution occurs.

The first of these was most universally and readily accepted during Darwin's time, because the idea (if not the evidence) was current in biological thought in the preceding generation. The second was differently conceptualized before and after Darwin, the result of fundamental differences in prior assumptions about species: Variation *between* species was readily acknowledged as part of the very definition, because species were long seen as fixed and essential types—whether the result of divine creation or of natural law. Darwin emphasized variations among individuals *within* species, which variations were themselves the basis for the divergence of populations into new species. To those who clung to the typological tradition, variations within a species were trivial and unimportant.

This major incompatibility in metaphysical thought naturally made natural selection itself unacceptable, because selection worked on individuals and not species directly. Hence, long before Mendel, Darwinian evolutionary theory was at odds with evolutionary theories based on essentialist thought, among biologists and nonbiologists, and scientists and philosophers, alike.

The coming of genetics, in this perspective, did little to change this, as is amply explained by Mayr (1980, 1–48) and Lewontin (1980, 58–68). For geneticists, change in species could only occur through mutation, because species tend to be uniformly pure lines; natural selection was at best a filter that eliminated bad mutations. For geneticists, variation was discontinuous, without transitions, as Morgan's fruit flies demonstrated; for

Darwinists, variation was continuous, often grading imperceptibly across population and individuals within a species. The Galapagos finches were good examples. For geneticists, evolution could only occur abruptly; for Darwinists, evolution was gradual and not *per saltum* (in sudden leaps).

So the Mendelians and the Darwinians took a long time to achieve common understanding, where the genotypes studied in the laboratory became part of the same construct as the phenotypes studied in nature. It was only when variations measured in gene frequencies, manifested and subject to selection in their phenotypes, became common ground for naturalists and experimentalists alike that genetics became part of the solution to the evolution question.

---

52. Gregor Mendel

Johann Mendel was born in the Austrian Empire, in what is now Hynčice, Czechoslovakia (about eighty miles northeast of Prague), on July 20, 1822. He died on January 6, 1884—two years after Darwin's death. His German family—he had two sisters—were traditional farmers, and Mendel as a youth was a gardener and beekeeper, stimulating an interest that would later bring his name to the forefront of science. He entered monastic life in 1843, at the Augustinian Abbey of St. Thomas, in Brno (center of the Province of Moravia), and took the name Gregor.

In 1851, he attended the University of Vienna, where he studied physics and astronomy (under Christian Doppler), as well as meteorology. Though most of Mendel's professional publications were on the last of these three, he became posthumously famous for his horticultural work on inheritance. The name *Mendel,* in the popular mind, is obviously associated with his inheritance experiments. This would curiously surprise Mendel today, as his two conference papers and his publication on these experiments were met with either criticism or less than enthusiastic responses.

The monastery's five-acre experimental garden afforded Mendel the opportunity to experiment with hybridizing pea plants (*Pisum* sp.)—which were not only easy to pollinate, but whose reproductive anatomy made wild cross-pollination very difficult; hence allowing excellent control of the variables involved. From 1856 to 1863, Mendel propagated and analyzed 29,000 pea plants, and where a particular experiment gave unexpected results, he always cross-checked and at times repropagated.

In addition to the statistical importance of his offspring ratios in these long experiments—and most certainly, the resulting concepts of *dominance* and *recessiveness* were revolutionary in importance—there was a more fundamental contribution. This was the shift from the *phenotypic* level of analysis to the *genotypic.* In the absence of Mendelian information, the only approach to studying inherited variation after Darwin was the measurement of the several alternative *manifestations* of simple traits in offspring. These manifestations constituted the basis for phenotypic analysis. One could not be faulted for one's ignorance of the fact that a particular phenotype might include *either* paired information identical in each parent's contribution (what is now known as a homozygous dominant) *or* a dominant contribution from one and a recessive from the other (now known as a heterozygous dominant).

This phenotypic approach had, in fact, developed some rigorous biometric analytic tools in its methodology—particularly in the statistical approach of Karl Pearson—and phenotypic analysis provided good insights into inheritance. The clash with Mendelism in the early twentieth century was only one of several debates that ensued with Mendel's rediscovery, before Bateson, R. A. Fisher, and others helped negotiate the evolutionary synthesis.

This unit begins, then, with an extensive reproduction of Mendel's original paper. It is gateway to that synthesis.

Experiments in Plant Hybridization

[3] **Introduction**

Experience in artificial fertilization, such as is effected with ornamental plants in order to obtain new variations in color, has led to the experiments which will here be discussed. The striking regularity with which the same hybrid forms always reappeared whenever fertilization took place between the same species induced further experiments to be undertaken, the object of which was to follow up the developments of the hybrids in their progeny....

That, so far, no generally applicable law governing the formation and development of hybrids has been successfully formulated can hardly be wondered at by anyone who is acquainted with the extent of the task, and can appreciate the difficulties with which experiments of this class have to contend. A final decision can only be arrived at when we shall have before us the results of *detailed experiments* made on plants belonging to the most diverse orders....

[4] The paper now presented records the results of such a detailed experiment. This experiment was practically confined to a small plant group, and is now, after eight years' pursuit, concluded in all essentials. Whether the plan upon which the separate experiments were conducted and carried out was the best suited to attain the desired end is left to the friendly decision of the reader....

[5] At the very outset special attention was devoted to the *Leguminosae* on account of their peculiar floral structure. Experiments which were made with several members of this family led to the result that the genus *Pisum* was found to possess the necessary qualifications....

In all, thirty-four more or less distinct varieties of Peas were obtained from several seedsmen and subjected to a two year's trial.... For fertilization twenty-two of these were selected and cultivated during the whole period of the experiments. They remained constant without any exception....

The seven characters Mendel chose for experiment included (1) seed form, either smooth or wrinkled; (2) color of the seed endosperm; (3) color of the seed coat; (4) seedpod form, smooth or constricted between seeds; (5) unripe seedpod color; (6) axial or terminal position of the flowers; and (7) stem length. On the latter, he writes, "In experiments with this character, in order to be able to discriminate with certainty, the long axis of 6 to 7 ft. was always crossed with the short one of ¾ ft. to 1½ ft."

In his identification of dominance and recessiveness in the results, Mendel recognizes a distinction between what is manifest in an offspring (phenotypes) and what precise paired genetic information the several offspring may actually possess (genotypes).

[9]... In the case of each of the seven crosses the hybrid–character resembles that of one of the parental forms so closely that the other either escapes observation completely or cannot be detected with certainty. This circumstance is of great importance in the determination and classification of the forms under which the offspring of the hybrids appear. Henceforth in this paper those characters which are transmitted entire, or almost unchanged in the hybridization, and therefore in themselves constitute the characters of the hybrid, are termed the *dominant*, and those which become latent in the process *recessive*. The expression *recessive* has been chosen because the characters thereby designated withdraw or entirely disappear in the hybrids, but nevertheless reappear unchanged in their progeny, as will be demonstrated later on.

It was furthermore shown by the whole of the experiments that it is perfectly immaterial whether the dominant character belongs to the seed plant or to the pollen plant; the form of the hybrid remains identical in both cases. This interesting fact was also emphasized by Gärtner, with the remark that even the most practiced expert is not in a position to determine in a hybrid which of the two parental species was the seed or the pollen plant....

In the following paragraph, Mendel has unknowingly revealed the principle of heterosis, or hybrid vigor, in stem length. This principle will later become an important element in genetic exploration.

[10] With regard to this last character [*stem height*] it must be stated that the longer of the two parental stems is usually exceeded by the hybrid, a fact which is possibly only attributable to the greater luxuriance which appears in all parts of plants when stems of very different lengths are crossed. Thus, for instance, in repeated experiments, stems of 1 ft. and 6 ft. in length yielded without exception hybrids which varied in length between 6 ft. and 7½ ft....

The First Generation From the Hybrids

In this generation there reappear, together with the dominant characters, also the recessive ones with their peculiarities fully developed, and this occurs in the definitely expressed average proportion of three to one, so that among each four plants of this generation three display the dominant character and one the recessive....

[13] **The Second Generation From the Hybrids**

Those forms which in the first generation exhibit the recessive character do not further vary in the second generation as regards this character; they remain constant in their offspring.

It is otherwise with those which possess the dominant character in the first generation [bred from the hybrids—i.e., the F_2 in modern terminology]. Of these *two*-thirds yield offspring which display the dominant and recessive characters in the proportion of three to one, and thereby show exactly the same ratio as the hybrid forms, while only *one*-third remains with the dominant character constant.

Mendel continues here with experiments 1–7. Then, in the following, he reflects on the fact that a 2:1 ratio in phenotype and 3:1 in genotype (not his terminology) actually reflects three genotypes—the homozygous dominant in a 1:4 ratio, the heterozygous dominant in a 2:4 ratio, and the homozygous recessive in a 1:4 ratio. This gives us the 1:2:1, represented by Mendel below as a 2:1:1 ratio.

[15]... It is therefore demonstrated that, of those forms which possess the dominant character in the first generation, two-thirds have the hybrid–character, while one-third remains constant with the dominant character.

The ratio 3:1, in accordance with which the distribution of the dominant and recessive characters results in the first generation, resolves itself therefore in all experiments into the ratio of 2:1:1, if the dominant character be differentiated according to its significance as a hybrid–character or as a parental one. Since the members of the first generation spring directly from the seed of the hybrids, it is now clear that the hybrids form seeds having one or other of the two differentiating characters, and of these one-half develop again the hybrid form, while the other half yield plants which remain constant and receive the dominant or the recessive characters in equal numbers.

Current description of the "characters" described in the following refers to "alleles," and current notation would represent their genotypes as AA + 2Aa + aa.

The Subsequent Generations From the Hybrids

The proportions in which the descendants of the hybrids develop and split up in the first and second generations presumably hold good for all subsequent progeny....

If ***A*** be taken as denoting one of the two constant characters, for instance the dominant, ***a***, the recessive, and ***Aa*** the hybrid form in which both are conjoined, the expression

$$A + \mathbf{2}Aa + a$$

[16] shows the terms in the series for the progeny of the hybrids of two differentiating characters....

Mendel's law of independent assortment is expressed in the following paragraph, while the law of segregation—readily demonstrated in most of the foregoing—is explicated in the final paragraphs.

The Offspring of Hybrids in Which Several Differentiating Characters are Associated

[20]...There is therefore no doubt that for the whole of the characters involved in the experiments the principle applies *that the offspring of the hybrids in which several essentially different characters are combined exhibit the terms of a series of combinations, in which the developmental series for each pair of differentiating characters are united.* It is demonstrated at the same time that *the relation of each pair of different characters in hybrid union is independent of the other differences in the two original parental stocks....* [italics in original]

[22] **The Reproductive Cells of the Hybrids**

[26]....It remains, therefore, purely a matter of chance which of the two sorts of pollen will become united with each separate egg cell. According, however, to the law of probability, it will always happen, on the average of many cases, that each pollen form ***A*** and ***a*** will unite [27] equally often with each egg cell form ***A*** and ***a***, consequently one of the two pollen cells ***A*** in the fertilization will meet with the egg cell ***A*** and the other with the egg cell ***a***, and so likewise one pollen cell ***a*** will unite with an egg cell ***A***, and the other with the egg cell ***a***.

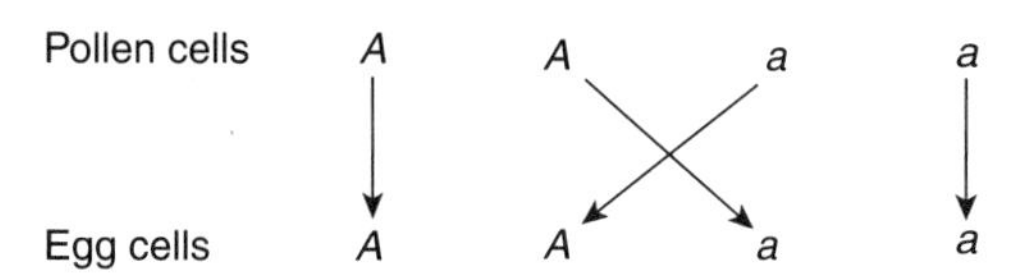

The result of the fertilization may be made clear by putting the signs for the conjoined egg and pollen cells in the form of fractions, those for the pollen cells above and those for the egg cells below the line. We then have

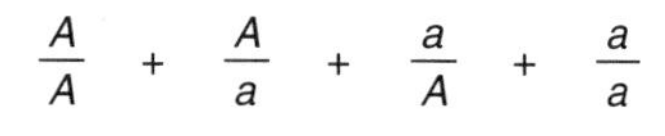

$$\frac{A}{A} + \frac{A}{a} + \frac{a}{A} + \frac{a}{a}$$

53. GEORGE H. HARDY

For all of its conceptual influence on genetics, the following paper by George Hardy should come later in this unit, for it was not until much closer to the evolutionary synthesis that its significance was truly recognized. Had it been seized upon in its time, as Dobzhansky notes (1980, 229–42), it would have reconciled the conceptual misunderstanding between the Mendelians and the Darwinians. This brief letter to the editor of Science laid the foundation for population genetics. Hardy was a mathematician, and this was his only published venture into genetics and evolution. Independently that same year, German physician Wilhelm Weinberg announced the same mathematical principle. This was not recognized for a quarter century, so the famous Hardy-Weinberg principle was first known as the Hardy principle. "I have never done anything 'useful,'" Hardy once wrote. "No discovery of mine has made, or is likely to make,... the least difference to the amenity of the world." How wrong he was.

Mendelion Prorpotions in a Mixed Population

To The Editor of Science: I am reluctant to intrude in a discussion concerning matters of which I have no expert knowledge, and I should have expected the very simple point which I wish to make to have been familiar to biologists. However, some remarks of Mr. Udny Yule, to which Mr. R. C. Punnett has called my attention, suggest that it may still be worth making.

In the *Proceedings of the Royal Society of Medicine* (Vol I., p. 165) Mr. Yule is reported to have suggested, as a criticism of the Mendelian position, that if brachydactyly is dominant "in the course of time one would expect, in the absence of counteracting factors, to get three brachydactylous persons to one normal."

It is not difficult to prove, however, that such an expectation would be quite groundless. Suppose that *Aa* is a pair of Mendelian characters, *A* being dominant, and that in any given generation the numbers of pure dominants (*AA*), heterozygotes (*Aa*), and pure recessives (*aa*) are as $p{:}2q{:}r$. Finally, suppose that the numbers are fairly large, so that the mating may be regarded as random, that the sexes are evenly distributed among the three varieties, and that all are equally fertile. A little mathematics of the multiplication-table type is enough to show that in the next generation the numbers will be as

$$(p + q)^2 : 2(p + q)(q + r) : (q + r)^2,$$

or as $p_1{:}2q_1{:}r_1$, say.

NOTE: The more conventional Hardy-Weinberg equation used today, demonstrating the equilibrium in genotypes at the p-q locus in a random mating population, is

$$(p + q)^2 = p^2 + 2pq + q^2.$$

The interesting question is—in what circumstances will this distribution be the same as that in the generation before? It is easy to see that the condition for this is $q^2 = pr$. And since $q_1^2 = p_1r_1$, whatever the values of *p*, *q*, and *r* may be, the distribution will in any case continue unchanged after the second generation.

Suppose, to take a definite instance, that *A* is brachydactyly, and that we start from a population of pure brachydactylous and pure normal persons, say in the ratio of 1:10,000. Then $p = 1$, $q = 0$, $r = 10{,}000$ and $p_1 = 1$, $q_1 = 10{,}000$, $r_1 = 100{,}000{,}000$. If brachydactyly is dominant, the proportion of brachydactylous persons in the second generation is 20,001:100,020,001, or practically 2:10,000, twice that in the first generation; and this proportion will afterwards have no tendency whatever to increase. If, on the other hand, brachydactyly were recessive, the proportion in the second generation would be 1:100,020,001, or practically 1:100,000,000, and this proportion would afterwards have no tendency to decrease.

In a word, there is not the slightest foundation for the idea that a dominant character should show a tendency to spread over a whole population, or that a recessive should tend to die out.

NOTE: It was subsequently shown that in the absence of selection, genetic drift would eventually eliminate the recessive. This "decay of variability" was demonstrated by Sewell Wright and others in the development of mathematical population genetics.

I ought perhaps to add a few words on the effect of the small deviations from the theoretical proportions which will, of course, occur in every generation. Such a distribution as $p_1:2q_1:r_1$, which satisfies the condition $q_1^2 = p_1r_1$, we may call a *stable* distribution. In actual fact we shall obtain in the second generation not $p_1:2q_1:r_1$ but a slightly different distribution $p:2q:r$, which is not "stable." This should, according to theory, give us in the third generation a "stable" distribution $p_2:2q_2:r_2$, also differing from $p_1:2q_1:r_1$; and so on. The sense in which the distribution $p_1:2q_1:r_1$ is "stable" is this, that if we allow for the effects of casual deviations in any subsequent generation, we should, according to theory, obtain at the next generation a new "stable" distribution differing but slightly from the original distribution.

I have, of course, considered only the very simplest hypotheses possible. Hypotheses other that [*sic*] that of purely random mating will give different results, and, of course, if, as appears to be the case sometimes, the character is not independent of that of sex, or has an influence on fertility, the whole question may be greatly complicated. But such complications seem to be irrelevant to the simple issue raised by Mr. Yule's remarks.

G. H. Hardy
Trinity College, Cambridge,
April 5, 1908

P. S. I understand from Mr. Punnett that he has submitted the substance of what I have said above to Mr. Yule, and that the latter would accept it as a satisfactory answer to the difficulty that he raised. The "stability" of the particular ratio 1:2:1 is recognized by Professor Karl Pearson (*Phil. Trans. Roy. Soc.* (A), vol. 203, p. 60).

54. Thomas Hunt Morgan

One of the first geneticists to emerge after the rediscovery of Mendel's work, Thomas H. Morgan was also one of the most influential. Known for his study of mutations in the fruit fly, *Drosophilia melanogaster*, his contributions to understanding the role of chromosomes in heredity revolutionized the field of genetics.

Born in Lexington, Kentucky, on September 25, 1866, Morgan received his doctorate from Johns Hopkins in 1890, having studied embryology at the Marine Biology Laboratory in Woods Hole, Massachusetts. He continued this research interest when he accepted, the following year, an appointment as department chair in biology at Bryn Mawr.

With the rediscovery of Mendel in 1900, Morgan became interested in the mechanisms of heredity. By 1903, he had published a book on evolution. Though he fully accepted evolution as "descent from common ancestors," he did not accept Darwin's natural selection as a valid mechanism for producing new species.

In 1904, Morgan was appointed professor of experimental zoology at Columbia University and began his research in earnest on heredity. From his "fly lab" at Columbia University, he trained many of the brightest in the new generation of geneticists, including Theodosius Dobzhansky. In 1928, Morgan accepted an offer to establish a similar program in biology at Caltech, where he remained until his death in 1945. His division of biology there attracted numerous esteemed scientists, including Linus Pauling, Edward Tatum, and George W. Beadle. Pauling was awarded the Nobel Prize in Chemistry in 1954, and Tatum and Beadle shared the Nobel Prize for Medicine or Physiology in 1958 for their work in genetics.

Morgan was the first geneticist to receive the Nobel Prize in Medicine or Physiology, in 1933. He was elected to the National Academy of Sciences and was a foreign member of the Royal Society of London. Ironically, in view of his great skepticism over the role of natural selection in evolution, he was awarded the Royal Society's Darwin Medal in 1924.

The following selections focus on his positions on natural selection, and the role of genetics in evolution.

The Scientific Basis of Evolution

The following passages illustrate the fundamental differences between the early geneticists and the Darwinian naturalists: To Morgan, as one of the former, only experimental methods constitute true science because these are more objective and capable of hypothesis testing. The historical evolutionary approach of the naturalists—based as it is only on observation, and hence description—is, in this regard, mostly speculative.

[13] With the opening years of the present century evolution began to be seriously studied by experimental methods for the first time in its long history. In 1900, de Vries described the mutations of the evening primrose and in 1901 he brought forward a theory of evolution

by mutation—a theory based on experimental procedure. In 1900, Mendel's paper on heredity came to light. Three years later (1903) Johannsen's work on pure lines appeared. It, too, was based on experiment. In the same year, Button first pointed out that the chromosome mechanism of the maturation stages of the egg and sperm-cell supply a mechanism that accounts for the two laws of heredity that Mendel had discovered. While the information relating to these stages of the maturation of the germ-cells was not based on experiment but was largely the result of direct observation, nevertheless some experimental work had been carried out that had made highly probable the view that the chromosomes are the bearers of the hereditary units. The evidence from these four sources and the subsequent developments furnish us today with ideas for an objective discussion of the theory of evolution in striking contrast to the older speculative method of treating evolution as a problem of history.

[14] As a result of this work it is now realized that the most promising method for the interpretation of evolution is through an appeal to experiment. By an appeal to experiment is meant the application of the same kind of procedure that has long been recognized in the physical sciences as the most dependable one in formulating an interpretation of the outer world.

[15] Until the present century the evidence on which the theory of descent rested was, as I have said, derived largely from observation, and was therefore for the most part descriptive. The theory was more of the order o[f] a broad generalization than of a scientific theory based on controlled experimental data.

[168] The sharp contrast between the older speculative procedure, that tried to explain secondary sexual characters, and the new experimental way of studying the same problem, is evident from what has been said. There is no need to claim that the modern method has solved the historical problem of the origin of secondary sexual differences, but it is clear that the more precise and verifiable work of recent times may in the end help towards a better understanding of one of the problems of evolution.

In considering mutation the major directive force in evolution and selection only a force that does some fine-tuning and trimming, the early geneticists misunderstood not only how naturalists viewed evolution but also the nature of variation in that process. The focus on species as populations, rather than species as "types" allowed the naturalist to examine the intergrading of variations within species in light of variation boundaries between species. In this regard, variation was continuous and evolution resulted from the gradual process of selection from generation to generation. As is apparent from the following passages, geneticists viewed both species and variations as discontinuous and thus evolution as a dramatic leap from type to type—not necessarily in a progressive way.

[109]... Of course no mutationist would deny that a new type must be able to survive and perpetuate itself if it is to take part in evolution; but this might be said to be only a commonplace. A mutationist might well insist that the essential part of Darwin's theory of natural selection is not survival, but Darwin's [110] postulate that the individual variations, everywhere present, furnish the raw materials for evolution. This the mutationist would deny....

If we think of evolution as an active process, is natural selection an agency capable of bringing about progressive changes, or does it not rather direct attention away from the real phenomenon, and offer at most only an explanation of the presence of certain types and the absence of others at any one period of geological history? The origin of these types—the real creative steps—not the preservation of certain of them after they have appeared, might rather be regarded as the essential phenomenon of evolution. If so, "the struggle for existence" and "the survival of the fittest" may express only a sort of truism or metaphor, and have nothing to do with the origination of new types out of antecedent ones.

In the following selection, Morgan's argument contrasts mutation as a creative evolutionary force in the absence of selection and natural selection as a minor role-player. In the final paragraph, Morgan makes a false distinction: Normal variations inherent in any population cannot produce a new species through natural selection, whereas rare variations occurring by mutation can. The focus on mutants and their abrupt effect kept geneticists from understanding the profound role that genetic recombination could have in evolutionary change.

[130] If all the new mutant types that have ever appeared had survived and left offspring like themselves, we [131] should find living today all the kinds of animals and plants now present, and countless others. This consideration shows that even without natural selection evolution might have taken place. What the theory does account for is the absence of many kinds of living things that could not survive, partly because they could not meet the conditions of the inorganic world, partly because they found no new environment suitable to their needs, partly because they were destroyed by other animals or plants, and partly because they could not compete with the original type. Natural selection may then be invoked to explain the absence of a vast array of forms that have appeared, but this is saying no more than that most of them have not had a survival value. The argument shows that natural selection does not play the role of a creative principle in evolution. [148]

As has been explained, the kind of variability on which Darwin based his theory of natural selection can no longer [149] be used in support of that theory, because, in the first place, in so far as fluctuating variations are due to environmental effect, these differences are now known not to be inherited, and because, in the second place, selection of the differences between individuals, due to the then existing genetic variants, while changing the number of individuals of a given kind, will not introduce anything new. The essential of the evolutionary process is the occurrence of new characteristics.

55. Vitalist Theories

Though Mendelism offered a new challenge to Darwinism in its saltational view of change, there remained other less scientifically based challenges as well. Most of these sought to remedy the purely mechanistic, selfish, and purposeless view of evolution fostered by natural selection. Vitalist theories, such as those of Hans Driesch (1867–1941) and the philosopher Henri Bergson (1859–1941), attempted to retain the spirit of "natural theology" without the theology.

Driesch gave the Gifford Lectures in 1908 (*The Science and Philosophy of the Organism*, vols. I and II. London: Adam and Charles Black, 1908). He argued that many of life's evolutionary questions could not be solved by any purely mechanistic view of nature. Instead, he argued, life is governed by entelechy, an inner and inherent organization of purpose.

In his *Creative Evolution* (1907), Bergson claimed that the apparent purposive evolution of complex beings and complex organs is a directional impetus provided by an inner force, an "Ėlan vital," and it is this that turns randomness into progressive change.

Opposed on both scientific and philosophical grounds, then, the Darwinist's reliance on the gradual, adaptive responses of random variations as the result of natural selection awaited some unifying idea that would improve upon unsatisfactory Lamarckian and other alternatives. The solution—the evolutionary synthesis of the 1940s–1960s—is described in the following two selections.

56. George Gaylord Simpson

Born in Chicago on June 16, 1902, Simpson spent his growing years in Denver. He attended the University of Colorado, but received his college degrees from Yale. Clearly one of the foremost paleontologists of his time, Simpson was also a major influence in fostering the modern evolutionary synthesis.

In his *Tempo and Mode in Evolution* (1944), he provided some early ideas on punctuated equilibrium. Simpson also considered this book his major influence on evolutionary theory. His work was instrumental in reconciling fossil evidence with genetics.

Simpson was on the staff of the American Museum of Natural History from 1927 to 1959 and was professor of geology at Columbia University from 1945 to 1959. From 1959 to 1970, he was Alexander Agassiz Professor, Museum of Comparative Zoology, at Harvard University.

He retired at the University of Arizona in 1982 and died in October 1984.

The Meaning of Evolution

[271] During the latter part of the nineteenth century and into the twentieth these two opposing materialist schools of evolutionary theory lambasted each other, the neo-Lamarckians demonstrating that Darwinian natural selection cannot be the whole story, the neo-Darwinians demonstrating that it must be part, at least, of the story and that acquired characters are not inherited as such. Each side, of course, had part of the truth and between them they had the whole truth almost within their grasp had they only been able to see this. Much of the trouble was that they fought on the wrong issues and asked each other the wrong questions. The correct issue on natural selection was not whether this is or is not the cause of adaptation, but how adaptation arises in the interplay of multiple forces of which Darwinian natural selection is obviously one, but obviously only one of many. Similarly the issue on inheritance of acquired characters was put in the wrong terms. Characters, as such, are not inherited, whether they be acquired characters or not. It is a series of determiners for a developmental system that is inherited. What characters result from this depends on the interplay of the inherited determiners, the activities of the organism, and the environment during development. Thus the neo-Lamarckians were right that there is a functional and causal relationship here, but the neo-Darwinians were right that its results are not inherited *as such*.

Failure to grasp the essence of these issues and failure to put together their pieces of the truth into a whole was largely caused by the fact that a large piece was still missing. This was later to be supplied by the geneticists.

[272] In the meantime, many students of evolution who were not rabid devotees of one of these two schools or the other fell into moods of suspicion or despair. There were two main materialistic theories and you could take the proof supplied by each that the other was untenable. What, then to do? One could gather more facts and suspend judgment as to what meaning they might eventually have. This sounds like a fine idea and it has even been hailed (following Bacon, who never succeeded in making a discovery by his famed inductive method) as the proper procedure in science. In reality, gathering facts, without a formulated reason for doing so and a pretty good idea as to what the facts may mean, is a sterile occupation and has not been the method of any important scientific advance.

Indeed facts are elusive and you usually have to know what you are looking for before you can find one.

If unwilling merely to suspend judgment, one could try to find an acceptable alternative to the two main materialistic theories of evolution. The trouble with this was that both were about as much right as they were wrong and efforts to avoid them both tended merely to be all wrong. There was no real success, even partial, in this effort until the science of genetics came along and even this, as we will see, confused the issue still more before it clarified it.

Or one could throw over materialistic and causalistic theory altogether. It was in this atmosphere that most of the extremely diverse vitalistic and finalistic theories, a veritable spate of them, were advanced. In almost all of these a sense of despair or of hope, an emotion even more blinding than despair, is evident. The relatively few firsthand investigators of evolution who abandoned causalism did so, for the most part, because they despaired of finding an adequate materialistic theory and could not endure the void of having no theory at all. Others, among them a number of professional and amateur philosophers, sounded a [273] note of hope which was often quite plainly hope of drawing meaning from something not understood or, and this is particularly striking, hope of finding that science did, after all, confirm what were in reality their intuitive or inherited and popular prejudices. This tendency to confirm prejudice accounts for the great popularity that finalist theories have sometimes enjoyed among those incompetent to judge them adequately from either a scientific or a philosophical point of view.

All these theories, vitalist, finalist, or both, involved some degree of abandonment of causalism. They did not explain evolution, but claimed that it is inexplicable and then gave a name to its inexplicability: "elan vital" (Bergson), "cellular consciousness" (Buis, under the pseudonym "Pierre-Jean"), "aristogenesis" (Osborn), "nomogenesis" (Berg), "holism" (Smuts), "hologenesis" (Rosa), "entelechy" (Driesch), "telefinalism" (du Noiiy)—the list could be greatly extended. As Huxley has remarked, the vitalists' ascribing evolution to an *elan vital* no more explained the history of life than would ascribing its motion to an *elan locomotif* explain the operation of a steam engine. The finalist left explanation still further behind, for he did not render even such lip service to causality as was often made by the nonfinalistic vitalists. In many cases the finalist merely viewed the phenomena of life with the unreasoning wonder of a child and decided that they happened simply because they were *meant* to happen. In other cases, as an eminent student has remarked in a different context, "The finalist was often the man who made a liberal use of the *ignava ratio,* or lazy argument: when you failed to explain a thing by the ordinary process of causality, you could 'explain' it by reference to some purpose of nature or of its Creator."[4]

The general features of vitalist and finalist philosophy [274] and theory have already been outlined on previous pages Once causalism is abandoned, there are no limitations on nights of the imagination and there are about as many separate vitalist and finalist theories as there have been vitalists and finalists. A review of each of these individually would be merely tedious and is not necessary. It has been shown that although the basic propositions common to all of them are nonscientific, that is, defined as outside the limits of scientific investigation, these would necessarily involve phenomena that can be investigated. It has also been pointed out that these diagnostic phenomena are in fact absent in the history of life. The mere fact that vitalism and finalism do not *explain* evolution would not warrant concluding that they therefore are not true: one cannot logically exclude a priori the possibility that evolution might really be noncausalistic and inexplicable. But the fact that the history of life is flatly inconsistent with their basic propositions does warrant the conclusion that vitalism and finalism are untenable.[5]

It must be concluded that the tendency of the vitalist and finalist theories was rather to obfuscate than to ad- [275] vance the study of evolution. They did, nevertheless, make some contribution to this. They emphasized and firmly established the fact that evolution involves forces that are directional in nature and creative in aspect and they exposed the weaknesses of earlier attempts to identify these forces. To this extent they had an influence in the synthesis that achieved the present causalistic explanation of this particular aspect of evolution. They have also had another effect not in the advance but in the acceptance of knowledge, which should not be underestimated. To great numbers of people unlearned in the subject, the fact of evolution has been emotionally distasteful and has been rejected on this basis alone, although the rejection has often been rationalized in other terms. Vitalistic and, more particularly, finalistic theories have persuaded many of these wishful thinkers that evolution is, after all, consonant with their emotions and prejudices. For them the pill has

[4]P. G. Galloway, *Philosophy of Religion* (New York, Scribner, 1914).

[5]And the previous demonstration of this fact warrants the present unfavorable review of the position of vitalism and finalism in the history of the study of evolution. Most of the literature of this subject is either ingenuous or ingenious special pleading and little of it can be recommended to a reader who may want to read a more favorable and still a judicious review. Fortunately there is one such work, although it is not available in English: L. Cuenot, *Invention et finaliti en biologie* (Paris, Flammarion, 1941).

Although Cuenot here avows himself a finalist, he still retains the charm and sobriety that mark the long sequence of his earlier work. He remains critical of the various finalist theories, and he is not blind to the scientific, logical, and philosophical weaknesses of the position in which he nevertheless finds himself. The reader should, however, note and bear in mind that when Cuenot wrote this book he was evidently completely unaware of the now current materialistic theory of evolution or of any of the crucial studies on population genetics and selection that gave impetus to this. The book does not mention even his fellow Parisian leader in this field, Teissier.

been sugar coated. In this respect, even du Noiiy's *Human Destiny*, the most popular but even within the finalist fold one of the least logical or reliable of such efforts, may have rendered a real service. The danger is that the sugar coating may be mistaken for the remedy. Any value in it will fail of realization unless it is a step in the direction of taking truth straight.

It was during the period around the turn of the century, when the conflicts of neo-Darwinism, neo-Lamarckism, vitalism, and finalism had thrown the study of evolution into great confusion, that the science of genetics was born. Even before the rediscovery of Mendelism de Vries had noted the sudden and random appearance of new varieties of plants. (His plants were primroses and we now know that his new varieties arose from chromosome mutations.) He decided that this, at last, was the real basis of evolution and he generalized it in his "mutation theory," published in 1901.

Scientists often display a human failing: whenever they [276] get hold of some new bit of truth they are inclined to decide that it is the whole truth. Thus the neo-Darwinians insisted that natural selection was the whole truth of evolution; the neo-Lamarckians held that interaction of structure-function-environment was the whole truth; the vitalists saw the whole truth in the creative aspect of life processes; and the finalists found all basic truth in the directional nature of evolution. Similarly, many of the early geneticists, although they soon learned far more about the mechanism involved, accepted de Vries' thesis and concluded that mutation was the whole truth of evolution. Mutations are random, so it was decided that evolution is random. The problem of adaptation was, in their opinion, solved by abolishing it: they proclaimed that there is no adaptation, only chance preadaptation.

Other theories had often stumbled over the fact that there is quite plainly a random element in evolution, the nature of which had been unknown. Now the mutationists had identified the source of this random element, but their theory stumbled over the fact that evolution is not wholly random. The vitalists and finalists were right in continuing to insist on this point, although they were wrong in their own overgeneralization of insisting that the directional element is universal and in maintaining that this element is inherent in life or in its goal. The mutationist discoveries were bewildering to many field naturalists and paleontologists, because they in particular were well aware that evolution *cannot* be a purely random process and that progressive adaptation certainly does occur. For a time the discoveries of the geneticists seemed only to make confusion worse confounded. Defeatism and escapism spread among many students of evolution. One very eminent vertebrate paleontologist ended a lifetime of study of evolution with the conclusion that he did not, after all, know anything about its causes; another decided in the declining years of [277] his prolonged and exceptionally fertile studies of the subject that good and bad angels must be directing evolution!

In fact, as the geneticists' studies progressed they were providing the last major piece of the truth so long sought regarding the causes of evolution. As this went on, it naturally began to dawn on students of the subject that each of the conflicting schools of theory had part of the truth and that none had all of it. The movement began with geneticists who wondered what effect natural selection, if it should really be a guiding force in evolution, would have on the genetic factors that they had discovered in individuals and in populations. They found that it would have a profound effect and that this effect was not exactly that predicted either by the neo-Darwinians or by the mutationists. These initial successes intensely stimulated study of evolution, which quickly regained its slipping position as the focal point of all the life sciences. Students of all the different aspects of life began to unite, each contributing from his field its special bit of evolutionary fact to add to the growing synthesis. The resulting synthetic theory[6] need not here be summarized, because it is the theory already presented, in broad outlines, in the preceding chapters.

The synthetic theory has no Darwin, being in its nature the work of many different hands. To mention any of these is to be culpable of important omissions, but if only to indicate the breadth of the synthesis it may be noted that [278] among the many contributors have been: in England, Fisher, Haldane, Huxley, Darlington, Waddington, and Ford; in the United States, Wright, Muller, Dobzhansky, Mayr, Dice, and Stebbins; in Germany, Timoféeff-Rassovsky and Rensch; in the Soviet Union, Chetverykov [*sic*]and Dubinin; in France, Teissier; in Italy, Buzzati-Traverso. I do not, of course, mean to say that these students all hold opinions identical in detail. Their fields of work, not to mention other personal variables, are so diverse that this would be a miracle. All, however, have made outstanding contributions to the modern synthesis and all seem to be agreed as to its most essential features.

This work has placed the study of the causes of evolution on a new and firm footing and has produced a degree of agreement as to these causes never before approached. We seem at last to have a unified theory—although a complex one inevitably, as evolution itself is a complex interaction of different processes—which is capable of facing all

[6]The theory has often been called neo-Darwinian, even by those who have helped to develop it, because its first glimmerings arose from confrontation of the Darwinian idea of natural selection with the facts of genetics. The term is, however, a misnomer and doubly confusing in this application. The full-blown theory is quite different from Darwin's and has drawn its materials from a variety of sources largely non-Darwinian and partly anti-Darwinian. Even natural selection in this theory has a sense distinctly different, although largely developed from, the Darwinian concept of natural selection. Second, the name "neo-Darwinian" has long been applied to the school of Weismann and his followers, whose theory was radically different from the modern synthetic theory and certainly should not be confused with it under one name.

the classic problems of the history of life and of providing a causalistic solution of each.

This is not to say that the whole mystery has been plumbed to its core or even that it ever will be. The ultimate mystery is beyond the reach of scientific investigation, and probably of the human mind. There is neither need nor excuse for postulation of nonmaterial intervention in the origin of life, the rise of man, or any other part of the long history of the material cosmos. Yet the origin of that cosmos and the causal principles of its history remain unexplained and inaccessible to science. Here is hidden the First Cause sought by theology and philosophy. The First Cause is not known and I suspect that it never will be known to living man. We may, if we are so inclined, worship it in our own ways, but we certainly do not comprehend it.

Within the realm of what is clearly knowable, the main [279] problem seems to me and many other investigators to be solved, but much still remains to be learned. Our knowledge of the material history of life is considerable, but it is only a tithe of what we should and can know. Study of the intricate dynamics of natural populations and their inter-relationships is well begun, but not much more. Study of the evidently very important psychological factors in evolution seems hardly begun. In the study of individual life processes are gaps that are probably the most serious remaining in general knowledge of evolution. Precisely what is a gene and how does it act? The path to the answer is glimpsed in some recent work—that of Beadle, for instance—but the end is not yet reached. How is a genetic system translated into completed organic form? Unless this ignorance is personal, even the path to the answer is not yet really evident, as recent experimental embryology seems rather to skirt than to follow such a path.

It is to be expected that these probable future discoveries will not only greatly deepen but also modify our current ideas of evolutionary processes. It is, however, improbable that these ideas will be vitiated in an essential way. If we do not know each single step from reptilian ear to mammalian ear, we know a multitude of steps along the line. If we do not know how genetic system leads to organic system, we know that it does and we know, broadly, the equivalents in the two. Any report on scientific enquiry or on its human import will be an interim report, as long as our species lasts and continues to value truth, but in the current knowledge of evolution we have an excellent basis for such an interim report.

---

57. ERNST MAYR

Ernst Mayr's contributions to modern evolutionary science are immense. His focus on species populations and speciation was both pragmatic and ontological, and he is widely accepted as the father of the modern philosophy of biology.

Furthermore, and more to our purpose, he was a major craftsman of the evolutionary synthesis of the 1930–40 era. Erudite and clear in both thought and writing, Mayr gives us a privileged insight into the numerous conflicting ideas and competing theories that tangled the web of evolution in its post-Mendelian years.

Born in Kempten, Germany, on July 5, 1904, Mayr was initially drawn to ornithology and taxonomy, completing a doctorate in the former at the University of Berlin in1926 at the young age of twenty-one. After three years of fieldwork in New Guinea, Mayr accepted, in 1931, a position at the American Museum of Natural History. His first book, published in 1942, was groundbreaking: *Systematics and the Origin of Species* introduced the biological species concept and put the final conceptual touches on the evolutionary synthesis concept of peripatric speciation.

In 1953, Mayr joined the faculty at Harvard, serving as director of the Museum of Comparative Zoology from 1961 to 1970, where he enjoyed a close colleagueship with Stephen Jay Gould and Niles Eldridge. Mayr's idea of peripatric speciation formed a theoretical basis for the Gould and Eldridge theory of punctuated equilibrium.

In lieu of additional original selections from classical genetics, I submit Mayr's personal reflections of the evolutionary synthesis and the dramatis personae involved. Mayr died in February 2005, in Bedford, Massachusetts, at age 101!

80 Years of Watching the Evolutionary Scenery

Having reached the rare age of 100 years, I find myself in a unique position: I'm the last survivor of the golden age of the Evolutionary Synthesis. That status encourages me to present a personal account of what I experienced in the years (1920s to the 1950s) that were so crucial in the history of evolutionary biology.

Evolutionary biology in its first 90 years (1859 to the 1940s) consisted of two widely divergent fields: evolutionary change in populations and biodiversity, the domains of geneticists and naturalists (systematicists), respectively. Histories covering this period were usually written by geneticists, who often neglected the evolution of biodiversity. As I am a naturalist, I consider this neglect to be a grave deficiency of most historical treatments.

Curiously, I cannot pinpoint the age at which I became an evolutionist. I received all of my education in Germany, where evolution was not really controversial. In the gymnasium (equivalent to a U.S. high school), my biology teacher took evolution for granted. So, I am quite certain, did my parents—who, to interest their three teenage sons, subscribed to a popular natural history journal that accepted evolution as a fact. Indeed, in Germany at that time there was no Protestant fundamentalism. And after I had entered university, no one raised any questions about evolution, either in my medical curriculum or in my preparations for the Ph.D. Those who were unable to adopt creation as a plausible solution for biological diversity concluded that evolution was the only rational explanation for the living world.

Even though creationism was not a major issue, evolutionary biology was nonetheless badly split by controversies. The disagreements concerned the causation of evolutionary change and the validity of various theories of evolution. I was a beneficiary of the furor, a bystander during the numerous clashes between supporters of the opposing theories favored during the first decades of the 20th century.

There are [*sic*] a multitude of reasons why so many controversies about evolution emerged during that period. For instance, the philosophy of science at that time was totally dominated by physics and by typology (essentialism). This philosophy was appropriate for the physical sciences but entirely unsuitable as a foundation for theories dealing with biological populations (see below). Perhaps more important was the fact that the paradigm of Darwinian evolution was not a single theory, as Darwin always insisted, but was actually composed of five quite independent theories. Two of these were readily accepted by the Darwinians: the simple fact of evolution (the "non-constancy of species" as Darwin called it) and the branching theory of common descent. The other three—gradual evolution, the multiplication of species, and natural selection—were accepted by only a minority of Darwin's followers. Indeed, these three theories were not universally accepted until the so-called Evolutionary Synthesis of the 1940s.

Superimposed on these conceptual differences were others that arose because of the preferences of evolutionists in different countries. The evolutionary theories considered valid in England or in France were rejected in Germany or the United States. One powerful author in a particular country often could determine the thinking of all his fellow scientists. Finally, different evolutionary theories were often favored by scholars in different branches of biology—say, genetics, or developmental biology, or natural history. To understand what happened during the Evolutionary Synthesis, one must be aware of the sources of disagreement during that earlier period. It was my good fortune that I became acquainted with most of the major schools of evolutionary thought of the first third of the 20th century and was therefore able to compare their claims.

In 1925–26, I prepared for my Ph.D. at the Natural History Museum of Berlin. Even though it was part of Berlin's Humboldt University, this museum had its own faculty and student body. This separation resulted in the development of two branches of evolutionary biology. The laboratory (experimental) geneticists in the University studied the processes taking place in a single population—in a single gene pool. Their emphasis was on gradual evolutionary change in a phyletic sequence of populations. The museum naturalists, including myself, were much more interested in a different aspect of evolution, the origin of biodiversity, and particularly the origin of new species and higher taxa. We studied the process of speciation and the transition from species level evolution to macroevolution—in other words, the origin of new types of organisms and the evolution of higher taxa.

We naturalists thought that evolution was indeed a gradual process, as Darwin had always insisted. Our material provided hundreds of illustrations of widespread species that gradually changed throughout their geographic range. By contrast, most early Mendelians, impressed by the discontinuous nature of genetic changes ("mutations"), thought that these mutations provided evidence for a saltational origin of new species.

When Mendel's laws were rediscovered in 1900, there was widespread hope that they would lead to a unification of the conflicting theories on speciation. Unfortunately, it turned out that the three geneticists most interested in evolution—Bateson, DeVries, and Johannsen—were typologists and opted for a mutational origin (by saltation) of new species. Worse, they rejected gradual evolution through the natural selection of small variants. For their part, the naturalists erroneously thought that the geneticists had achieved a consensus based on saltational speciation, and this led to a long-lasting controversy between the naturalists and the early Mendelians.

The naturalists were unaware that there were other geneticists—like East, Castle, Baur, and Chetverikov—who may have been a majority. Their interpretation of small mutations and gradual evolution was completely compatible with the theories of the naturalists. Beginning in 1910, the work of the Columbia University group in New York under T. H. Morgan led to a refutation of the theories of the saltational Mendelians, and established the basis for the origin of a rigorous school of mathematical population genetics, culminating in the work of R. A. Fisher, J. B. S. Haldane, and Sewall Wright. Most important, this new school of population geneticists fully accepted natural selection—and that permitted a synthesis.

Several historians have mistakenly thought that this synthesis within genetics had solved all the problems of Darwinism. That assumption, however, failed to take account of an important gap. One of the two major branches of evolutionary biology, the study of the origin of biodiversity, had been left out of the major treatises of Fisher, Haldane, and Wright. Actually, unknown to these geneticists, the problems of the origin of biodiversity had already been solved in the 1920s by several European naturalists, most important among them, Moritz Wagner, Karl Jordan, Edward Poulton, Chetverikov, and Erwin Stresemann.

Thus, evolutionary biology around 1930 found itself in a curious position. It faced two major seemingly unsolved problems: the adaptive changes of populations and the origin of biodiversity. Two large and very active groups of evolutionists worked on these problems. One of these groups consisted of the population geneticists. As summarized in the works of Fisher, Haldane, and Wright, this group had solved the problem of gradual evolution of populations through natural selection. But they had not made any contribution to the problem of how species arise (speciation)—that is, to the problem of the origin of biodiversity. The other group of evolutionists consisted of the naturalists (taxonomists). Although unaware of the solution to the problem of gradual adaptive evolution, they had solved the open problems of the evolution of biodiversity through the contributions of the European naturalists Wagner, Jordan, Poulton, Stresemann, and Chetverikov. Thus, by 1930, the two great problems of evolutionary biology had been solved, but by different groups whose accomplishments were unknown to one another.

As a student in Germany in the 1920s, I belonged to a German school of evolutionary taxonomists that was unrepresented in the United States. Our tradition placed great stress on geographic variation within species, and particularly on the importance of geographic isolation and its role in leading to the origin of new species. It accepted a Lamarckian inheritance of newly acquired characters but simultaneously accepted natural selection as facilitating gradual evolution. We decisively rejected any saltational origin of new species, as had been postulated by DeVries.

Fortunately, there was one evolutionist who had the background to be able to resolve the conflict between the geneticists and the naturalists. It was Theodosius Dobzhansky. He had grown up in Russia as a naturalist and beetle taxonomist, but, in 1927, he joined Morgan's laboratory in America where he became thoroughly familiar with population genetics. He was ideally suited to show that the findings of the population geneticists and those of the European naturalists were fully compatible and that a synthesis of the theories of the two groups would provide a modern Darwinian paradigm, subsequently referred to as the "Evolutionary Synthesis."

What is particularly remarkable about this new paradigm is its stability. Dobzhansky's first approach was elaborated and modified in the ensuing years in the publications of Ernst Mayr, Julian Huxley, and Bernhard Rensch. I owe it to Dobzhansky that I played such an important role in the Synthesis. He knew that his own treatment of the evolution of biodiversity was insufficient. It was my task to fill the gaps in his 1937 account. Dobzhansky even provided the inspiration for the title of my book, *Systematics and the Origin of Species*—a title chosen deliberately as an equivalent to his *Genetics and the Origin of Species*. Curiously, there is no chapter on speciation in Dobzhansky's book. His description of the isolating mechanisms was erroneous. Isolating mechanisms are genetic properties of individuals, yet he included geographic barriers among them. In addition, his putative species definition refers to the stage of a process, but is not the description of a (species) population. I was able to correct this in my 1942 book.

At a meeting in Princeton in 1947, the new paradigm was fully acknowledged and it was confirmed again and again in the next 60 years. Whenever an author claimed to have found an error in the Synthesis, his claim was rapidly refuted. The two belief systems had only one inconsistency—the object of natural selection. For the geneticists the object of selection had been the gene since the 1920s, but for most naturalists it was the individual.

Elliot Sober showed how one could resolve this conflict. He pointed out that one must discriminate between selection of an object and selection for an object. The answer to the question of what is being selected for specifies the particular properties for which a given object of selection is favored. However, a particular gene can favor an individual without being the object of selection because it gives properties to the individual that favor its selection. It is a selection for these properties.

By the end of the 1940s the work of the evolutionists was considered to be largely completed, as indicated by the robustness of the Evolutionary Synthesis. But in the ensuing decades, all sorts of things happened that might have had a major impact on the Darwinian paradigm. First came Avery's demonstration that nucleic acids and not proteins are the genetic material. Then in 1953, the discovery of the double helix by Watson and Crick increased the analytical capacity of the geneticists by at least an order of magnitude. Unexpectedly, however, none of these molecular findings necessitated a revision of the Darwinian paradigm—nor did the even more drastic genomic revolution that has permitted the analysis of genes down to the last base pair.

It would seem justified to assert that, so far, no revision of the Darwinian paradigm has become necessary as a consequence of the spectacular discoveries of molecular biology. But there is something else that has indeed affected our understanding of the living world: that is its immense diversity. Most of the enormous variation of kinds of organisms has so far been totally ignored by the students of speciation. We have studied the origin of new species in birds, mammals, and certain genera of fishes, lepidopterans, and molluscs, and speciation has been observed to be allopatric (geographical) in most of the studied groups. Admittedly, there have been a few exceptions, particularly in certain families, but no exceptions have been found in birds and mammals where we find good biological species, and speciation in these groups is always allopatric. However, numerous other modes of speciation have also been discovered that are unorthodox in that they differ from allopatric speciation in various ways. Among these other modes are sympatric speciation, speciation by hybridization, by polyploidy and other chromosome rearrangements, by lateral gene transfer, and by symbiogenesis. Some of these nonallopatric modes are quite frequent in certain genera of cold-blooded vertebrates, but they may be only the tip of the iceberg. There are all the other phyla of multicellular eukaryotes, the speciation of most of them still quite unexplored. This is even truer for the 70-plus phyla of unicellular protists and for the prokaryotes. There are whole new worlds to be discovered with, perhaps, new modes of speciation among the forthcoming discoveries.

The new research has one most encouraging message for the active evolutionist: it is that evolutionary biology is an endless frontier and there is still plenty to be discovered. I only regret that I won't be present to enjoy these future developments.

Ideas to Think About

1. The brilliant geneticist R. A. Fisher, in 1935, cast doubt on the validity of Mendel's original ratios by claiming them too good to be true. Subsequent study suggests that, as Mendel began to approach a 3:1 ratio in his results, he continued to count the crosses until that ratio was confirmed. Others have suggested that he threw out the outliers when it became obvious that they were anomalies. Given the level of statistical knowledge in the 1860s, how would you assess these claims, if supported?
2. Hardy's equation demonstrated that given normal conditions, gene proportions in a population with genetic variability would remain unchanged through generations. How significant is this in the study of directional change through natural selection? What is the value of the equation?
3. Consider T. H. Morgan's claim that the historical approach of the naturalists toward explaining evolution is speculation, whereas the experimental approach of lab geneticists is scientific. This has long been claimed as a failure of the historical sciences. Can you counter this argument?
4. The experimentalists focused on discontinuous variation in the genetic process, whereas the naturalists focused on continuous variation among intergrading phenotypes. How does each emphasis differently treat the relationship between variation and its sources?
5. Resolving the apparent paradox of evolutionary continuity through systematic adaptive responses to the environment and the discontinuity sometimes apparent in the generation of biodiversity is aptly recognized by Mayr. What events in the 1930s resolved this and completed the synthesis?

For Further Reading

The Causes of Evolution, J. B. S. Haldane. Princeton, NJ: Princeton University Press (1990). This is a thoroughly captivating popularization of evolution in its transition from Darwin's era to the modern synthesis. Haldane was one of the principals in that transition, and in this work provides, as Harry Horn puts it, a "balance between gentle rhetoric and logical rigor."

Shaping Science with Rhetoric: The Cases of Dobzhansky, Schrodinger, and Wilson, Leah Ceccarelli. Chicago: University of Chicago Press (2001). This is a brief work of fewer than two hundred pages, but it is comprehensive in its purpose of galvanizing the imagination and stimulating interdisciplinary research. Dobzhansky's *Genetics and the Origin of Species* admirably served this purpose for the evolutionary synthesis. Schrodinger's *What is Life?* catalyzed molecular biology, which warrants reading this again following Unit 15 of this volume. E. O. Wilson's controversial *Consilience* merits further reading following Unit 14 of this volume.

UNIT 12

Beyond Biology: Social Darwinism and Eugenics

> While the law [of competition] may be sometimes hard for the individual, it is best for the race, because it insures the survival of the fittest in every department. We accept and welcome, therefore, as conditions to which we must accommodate ourselves, great inequality of environment, the concentration of business, industrial and commercial, in the hands of a few, and the law of competition between these, as being not only beneficial, but essential for the future progress of the race.
>
> ANDREW CARNEGIE, *Wealth (1899)*

The influence of the Darwinian spectacle—and one may justifiably call it that, given the circus of rhetoric that followed the 1859 publication—was far-reaching, even if we limit our attention to the nineteenth century. The affect on all cultural forms, political, literary, poetic, and dramatic, was, well, dramatic. If we extend this into the twentieth century (and, indeed, to the continuing influence in the twenty-first), we find an inventory of inspiration that crosscuts broad vistas of creativity. Alas, it has not been comprehensively assembled, and only the inveterate spelunker might yet probe its hidden intellectual chambers. The broad reach of biology into the social order—actually and by analogy—was a major source of what we might call "applied Darwinism."

The idea that progress is a natural condition of the social order did not await Darwin for its expression. It was present at the Enlightenment. Natural selection simply gave it a sense of scientific authenticity. The role of government in society, as defined in the Enlightenment philosophy of Rousseau, ought to be as protector of individual liberty, in contrast to the ideas of Hobbes a century earlier. This was echoed by John Stuart Mill. The idea, drawn from Enlightenment optimism for social improvement following the Industrial Revolution, readily led to the claim for the progress of humankind as part of the natural order. In its most developed form, however, this philosophy rejected Rousseau's call for government to help guarantee equality as well as liberty.

So, if in nature we find abundant evidence for the survival of the fittest in competition for resources, and if the transformation of humankind into the framework of nature was a logical result of the new natural order, then human society has progressed according to the same protocol for advancement. Civilization is thus destiny working its way to perfection.

Enter Herbert Spencer. Progress, he argued, is part of the natural order—alike for the physical universe, biological species, and human society. Its fundamental quality was the transformation from simplicity to complexity, the development from a homogeneous organization to a heterogeneous one. For Spencer, civil society was ever an exercise in allowing this process to take place without undue interference. The dialectic always involved tensions between the inherent elements of simple society and the progressive elements of complexity. As these tensions were naturally reconciled in nature, they would equally work themselves out in society, with a resultant increase in individual freedom.

Spencer spanned the Darwinian era. His ideas were not drawn from the Darwinian model, but they benefited vastly from it. The lesson to be learned quickly became: social, like biological, competition would result in the survival of the fittest. Trying to support the less fit at the expense of the fit was government interference. During the nineteenth and early twentieth centuries, and particularly in America, this concept resonated well. A philosophy of Social Darwinism—and its more active extensions, including eugenics—enjoyed a substantial period of popular support.

William Graham Sumner, father of sociology, as it was initially defined in the United States, was influenced by and followed the sociopolitical thoughts of Spencer. He was strongly influenced, as well, by Darwin. We see in his philosophy the liberal approach to capitalist competition in a minimalist government plus a strong isolationist approach to foreign affairs.

The reliance on self-improvement that focused on individualism during the late nineteenth and early twentieth centuries, both here and abroad, emphasized a laissez-faire role of the state that was vehemently antisocialist. This fear of government interference in both the economic and political spheres, however, did not entirely limit the state's role in actively ensuring a progressive result. The state should also remove barriers that could threaten achievements of hard work and diligent competitive success. In its most extreme form, this led to efforts to control the reproduction of those segments of society considered unfit, whose inherited inabilities encouraged welfare. They were a burden to the state, holding back its inexorable march toward excellence. This attitude resulted in a strong eugenics movement that, in the United States during the early part of the twentieth century, fostered a number of compulsory sterilization laws. Unfortunately, court challenges to these were not always successful. The notorious *Buck* v. *Bell* Supreme Court decision is included here as an example.

58. Herbert Spencer

Born in Derby, England, on April 27, 1820, Herbert Spencer inherited from his father, William George Spencer, both a religious deism and a commitment to Enlightenment values. The elder Spencer was active in the intellectual circles of the Derby-Birmingham communities, where Boulton, Priestly, Watt, and Wedgwood transformed science into industry, and was at one time secretary of the Derby Philosophical Society, founded by Erasmus Darwin.

Herbert Spencer acquired a strong belief in the natural progress of society and the value of unbridled human freedom in the competitive struggle for success. This would lead, he believed, not only to progressive social evolution, but also to a future perfection of the human social condition. Surrounded by the vast potential benefits of science and technology, Spencer applied the concept of natural law and the evolutionary success it made possible to human societies. He saw no fundamental break between the biological and the social in terms of basic principles of cause and effect and the innate progress towards a final "equilibrium." Like Comte, he sought a unity in all sciences, an ultimate scientific truth, and the universality of natural law as it applied to all things—physical, biological, social, intellectual, and emotional.

He read Darwin after he had already published his essay on "Progress: Its Law and Cause" outlining his philosophy. He accepted Darwin's argument, coining the term *survival of the fittest* to represent natural selection. Spencer favored Lamarckian evolution over Darwinian, however, as its "use-disuse" concept of acquired traits better fit Spencer's notion of the improvement of humankind through struggle.

Progress: Its Law and Cause

[445] The current conception of Progress is somewhat shifting and indefinite. Sometimes it comprehends little more than simple growth—as of a nation in the number of its members and the extent of territory over which it has spread. Sometimes it has reference to quantity of material products—as when the advance of agriculture and manufactures is the topic. Sometimes the superior quality of these products is contemplated; and sometimes the new or improved appliances by which they are produced. When again we speak of moral or intellectual progress, we refer to the state of the individual or people exhibiting it; whilst, when the progress of Knowledge, of Science, of Art, is commented upon, we have in view certain abstract results of human thought and action. Not only, however, is the current conception of Progress more or less vague, but it is in great measure erroneous. It takes in not so much the reality of Progress as its accompaniments—not so much the substance as the shadow. That progress in intelligence which takes place during the evolution of the child into the man, or the savage into the philosopher, is commonly regarded as consisting in the greater number of facts known and laws understood: whereas the actual progress consist in the produce of a greater quantity and variety of articles for the satisfaction of men's wants; in the increasing security of person and property; in the widening freedom of action enjoyed whereas, rightly understood, social progress consists in those changes of structure in the social organism which have entailed [446] these consequences. The current conception is a teleological one. The phenomena are contemplated solely as bearing on human happiness. Only those changes are held to constitute progress which directly or indirectly tend to heighten human happiness. And they are thought to constitute progress simply *because* they tend to heighten human happiness. But rightly to understand Progress, we must inquire what is the nature of these changes, considered apart from our interests. Ceasing, for example, to regard the successive geological modifications that have taken place in the Earth, as modifications that have gradually fitted it for the habitation of Man, and as therefore a geological progress, we must seek to determine the character common to these modifications—the law to which they all conform. And similarly in every other case. Leaving out of sight concomitants and beneficial consequences, let us ask what Progress is in itself.

In respect to that progress which individual organisms display in the course of their evolution, this question has been answered by the Germans. The investigations of Wolff, Goethe, and Van Baer have established the truth that the series of changes gone through during the development of a seed into a tree, or an ovum into an animal, constitute an advance from homogeneity of structure to heterogeneity of structure. In its primary stage, every germ consists of a substance that is uniform throughout, both in texture and chemical composition. The first step in its development is the appearance of a difference between two parts of this substance; or, as the phenomenon is described in physiological language—a differentiation. Each of these differentiated divisions presently begins itself to exhibit some contrast of parts; and by these secondary differentiations become as definite as the original one. This progress is continuously repeated—is simultaneously going on in all parts of the growing embryo; and by endless multiplication of these differentiations there is ultimately produced that complex combination of tissues and organs constituting the adult animal or plant. This is the course of evolution followed by all organisms whatever. It is settled beyond dispute that organic progress consists in a change from the homogeneous to the heterogeneous.

Now, we propose in the first place to show, that this law of organic progress is the law of all progress. Whether it be in the development of the Earth, in the development of Life upon its surface, the development of Society, of Government, of Manufactures, of Commerce, of Language, Literature, Science, Art, this same evolution of the simple into the complex, through a process of continuous differentiation, holds throughout. From the earliest traceable cosmical changes down to the latest results [447] of civilization, we shall find that the transformation of the homogeneous into the heterogeneous, is that in which Progress essentially consists....

[451] Whether an advance from the homogeneous to the heterogeneous is or is not displayed in the biological history of the globe, it is clearly enough displayed in the progress of the latest and most heterogeneous creature—Man. It is alike true that, during the period in which the Earth has been peopled, the human organism has become more heterogeneous among the civilized divisions of the species and that the species, as a whole, has been growing more heterogeneous in virtue of the multiplication of races and the differentiation of these races from each other....

[454] In the course of ages, there arises, as among ourselves, a highly complex political organization of monarch, ministers, lords and commons, with their subordinate administrative departments, courts of justice, revenue offices, &c., supplemented in the provinces by municipal governments, county governments, parish or union governments—all of them more or less elaborated. By its side there grows up a highly complex religious organization, with its various grades of officials from archbishops down to sextons, its colleges, convocations, ecclesiastical courts, &c.; to all which must be added the ever-multiplying independent sects, each with its general and local authorities. And at the same time there is developed a highly complex aggregation of customs, manners, and temporary fashions, enforced by society at large, and serving to control those minor transactions between man and man which are not regulated by civil and religious law. Moreover it is to be observed that this [455] ever-increasing heterogeneity in the governmental appliances of each nation, has been accompanied by an increasing heterogeneity in the governmental appliances of different nations all of which are more or less unlike in their political systems and legislation in their creeds and religious institutions, in their customs and ceremonial usages.

Simultaneously there has been going on a second differentiation of a still more familiar kind; that, namely, by which the mass of the community has become segregated into distinct classes and orders of workers. While the governing part has been undergoing the complex development above described, the governed part has been undergoing an equally complex development, which has resulted in that minute division of labour characterizing advanced nations. It is needless to trace out this progress from its first stages, up through the caste divisions of the East and the incorporated guilds of Europe, to the elaborate producing and distributing organization existing among ourselves. Political economists have made familiar to all, the evolution which, beginning with a tribe whose members severally perform the same actions each for himself, ends with a civilized community whose members severally perform different actions for each other; and they have further explained the evolution through which the solitary producer of any one commodity, is transformed into a combination of producers who united under a master, take separate parts in the manufacture of such commodity. But there are yet other and higher phases of this advance from the homogeneous to the heterogeneous in the industrial structure of the social organism. Long after considerable progress has been made in the division of labour among different classes of workers, there is still little or no division of labour among the widely separated parts of the community: the nation continues comparatively homogeneous in the respect that in each district the same occupations are pursued. But when roads and other means of transit become numerous and good, the different districts begin to assume different functions, and to become mutually dependent....

[456] Not only is the law thus clearly exemplified in the evolution of the social organism, but it is exemplified with equal clearness in the evolution of all products of human thought and action; whether concrete or abstract, real or ideal....

[483] Without further accumulation of evidence, we venture to think our case is made out. The many imperfections of statement which brevity has necessitated, do not, we believe, militate [484] against the truth of the propositions enunciated. The qualifications here and there demanded would not, if made, affect the inferences. Though in one instance, in which sufficient evidence is not attainable, we have been unable to show that the law of Progress applies; yet there is high probability that the same generalization holds which holds throughout the rest of creation.

59. William Graham Sumner

Born in Boston on October 30, 1840, Sumner is best known as a sociologist interested in folkways and cultural diffusion. Three years in Europe following graduation from Yale in 1863 acquainted him with the literature and philosophy of history. He spent much of 1866 at Oxford, where discussions with students led to his advocacy of free markets and a laissez-faire approach to economics and government.

In 1872, Sumner accepted an appointment at Yale. He was the first to teach a course called "sociology." He read and reread Spencer, and read Darwin for the first time, along with other natural scientists. "I greatly regretted that I had no education in natural science," Sumner later wrote, "especially in biology; but I found that the 'philosophy of history' and the 'principles of philology,' as I had learned them, speedily adjusted themselves to the new conception, and won a new meaning and power from it." Sumner soon recognized the theoretical value of joining the two. "I was constantly getting evidence that sociology, if it borrowed the theory of evolution in the first place, would speedily render it back again enriched by new and independent evidence."

Sumner helped to found the American Sociological Association in 1905 and served as its president in 1908–9. Among his students were the economist Thorstein Veblen and anthropologist Albert Galloway Keller. He and Keller collaborated in numerous publications. The collection of essays from which the following is taken was edited by Keller following Sumner's death in 1910. There is a persuasive argument that Sumner in his later career rejected Social Darwinism, whereas Keller continued to strongly support it (Smith 1979).

The Challenge of Facts and Other Essays

[17] Socialism is no new thing. In one form or another it is to be found throughout all history. It arises from all observation of certain harsh facts in the lot of man on earth, the concrete expression of which is poverty and misery. These facts challenge us. It is folly to try to shut our eyes to them. We have first to notice what they are, and then to face them squarely.

Man is born under the necessity of sustaining the existence he has received by all onerous struggle against nature, both to win what is essential to his life and to ward off what is prejudicial to it. He is born under a burden and a necessity. Nature holds what is essential to him, but she offers nothing gratuitously. He may win for his use what she holds, if he can. Only the most meager and inadequate supply for human needs can be obtained directly from nature. There are trees which may be used for fuel and for dwellings, but labor is required to fit them for this use. There are ores in the ground, but labor is necessary to get out the metals and make tools or weapons. For any real satisfaction, labor is necessary to fit the products of nature for human use. In this struggle every individual is under the pressure of the necessities for food, clothing, shelter, fuel, and every individual brings with him more or less energy for the conflict necessary to supply his needs. The relation, therefore, between each man's needs and each man's energy, or "individualism," is the first fact of human life.

[18] The history of the human race shows a great variety of experiments in the relation of the sexes and in the organization of the family. These experiments have been controlled by economic circumstances, but, as man has gained more and more control over economic circumstances, monogamy and the family education of children have been more and more sharply developed. If there is one thing in regard to which the student of history and sociology can affirm with confidence that social institutions have made "progress" or grown "better," it is in this arrangement of marriage and the family. All experience proves that monogamy, pure and strict, is the sex relation which conduces most to the vigor and intelligence of the race, and that the family education of children is the institution by which the race as a whole advances most rapidly, from generation to generation, in the struggle with nature. Love of man and wife, as we understand it, is a modern sentiment. The devotion and sacrifice of parents for children is a sentiment which has been developed steadily and is now more intense and far more widely practiced throughout society than in earlier times. The relation is also coming to be regarded in a light quite different from that in which it was [19] formerly viewed. It used to be believed that the parent had unlimited claims on the child and rights over him. In a truer view of the matter, we are coming to see that the rights are on the side of the child and the duties on the side of the parent....

[20] The next great fact we have to notice in regard to the struggle of human life is that labor which is spent in a direct struggle with nature is severe in the extreme and is but slightly productive. To subjugate nature, man needs weapons and tools. These, however, cannot be won unless the food and clothing and other prime and direct necessities are supplied in such amounts that they can be consumed while tools and weapons are being made, for the tools and weapons themselves satisfy no needs directly. A man who tills the ground with his fingers or with a pointed stick picked up without labor will get a small crop. To fashion even the rudest spade or hoe will cost time, during which the laborer must still eat and drink and wear, but the tool when obtained, will multiply immensely the power to produce. Such products of labor, used to assist production, have a function so peculiar in the nature of things that we need to distinguish them. We call them capital. A lever is capital, and the advantage of lifting a weight with a lever over lifting it by direct exertion is only a feeble illustration of the power of capital in production. The origin of capital lies in the darkness before history, and it is probably impossible for us to imagine the slow and painful steps by which the race began the formation of it. Since then it has gone on rising to higher and [21] higher powers by a ceaseless involution, if I may use a mathematical expression. Capital is labor raised to a higher power by being constantly multiplied into itself. Nature has been more and more subjugated

by the human race through the power of capital, and every human being now living shares the improved status of the race to a degree which neither he nor any one else can measure, and for which he pays nothing....

[23] The constant tendency of population to outstrip the means of subsistence is the force which has distributed population over the world, and produced all advance in civilization. To this day the two means of escape for an overpopulated country are emigration and an advance in the arts. The former wins more land for the same people; the latter makes the same land support more persons. If, however, either of these means opens a chance for an increase of population, it is evident that the advantage so won may be speedily exhausted if the increase takes place. The social difficulty has only undergone a temporary amelioration, and when the conditions of pressure and competition are renewed, misery and poverty reappear. The victims of them are those who have inherited disease and depraved appetites, or have been brought up in vice and ignorance, or have themselves yielded to vice, extravagance, idleness, and imprudence. In the last analysis, therefore, we come back to vice, in its original and hereditary forms, as the correlative of misery and poverty.

The condition for the complete and regular action of the force of competition is liberty. Liberty means the security given to each man that, if he employs his energies to sustain the struggle on behalf of himself and those he cares for, he shall dispose of the produce exclusively as he chooses. It is impossible to know whence any definition or criterion of justice can be derived, if it is not deduced from this view of things; or if it is not the definition of justice that each shall enjoy the fruit of [24] his own labor and self-denial, and of injustice that the idle and the industrious, the self-indulgent and the self-denying, shall share equally in the product....

[25] Private property, also, which we have seen to be a feature of society organized in accordance with the natural conditions of the struggle for existence produces inequalities between men. The struggle for existence is aimed against nature. It is from her niggardly hand that we have to wrest the satisfaction for our needs, but our fellow-men are our competitors for the meager supply. Competition, therefore, is a law of nature. Nature is entirely neutral; she submits to him who most energetically and resolutely assails her. She grants her rewards to the fittest, therefore, without regard to other considerations of any kind. If, then, there be liberty, men get from her just in proportion to their works, and their having and enjoying are just in proportion to their being and their doing. Such is the system of nature. If we do not like it, and if we try to amend it, there is only one way in which we can do it. We can take from the better and give to the worse. We can deflect the penalties of those who have done ill and throw them on those who have done better. We can take the rewards from those who have done better and give them to those who have done worse. We shall thus lessen the inequalities. We shall favor the survival of the unfittest, and we shall accomplish this by destroying liberty. Let it be understood that we cannot go outside of this alternative; liberty, inequality, survival of the fittest; not-liberty, equality, survival of the unfittest. The former carries society forward and favors all its best members; the latter carries society downwards and favors all its worst members....

[26] What we mean by liberty is civil liberty, or liberty under law; and this means the guarantees of law that a man shall not be interfered with while using his own powers for his own welfare. It is, therefore, a civil and political status; and that nation has the freest institutions in which the guarantees of peace for the laborer and security for the capitalist are the highest. Liberty, therefore, does not by any means do away with the struggle for existence. We might as well try to do away with the need of eating, for that would, in effect, be the same thing. What civil liberty does is to turn the competition of man with man from violence and brute force into an industrial competition under which men vie with one another for the acquisition of material goods by industry, energy, skill, frugality, prudence, temperance, and other industrial virtues. Under this changed order of things the inequalities are not done away with. Nature still grants her rewards of having and enjoying, according to our being and doing, but it is now the man of the highest training and not the man of the heaviest fist who gains the highest reward.

[35] A century ago there were very few wealthy men except owners of land. The extension of commerce, manufactures, and mining, the introduction of the factory system and machinery, the opening of new countries, and the great discoveries and inventions have created a new middle class, based on wealth, and developed out of the peasants, artisans, unskilled laborers, and small shopkeepers of a century ago. The consequence has been that the chance of acquiring capital and all which depends on capital has opened [36] before classes which formerly passed their lives in a dull round of ignorance and drudgery.... The appetite for enjoyment has been awakened and nourished in classes which formerly never missed what they never thought of, and it has produced eagerness for material good, discontent, and impatient ambition. This is the reverse side of that eager uprising of the industrial classes which is such a great force in modern life....

The socialist regards this misery as the fault of society. He [37] thinks that we can organize society as we like and that an organization can be devised in which poverty and misery shall disappear. He goes further even than this. He assumes that men have artificially organized society as it now exists. Hence if anything is disagreeable or hard in the present state of society, it follows, on that view, that the task of organizing society has been imperfectly and badly performed, and that it needs to be done over again....

The truth is that the social order is fixed by laws of nature precisely analogous to those of the physical order. The most that man can do is by ignorance and self-conceit to mar the operation of social laws. The evils of society are to a great extent the result of the dogmatism and self-interest of statesmen, philosophers, and ecclesiastics who in past time have done just what the socialists now want to do. Instead of studying the natural laws of the social order, they assumed that they could organize society as they chose, they made up their minds what kind of a society they wanted to make, and they planned their little measures for the ends they had resolved upon.

[39] The socialist looking at these facts says that it is capital which produces the inequality. It is the inequality of men in what they get out of life which shocks the socialist. He finds enough to criticize in the products of past dogmatism and bad statesmanship to which I have alluded, and the program of reforms to be accomplished and abuses to be rectified which the socialists have set up have often been admirable. It is their analysis of the situation which is at fault. Their diagnosis of the social disease is founded on sectarian assumptions, not on the scientific study of the structure and functions of the social body. In attacking capital they are simply attacking the foundations of civilization, and every socialistic scheme which has ever been proposed, so far as it has lessened the motives to saving or the security of capital, is anti-social and anti-civilizing. Rousseau, who is the great father of the modern socialism, laid accusation for the inequalities existing amongst men upon wheat and iron. What he meant was that wheat is a symbol of agriculture, and when men took to agriculture and wheat diet they broke up their old tribal relations, which were partly communistic, and developed individualism and private property. At the same time agriculture called for tools and machines, of which iron is a symbol; but these tools and machines are capital. Agriculture, individualism, tools, capital were, according to Rousseau's ideas, the causes of inequality. He was, in a certain way, correct, as we have already seen by our own analysis of the facts of the social order.

60. The Rise of Eugenics

If society is susceptible to the same natural law as biology, and if stockbreeders can manipulate natural law for the betterment of domestic breeds, why should not government manipulate human breeding for the betterment of society? This is not a new suggestion, nor was it new in the late nineteenth and early twentieth centuries. It is, as a matter of fact, as old as Socrates's conversation with Glaucon (Unit 1) in the fourth century BC.

But with the hopeful optimism for Victorian Britain's future riding on both scientific and social reform, it was inevitable that the eugenic concept would rise again. Inadequacy of the preexisting Poor Laws of Great Britain was magnified in the early nineteenth century by rising unemployment and poverty and the increasing gap between the destitute and the growing middle class. Welfare reform engaged debate between those such as Jeremy Bentham, who espoused a punitive approach to indigence, and Thomas Malthus, who was more concerned with preventing overpopulation by controlling illegitimacy.

Both abroad and in the United States, support for some form of legal eugenics became one of the measures of welfare reform. Even after the Mendelian rediscovery, however, and most particularly in view of a strong Lamarckian sentiment on inheritance that continued through the second decade of the twentieth century, there was little understanding of and far less experimental research on inheritance factors in behavior, character, intelligence, and common pathological conditions.

Even as Galton produced mathematical formulae to measure probabilities of inheritance, therefore, the determination of exactly what could be inherited was being debated. Those who favored eugenic sterilization—and many did—as an effective method of control against undesirable inheritance could not agree on exactly what conditions this might legitimately include. Use of pedigrees and familial patterns as measures were as notoriously unreliable as they were insidiously compelling. If poverty runs in a family, can one eliminate it by reproductive control? If antisocial behavior characterizes several generations of a destitute family, can we control this through sterilization?

The will to believe is more powerful than skeptical dissent, particularly with the appearance of success. "Ignorance and credulous hope," wrote Samuel Hopkins Adams in 1905, "make the market for most proprietary remedies." And so compulsory eugenic sterilization laws began to crop up, here and there, in the shadows and with little fanfare. In 1897, the Michigan legislature passed the first compulsory sterilization law. The governor vetoed it. In 1907, an Indiana law forcing sterilization of the "feebleminded" was successfully passed. In 1909, both California and Washington passed such laws. Finally, in 1927, the U.S. Supreme Court ruled as constitutional state laws mandating sterilization on the basis of mental handicap. Nazi propaganda in 1936 defended its mandated sterilization program by citing the U.S. laws. It was not until 1981, when Oregon performed the final compulsory sterilization, that the practice in the United States formally ceased.

61. Francis Galton

Francis Galton (1822–1911), half-cousin of Charles Darwin (they shared grandfather Erasmus Darwin), was an accomplished statistician who pioneered the concepts and tests of correlation and regression toward the mean. His community sample surveys and psychometric tests helped pioneer rigorous methodologies in the social sciences. His interest in the inheritance of intelligence led him to the field of eugenics, a name he coined.

Darwin's *Origin of Species* influenced him greatly—particularly Darwin's first chapter, on variation under domestication. Although he disagreed that evolution could occur through slow, gradual change, he became a staunch supporter of Darwin. He was among those present at the famous Huxley-Wilberforce debate at Oxford in 1860.

The following article was originally an address given at the Sociological Society of London (at a meeting of the London School of Economics) on May 16, 1904, and chaired by Karl Pearson. The respondents were either skeptical or critical, which angered Galton.

Eugenics: Its Definition, Scope, and Aims

Eugenics is the science which deals with all influences that improve the inborn qualities of a race; also with those that develop them to the utmost advantage. The improvement of the inborn qualities, or stock, of some one human population will alone be discussed here.

What is meant by improvement? What by the syllable *eu* in "eugenics," whose English equivalent is "good"? There is considerable difference between goodness in the several qualities and in that of the character as a whole. The character depends largely on the *proportion* between qualities, whose balance may be much influenced by education. We must therefore leave morals as far as possible out of the discussion, not entangling ourselves with the almost hopeless difficulties they raise as to whether a character as a whole is good or bad. Moreover, the goodness or badness of character is not absolute, but relative to the current form of civilization. A fable will best explain what is meant. Let the scene be the zoological gardens in the quiet hours of the night, and suppose that, as in old fables, the animals are able to converse, and that some very wise creature who had easy access to all the cages, say a philosophic sparrow or rat, was engaged in collecting the opinions of all sorts of animals with a view of elaborating a system of absolute morality. It is needless [2] to enlarge on the contrariety of ideals between the beasts that prey and those they prey upon, between those of the animals that have to work hard for their food and the sedentary parasites that cling to their bodies and suck their blood, and so forth. A large number of suffrages in favor of maternal affection would be obtained, but most species of fish would repudiate it, while among the voices of birds would be heard the musical protest of the cuckoo. Though no agreement could be reached as to absolute morality, the essentials of eugenics may be easily defined. All creatures would agree that it was better to be healthy than sick, vigorous than weak, well-fitted than ill-fitted for their part in life; in short, that it was better to be good rather than bad specimens of their kind, whatever that kind might be. So with men. There are [*sic*] a vast number of conflicting ideals, of alternative characters, of incompatible civilizations; but they are wanted to give fulness [*sic*] and interest to life. Society would be very dull if every man resembled the highly estimable Marcus Aurelius or Adam Bede. The aim of eugenics is to represent each class or sect by its best specimens; that done, to leave them to work out their common civilization in their own way.

A considerable list of qualities can easily be compiled that nearly everyone except "cranks" would take into account when picking out the best specimens of his class. It would include health, energy, ability, manliness, and courteous disposition. Recollect that the natural differences between dogs are highly marked in all these respects, and that men are quite as variable by nature as other animals of like species. Special aptitudes would be assessed highly by those who possessed them, as the artistic faculties by artists, fearlessness of inquiry and veracity by scientists, religious absorption by mystics, and so on. There would be self-sacrificers, self-tormentors, and other exceptional idealists; but the representatives of these would be better members of a community than the body of their electors. They would have more of those qualities that are needed in a state—more vigor, more ability, and more consistency of purpose. The community might be trusted to refuse representatives of criminals, and of others whom it rates as undesirable.

[3] Let us for a moment suppose that the practice of eugenics should hereafter raise the average quality of our nation to that of its better moiety at the present day, and consider the gain. The general tone of domestic, social, and political life would be higher. The race as a whole would be less foolish, less frivolous, less excitable, and politically more provident than now. Its demagogues who "played to the gallery" would play to a more sensible gallery than at present. We should be better fitted to fulfil [*sic*] our vast imperial opportunities. Lastly, men of an order of ability which is now very rare would become more frequent, because, the level out of which they rose would itself have risen.

The aim of eugenics is to bring as many influences as can be reasonably employed, to cause the useful classes in the community to contribute *more* than their proportion to the next generation.

The course of procedure that lies within the functions of a learned and active society, such as the sociological may become, would be somewhat as follows:

1. Dissemination of a knowledge of the laws of heredity, so far as they are surely known, and promotion of their further study. Few seem to be aware how greatly the knowledge of what may be termed the *actuarial* side of heredity has advanced in recent years. The *average* closeness of kinship in each degree now admits of exact definition and of

being treated mathematically, like birth- and death-rates, and the other topics with which actuaries are concerned.

2.Historical inquiry into the rates with which the various classes of society (classified according to civic usefulness) have contributed to the population at various times, in ancient and modern nations. There is strong reason for believing that national rise and decline is closely connected with this influence. It seems to be the tendency of high civilization to check fertility in the upper classes through numerous causes, some of which are well known, others are inferred, and others again are wholly obscure. The latter class are apparently analogous to those which bar the fertility of most species of wild animals in zoological gardens. Out of the hundreds and thousands of species that [4] have been tamed, very few indeed are fertile when their liberty is restricted and their struggles for livelihood are abolished; those which are so, and are otherwise useful to man, becoming domesticated. There is perhaps some connection between this obscure action and the disappearance of most savage races when brought into contact with high civilization, though there are other and well-known concomitant causes. But while most barbarous races disappear, some, like the negro, do not. It may therefore be expected that types of our race will be found to exist which can be highly civilized without losing fertility; nay, they may become more fertile under artificial conditions, as is the case with many domestic animals.

3. Systematic collection of facts showing the circumstances under which large and thriving families have most frequently originated; in other words, the *conditions* of eugenics. The definition of a thriving family, that will pass muster for the moment at least, is one in which the children have gained distinctly superior positions to those who were their classmates in early life. Families may be considered "large" that contain not less than three adult male children. It would be no great burden to a society including many members who had eugenics at heart, to initiate and to preserve a large collection of such records for the use of statistical students. The committee charged with the task would have to consider very carefully the form of their circular and the persons intrusted [*sic*] to distribute it. They should ask only for as much useful information as could be easily, and would be readily supplied by any member of the family appealed to. The point to be ascertained is the *status* of the two parents at the time of their marriage, whence its more or less eugenic character might have been predicted, if the larger knowledge that we now hope to obtain had then existed. Some account would be wanted of their race, profession, and residence; also of their own respective parentages, and of their brothers and sisters. Finally the reasons would be required why the children deserved to be entitled a "thriving" family. This manuscript collection might hereafter develop into a " golden book" of thriving families.... [5] The act of systematically collecting records of thriving families would have the further advantage of familiarizing the public with the fact that eugenics had at length become a subject of serious scientific study by an energetic society.

4. Influences affecting marriage.... The passion of love seems so overpowering that it may be thought folly to try to direct its course. But plain facts do not confirm this view. Social influences of all kinds have immense power in the end, and they are very various. If unsuitable marriages from the eugenic point of view were banned socially, or even regarded with the unreasonable disfavor which some attach to cousin-marriages, very few would be made. The multitude of marriage restrictions that have proved prohibitive among uncivilized people would require a volume to describe.

5. Persistence in setting forth the national importance of eugenics. There are three stages to be passed through: (1) It must be made familiar as an academic question, until its exact importance has been understood and accepted as a fact. (2) It must be recognized as a subject whose practical development deserves serious consideration. (3) It must be introduced into the national conscience, like a new religion. It has, indeed, strong claims to become an orthodox religious tenet of the future, for eugenics co-operate with the workings of nature by securing that humanity shall be represented by the fittest races. What nature does blindly, slowly, and ruthlessly, man may do providently, quickly, and kindly. As it lies within his power, so it becomes his duty to work in that direction. The improvement of our stock seems to me one of the highest objects that we can [6] reasonably attempt. We are ignorant of the ultimate destinies of humanity, but feel perfectly sure that it is as noble a work to raise its level, in the sense already explained, as it would be disgraceful to abase it. I see no impossibility in eugenics becoming a religious dogma among mankind, but its details must first be worked out sedulously in the study. Overzeal leading to hasty action would do harm, by holding out expectations of a near golden age, which will certainly be falsified and cause the science to be discredited. The first and main point is to secure the general intellectual acceptance of eugenics as a hopeful and most important study. Then let its principles work into the heart of the nation, which will gradually give practical effect to them in ways that we may not wholly foresee.

62. *Buck v. Bell*

On March 20, 1924, Virginia's General Assembly passed a statute authorizing the involuntary sterilization of state institution inmates deemed to be feebleminded, including those diagnosed as "insane, idiotic, imbecile, or epileptic."

On September 10, 1924, Albert Priddy, superintendant of the Virginia Colony for Epileptics and Feeble-Minded—seeking a test of the law's constitutionality—asked the Colony's board of directors to order the sterilization of Carrie Buck, a seventeen-year-old inmate. She had recently given birth to an illegitimate daughter and was deemed promiscuous and feebleminded. A legal challenge was arranged and filed on Carrie's behalf.

Previous state laws had been legally flawed, but Virginia's was carefully crafted after a model law published by Harry Laughlin, director of the Eugenics Record Office in Cold Spring Harbor, New York. In the state court, a field worker from the ERO testified that Carrie's seven-month-old child, Vivian, "showed backwardness." Priddy testified that the members of the Buck family "belong to the shiftless, ignorant, and worthless class of anti-social whites of the South."

The Virginia Court of Appeals ruled for the Colony (143 Va. 310), and the case was further appealed to the U.S. Supreme Court. The famous opinion affirming the Virginia law was written by Chief Justice Oliver Wendell Holmes and is printed here. It has been subsequently shown that Carrie was not at all promiscuous, but had been raped by a nephew of Carrie's foster parents during their absence from the home. Furthermore, Vivian subsequently performed normally in school and even made the honor roll.

In 1942, the Supreme Court struck down an Oklahoma sterilization statute allowing the sterilization of criminals. Eugenic sterilization in principle, however, has never been tested in the courts and has thus not been deemed unconstitutional.

274 U.S. 200, Buck v. *Bell, Opinion of the Court*

[205] This is a writ of error to review a judgment of the Supreme Court of Appeals of the State of Virginia affirming a judgment of the Circuit Court of Amherst County by which the defendant in error, the superintendent of the State Colony for Epileptics and Feeble Minded, was ordered to perform the operation of salpingectomy upon Carrie Buck, the plaintiff in error, for the purpose of making her sterile. 143 Va. 310. The case comes here upon the contention that the statute authorizing the judgment is void under the Fourteenth Amendment as denying to the plaintiff in error due process of law and the equal protection of the laws.

Carrie Buck is a feeble minded white woman who was committed to the State Colony above mentioned in due form. She is the daughter of a feeble minded mother in the same institution, and the mother of an illegitimate feeble minded child. She was eighteen years old at the time of the trial of her case in the Circuit Court, in the latter part of 1924. An Act of Virginia, approved March 20, 1924, recites that the health of the patient and the welfare of society may be promoted in certain cases by the sterilization of mental defectives, under careful safeguard, &c.; that the sterilization may be effected in males by vasectomy and in females by salpingectomy, without serious pain or substantial danger to life; that the Commonwealth is supporting in various institutions many defective persons who, if now discharged, would become [206] a menace, but, if incapable of procreating, might be discharged with safety and become self-supporting with benefit to themselves and to society, and that experience has shown that heredity plays an important part in the transmission of insanity, imbecility, &c. The statute then enacts that, whenever the superintendent of certain institutions, including the above-named State Colony, shall be of opinion that it is for the best interests of the patients and of society that an inmate under his care should be sexually sterilized, he may have the operation performed upon any patient afflicted with hereditary forms of insanity, imbecility, &c., on complying with the very careful provisions by which the act protects the patients from possible abuse.

The superintendent first presents a petition to the special board of directors of his hospital or colony, stating the facts and the grounds for his opinion, verified by affidavit. Notice of the petition and of the time and place of the hearing in the institution is to be served upon the inmate, and also upon his guardian, and if there is no guardian, the superintendent is to apply to the Circuit Court of the County to appoint one. If the inmate is a minor, notice also is to be given to his parents, if any, with a copy of the petition. The board is to see to it that the inmate may attend the hearings if desired by him or his guardian. The evidence is all to be reduced to writing, and, after the board has made its order for or against the operation, the superintendent, or the inmate, or his guardian, may appeal to the Circuit Court of the County. The Circuit Court may consider the record of the board and the evidence before it and such other admissible evidence as may be offered, and may affirm, revise, or reverse the order of the board and enter such order as it deems just. Finally any party may apply to the Supreme Court of Appeals, which, if it grants the appeal, is to hear the case upon the record of the trial [207] in the Circuit Court, and may enter such order as it thinks the Circuit Court should have entered. There can be no doubt that, so far as procedure is concerned, the rights of the patient are most carefully considered, and, as every step in this case was taken in scrupulous compliance with the statute and after months of observation, there is no doubt that, in that respect, the plaintiff in error has had due process of law.

The attack is not upon the procedure, but upon the substantive law. It seems to be contended that in no circumstances could such an order be justified. It certainly is contended that the order cannot be justified upon the existing grounds. The judgment finds the facts that have been recited, and that Carrie Buck

> is the probable potential parent of socially inadequate offspring, likewise afflicted, that she may be sexually sterilized without detriment to her general health, and that her welfare and that of society will be promoted by her sterilization,

and thereupon makes the order. In view of the general declarations of the legislature and the specific findings of the Court, obviously we cannot say as matter of law that the grounds do not exist, and, if they exist, they justify the result. We have seen more than once that the public welfare may call upon the best citizens for their lives. It would be strange if it could not call upon those who already sap the strength of the State for these lesser sacrifices, often not felt to be such by those concerned, in order to prevent our being swamped with incompetence. It is better for all the world if, instead of waiting to execute degenerate offspring for crime or to let them starve for their imbecility, society can prevent those who are manifestly unfit from continuing their kind. The principle that sustains compulsory vaccination is broad enough to cover cutting the Fallopian tubes. *Jacobson v. Massachusetts,* 197 U.S. 11. Three generations of imbeciles are enough. [208]

But, it is said, however it might be if this reasoning were applied generally, it fails when it is confined to the small number who are in the institutions named and is not applied to the multitudes outside. It is the usual last resort of constitutional arguments to point out shortcomings of this sort. But the answer is that the law does all that is needed when it does all that it can, indicates a policy, applies it to all within the lines, and seeks to bring within the lines all similarly situated so far and so fast as its means allow. Of course, so far as the operations enable those who otherwise must be kept confined to be returned to the world, and thus open the asylum to others, the equality aimed at will be more nearly reached.

Judgment affirmed.

MR. JUSTICE BUTLER dissents.

Ideas to Think About

1. Analyze Spencer's extension of the concept of "progress": first, as applied to biology, is the homogeneity-to-heterogeneity claim a valid one (e.g., are humans more heterogeneous than other forms, because some possess more chromosomes and genes)? Second, are there flaws in asserting the human social condition as analogous? Finally, compare "progress" in Spencer's sense to "complexity."
2. In what way does Sumner extend this concept of "progress" and assign value to it? In his use of the terms *labor, "capital,* and *wealth,* what legacy does he owe to Adam Smith? How does he go beyond this?
3. Is there any logical connection between the Social Darwinian concepts of Spencer and Sumner and the eugenics argument of Galton? Is there any inherent contradiction (e.g., in the antisocialist position)? Can you find in Galton any support from the classical Greek position on man and society?
4. How much of the opprobrium directed toward eugenics is based on the unsubstantiated or faulty genetic argument? Much of the modern eugenics movement has been directed toward voluntary reproductive restraint. How do you respond to this?

For Further Reading

Social Darwinism in American Thought, Richard Hofstadter. Boston: Beacon Press (1983). First published in 1944, this doctoral dissertation remains the definitive work on the history of movements to apply biological evolutionary principles to human society.

Darwin's Coat-Tails: Essays on Social Darwinism, Paul Crook. New York: Peter Lang Publishing Group (2007). The author's long-term research into the extensions of Darwinism is captured in this series of essays. The subject matter of Social Darwinism is unusually broad here, as it attempted to justify capitalism and socialism, war and peace, and racism and the oppressive philosophy of the Third Reich.

UNIT 13

Evolution and Religion Revisited: Creationism

> Our willingness to accept scientific claims that are against common sense is the key to an understanding of the real struggle between science and the supernatural. We take the side of science in spite of the patent absurdity of some of its constructs, in spite of its failure to fulfill many of its extravagant promises of health and life, in spite of the tolerance of the scientific community for unsubstantiated just-so stories, because we have a prior commitment, a commitment to materialism.
>
> —RICHARD C. LEWONTIN, *1997*

The creationists are slack-jawed in amazement at Lewontin's admission. The admission, they say, confirms professional doubts about the validity of science, and consequently about the validity of its theories and assertions (see Phil Johnson's article, *infra*). However, many scientific claims—as far back as those of Copernicus—were "against common sense" because they were counterintuitive. Some scientific constructs are indeed "absurd," as was the oxygen theory of combustion in the face of the more logical phlogiston theory. The fact that intuitive reality can be trumped by experiment is testimony to the staying power of "a commitment to materialism." Science has its detractors, but Lewontin is not one of them.

Creationism—and particularly its more recent incarnation as intelligent design—itself makes two "scientific" assertions: (1) that the complexity of biological organization is the result of intentional design, and (2) that naturalism/materialism is inadequate to account for species emergence. On the surface, it would seem that these two assertions are unrelated, that neither has any necessary implication for the other.

A moment's reflection, however, reveals a strong epistemological connection. The former assertion seeks to affirm that logical analysis and systematic inquiry can demonstrate the scientific validity of the design argument, which is itself much more sophisticated than its earlier form in the seventeenth and eighteenth centuries. The latter assertion seeks to demonstrate that conventional science cannot completely close the knowledge gap between the known and the imagined.

Either of the two—confirmation of evidence entailing intelligent design, or disconfirmation of evidence supporting biological evolution—would open the pathway toward a redefinition of science, in which naturalism and supernaturalism become joined. The two are only partially evidence-based.

Much of the evolution-creation debate over the past four decades has consisted in exchange of claims and counterclaims on the evidence. Creationist claims that some organs or biochemical pathways are irreducibly complex—that incremental selective advantage of interdependent parts could not assemble such structures or processes—focused first on the mammalian eye (which both Paley and Darwin were concerned about), then on the physiological cascade in the pathways to blood coagulation, and most recently on the protein assembly line for the bacterial flagellum (see Behe 1996). Marine biologists and biochemists, respectively, have demonstrated that designating these pathways as irreducibly complex is invalid. Other efforts on the part of creationists have attempted to demonstrate that specified complexity characterizes many biological phenomena, particularly when biological functions are viewed as sophisticated information-processing systems, and that only an intelligent agent can be responsible. Dembski's "complex specified information" is a prime example of this (Dembski 1998, 1999). Evidentiary claims against evolution have included the sudden appearance of phyla in the Early Cambrian, long periods of stasis in many species, and lack of transitional forms in the fossil record—this latter addressed by Darwin in his initial publication.

To review the arguments on these factual and epistemological topics here would involve levels of arcane knowledge too detailed for our purpose. Furthermore, the evolution-creation debate has its recent history and its epistemological terms focused squarely on public education, not on philosophical preference, and this educational aspect is beyond our purpose, notwithstanding its critical importance. Perhaps most importantly, however, is the fact that intelligent design/creationism is not a scientific theory at all and does not merit any consideration here as such.

There is another—and more fundamental—issue to the evolution-creation debate that *is* worthy of our attention. It strikes at the historical core of evolution, indeed of science itself. This metaphysical issue has become central to intellectual exchange among the participants. It is the subject of the Lewontin quote ([1997], see this unit's epigraph) and of Johnson's response, and it asks the question: Is the fact that science is based on naturalism sufficiently arbitrary that supernatural cause might legitimately be included in a redefinition, or is the conduct of science so methodologically specific as to exclude the supernatural, and ought it to be so?

Although this is an issue seldom addressed by scientists, it is an important one for creationists, who seek the legitimacy of feeding at the same table. The issue lies below the surface—but causative—in the popular distrust of science and in the school debates rolling across the country on teaching evolution. Moreover, because scientists cannot afford to glibly dismiss negative public opinion of science and evolution as due to lack of education, it is important to focus on the more fundamental conceptual issues.

This focus is provided in the selections for this unit.

63. STEPHEN JAY GOULD

Nonoverlapping Magisteria

Stephen Jay Gould's untimely death, in 2002, at age sixty deprived the world of perhaps the greatest science popularizer since Thomas Huxley. A master essayist of science writing, Gould's "This View of Life" column in *Natural History* magazine began in 1974 and featured thought-provoking—and often iconoclastic—interpretations of life's history. He argued forcefully for the contingency, rather than the predictability, of evolution. He was convincing in denying evolution any progressive trend.

He is most famous for the theory of "punctuated equilibrium," coauthored with Niles Eldridge. This is the argument that most evolutionary sequences have consisted of long periods of stasis followed by short and abrupt periods of branching speciation. This theory, opposing the gradual model of Darwin, has been (mis-)used by creationists as evidence against macroevolution.

Born in Queens, New York, on September 10, 1941, Gould's career was predominantly spent at Harvard, where he held professorships in paleontology and geology, and where he was curator of invertebrate paleontology at Harvard's Museum of Comparative Zoology. He was also visiting professor of biology at New York University.

The following essay, widely reprinted, argues against the incompatibility of religious belief and acceptance of evolution. It firmly places Gould on the humanistic side of a contentious debate among evolutionists, with Oxford's Richard Dawkins representing the strict naturalist side.

Incongruous places often inspire anomalous stories. In early 1984, I spent several nights at the Vatican housed in a hotel built for itinerant priests. While pondering over such puzzling issues as the intended function of the bidets in each bathroom, and hungering for something other than plum jam on my breakfast rolls (why did the basket only contain hundreds of identical plum packets and not a one of, say, strawberry?), I encountered yet another among the innumerable issues of contrasting cultures that can make life so interesting. Our crowd (present in Rome for a meeting on nuclear winter sponsored by the Pontifical Academy of Sciences) shared the hotel with a group of French and Italian Jesuit priests who were also professional scientists.

At lunch, the priests called me over to their table to pose a problem that had been troubling them. What, they wanted to know, was going on in America with all this talk about "scientific creationism"? One asked me: "Is evolution really in some kind of trouble. and if so, what could such trouble be? I have always been taught that no doctrinal conflict exists between evolution and Catholic faith, and the evidence for evolution seems both entirely satisfactory and utterly overwhelming. Have I missed something?"

A lively pastiche of French, Italian, and English conversation then ensued for half an hour or so, but the priests all seemed reassured by my general answer: Evolution has encountered no intellectual trouble; no new arguments have been offered. Creationism is a homegrown phenomenon of American sociocultural history—a splinter movement (unfortunately rather more of a beam these days) of Protestant fundamentalists who believe that every word of the Bible must be literally true, whatever such a claim might mean. We all left satisfied, but I certainly felt bemused by the anomaly of my role as a Jewish agnostic, trying to reassure a group of Catholic priests that evolution remained both true and entirely consistent with religious belief.

Another story in the same mold: I am often asked whether I ever encounter creationism as a live issue among my Harvard undergraduate students. I reply that only once, in nearly thirty years of teaching, did I experience such an incident. A very sincere and serious freshman student came to my office hours with the following question that had clearly been troubling him deeply: "I am a devout Christian and have never had any reason to doubt evolution, an idea that seems both exciting and particularly well documented. But my roommate, a proselytizing Evangelical, has been insisting with enormous vigor that I cannot be both a real Christian and an evolutionist. So tell me, can a person believe both in God and evolution?" Again, I gulped hard, did my intellectual duty, and reassured him that evolution was both true and entirely compatible with Christian belief—a position I hold sincerely, but still an odd situation for a Jewish agnostic.

These two stories illustrate a cardinal point, frequently unrecognized but absolutely central to any understanding of the status and impact of the politically potent, fundamentalist doctrine known by its self-proclaimed oxymoron as "scientific creationism"—the claim that the Bible is literally true, that all organisms were created during six days of twenty-four hours, that the earth is only a few thousand years old, and that evolution must therefore be false. Creationism does not pit science against religion (as my opening stories indicate), for no such conflict exists. Creationism does not raise any unsettled intellectual issues about the nature of biology or the history of life. Creationism is a local and parochial movement, powerful only in the United States among Western nations, and prevalent only among the few sectors of American Protestantism that choose to read the Bible as an inerrant document, literally true in every jot and tittle.

I do not doubt that one could find an occasional nun who would prefer to teach creationism in her parochial school biology class or an occasional orthodox rabbi who does the same in his yeshiva, but creationism based on biblical literalism makes little sense in either Catholicism or Judaism for neither religion maintains any extensive tradition for reading the Bible as literal truth rather than illuminating literature, based partly on metaphor and allegory (essential components of all good writing) and demanding interpretation for proper understanding. Most Protestant groups, of course, take the same position—the fundamentalist fringe notwithstanding.

The position that I have just outlined by personal stories and general statements represents the standard attitude of all major Western religions (and of Western science) today. (I cannot, through ignorance, speak of Eastern religions, although I suspect that the same position would prevail in most cases.) The lack of conflict between science and religion arises from a lack of overlap between their respective domains of professional expertise—science in the empirical constitution of the universe, and religion in the search for proper ethical values and the spiritual meaning of our lives. The attainment of wisdom in a full life requires extensive attention to both domains—for a great book tells us that the truth can make us free and that we will live in optimal harmony with our fellows when we learn to do justly, love mercy, and walk humbly.

In the context of this standard position, I was enormously puzzled by a statement issued by Pope John Paul II on October 22, 1996, to the Pontifical Academy of Sciences, the same body that had sponsored my earlier trip to the Vatican. In this document, entitled "Truth Cannot Contradict Truth," the pope defended both the evidence for evolution and the consistency of the theory with Catholic religious doctrine. Newspapers throughout the world responded with frontpage headlines, as in the *New York Times* for October 25:

"Pope Bolsters Church's Support for Scientific View of Evolution."

Now I know about "slow news days" and I do admit that nothing else was strongly competing for headlines at that particular moment. (The *Times* could muster nothing more exciting for a lead story than Ross Perot's refusal to take Bob Dole's advice and quit the presidential race.) Still, I couldn't help feeling immensely puzzled by all the attention paid to the pope's statement (while being wryly pleased, of course, for we need all the good press we can get, especially from respected outside sources). The Catholic Church had never opposed evolution and had no reason to do so. Why had the pope issued such a statement at all? And why had the press responded with an orgy of worldwide, front-page coverage?

I could only conclude at first, and wrongly as I soon learned, that journalists throughout the world must deeply misunderstand the relationship between science and religion, and must therefore be elevating a minor papal comment to unwarranted notice. Perhaps most people really do think that a war exists between science and religion, and that (to cite a particularly newsworthy case) evolution must be intrinsically opposed to Christianity. In such a context, a papal admission of evolution's legitimate status might be regarded as major news indeed—a sort of modern equivalent for a story that never happened, but would have made the biggest journalistic splash of 1640: Pope Urban VIII releases his most famous prisoner from house arrest and humbly apologizes, "Sorry, Signor Galileo... the sun, er, is central."

But I then discovered that the prominent coverage of papal satisfaction with evolution had not been an error of non-Catholic Anglophone journalists. The Vatican itself had issued the statement as a major news release. And Italian newspapers had featured, if anything, even bigger headlines and longer stories. The conservative Il Giornale, for example, shouted from its masthead: "Pope Says We May Descend from Monkeys."

Clearly, I was out to lunch. Something novel or surprising must lurk within the papal statement but what could it be?—especially given the accuracy of my primary impression (as I later verified) that the Catholic Church values scientific study, views science as no threat to religion in general or Catholic doctrine in particular, and has long accepted both the legitimacy of evolution as a field of study and the potential harmony of evolutionary conclusions with Catholic faith.

As a former constituent of Tip O'Neill's, I certainly know that "all politics is local"—and that the Vatican undoubtedly has its own internal reasons, quite opaque to me, for announcing papal support of evolution in a major statement. Still, I knew that I was missing some important key, and I felt frustrated. I then remembered the primary rule of intellectual life: when puzzled, it never hurts to read the primary documents—a rather simple and self-evident principle that has, nonetheless, completely disappeared from large sectors of the American experience.

I knew that Pope Pius XII (not one of my favorite figures in twentieth-century history, to say the least) had made the primary statement in a 1950 encyclical entitled Humani Generis. I knew the main thrust of his message: Catholics could believe whatever science determined about the evolution of the human body, so long as they accepted that, at some time of his choosing, God had infused the soul into such a creature. I also knew that I had no problem with this statement, for whatever my private beliefs about souls, science cannot touch such a subject and therefore cannot be threatened by any theological position on such a legitimately and intrinsically religious issue. Pope Pius XII, in other words, had properly acknowledged and respected the separate domains of science and theology. Thus, I found myself in total agreement with Humani Generis—but I had never read the document in full (not much of an impediment to stating an opinion these days).

I quickly got the relevant writings from, of all places, the Internet. (The pope is prominently on-line, but a Luddite like me is not. So I got a computer-literate associate to dredge up the documents. I do love the fracture of stereotypes implied by finding religion so hep and a scientist so square.) Having now read in full both Pope Pius's Humani Generis of 1950 and Pope John Paul's proclamation of October 1996, I finally understand why the recent statement seems so new, revealing, and worthy of all those headlines. And the message could not be more welcome for evolutionists and friends of both science and religion.

The text of Humani Generis focuses on the magisterium (or teaching authority) of the Church—a word derived not from any concept of majesty or awe but from the different notion of teaching, for magister is Latin for "teacher." We may, I think, adopt this word and concept to express the central point of this essay and the principled resolution of supposed "conflict" or "warfare" between science and religion. No such conflict should exist because each subject has a legitimate magisterium, or domain of teaching authority—and these magisteria do not overlap (the principle that I would like to designate as NOMA, or "nonoverlapping magisteria").

The net of science covers the empirical universe: what is it made of (fact) and why does it work this way (theory). The net of religion extends over questions of moral meaning and value. These two magisteria do not overlap, nor do they encompass all inquiry (consider, for starters, the magisterium of art and the meaning of beauty). To cite the arch cliches, we get the age of rocks, and religion retains the rock of ages; we study how the heavens go, and they determine how to go to heaven.

This resolution might remain all neat and clean if the nonoverlapping magisteria (NOMA) of science and religion were separated by an extensive no man's land. But, in fact, the two magisteria bump right up against each other, interdigitating in wondrously complex ways along their joint border. Many of our deepest questions call upon aspects of both for different parts of a full answer—and the sorting of legitimate domains can become quite complex and difficult. To cite just two broad questions involving both evolutionary facts and moral arguments: Since evolution made us the only earthly creatures with advanced consciousness, what responsibilities are so entailed for our relations with other species? What do our genealogical ties with other organisms imply about the meaning of human life?

Pius XII's Humani Generis is a highly traditionalist document by a deeply conservative man forced to face all the "isms" and cynicisms that rode the wake of World War II and informed the struggle to rebuild human decency from the ashes of the Holocaust. The encyclical, subtitled "Concerning some false opinions which threaten to undermine the foundations of Catholic doctrine" begins with a statement of embattlement:

Disagreement and error among men on moral and religious matters have always been a cause of profound sorrow to all good men, but above all to the true and loyal sons of the Church, especially today, when we see the principles of Christian culture being attacked on all sides.

Pius lashes out, in turn, at various external enemies of the Church: pantheism, existentialism, dialectical materialism, historicism, and of course and preeminently, communism. He then notes with sadness that some well-meaning folks within the Church have fallen into a dangerous relativism—"a theological pacifism and egalitarianism, in which all points of view become equally valid"—in order to include people of wavering faith who yearn for the embrace of Christian religion but do not wish to accept the particularly Catholic magisterium.

What is this world coming to when these noxious novelties can so discombobulate a revealed and established order? Speaking as a conservative's conservative, Pius laments:

> Novelties of this kind have already borne their deadly fruit in almost all branches of theology.... Some question whether angels are personal beings, and whether matter and spirit differ essentially.... Some even say

> that the doctrine of Transubstantiation, based on an antiquated philosophic notion of substance, should be so modified that the Real Presence of Christ in the Holy Eucharist be reduced to a kind of symbolism.

Pius first mentions evolution to decry a misuse by overextension often promulgated by zealous supporters of the anathematized "isms":

> Some imprudently and indiscreetly hold that evolution... explains the origin of all things.... Communists gladly subscribe to this opinion so that, when the souls of men have been deprived of every idea of a personal God, they may the more efficaciously defend and propagate their dialectical materialism.

Pius's major statement on evolution occurs near the end of the encyclical in paragraphs 35 through 37. He accepts the standard model of NOMA and begins by acknowledging that evolution lies in a difficult area where the domains press hard against each other. "It remains for US now to speak about those questions which, although they pertain to the positive sciences, are nevertheless more or less connected with the truths of the Christian faith." [Interestingly, the main thrust of these paragraphs does not address evolution in general but lies in refuting a doctrine that Pius calls "polygenism," or the notion of human ancestry from multiple parents—for he regards such an idea as incompatible with the doctrine of original sin, "which proceeds from a sin actually committed by an individual Adam and which, through generation, is passed on to all and is in everyone as his own." In this one instance, Pius may be transgressing the NOMA principle—but I cannot judge, for I do not understand the details of Catholic theology and therefore do not know how symbolically such a statement may be read. If Pius is arguing that we cannot entertain a theory about derivation of all modern humans from an ancestral population rather than through an ancestral individual (a potential fact) because such an idea would question the doctrine of original sin (a theological construct), then I would declare him out of line for letting the magisterium of religion dictate a conclusion within the magisterium of science.]

Pius then writes the well-known words that permit Catholics to entertain the evolution of the human body (a factual issue under the magisterium of science), so long as they accept the divine Creation and infusion of the soul (a theological notion under the magisterium of religion):

> The Teaching Authority of the Church does not forbid that, in conformity with the present state of human sciences and sacred theology, research and discussions, on the part of men experienced in both fields, take place with regard to the doctrine of evolution, in as far as it inquires into the origin of the human body as coming from pre-existent and living matter—for the Catholic faith obliges us to hold that souls are immediately created by God.

I had, up to here, found nothing surprising in *Humani Generis*, and nothing to relieve my puzzlement about the novelty of Pope John Paul's recent statement. But I read further and realized that Pope Pius had said more about evolution, something I had never seen quoted, and that made John Paul's statement most interesting indeed. In short, Pius forcefully proclaimed that while evolution may be legitimate in principle, the theory, in fact, had not been proven and might well be entirely wrong. One gets the strong impression, moreover, that Pius was rooting pretty hard for a verdict of falsity. Continuing directly from the last quotation, Pius advises us about the proper study of evolution:

> However, this must be done in such a way that the reasons for both opinions, that is, those favorable and those unfavorable to evolution, be weighed and judged with the necessary seriousness, moderation and measure.... Some, however, rashly transgress this liberty of discussion, when they act as if the origin of the human body from pre-existing and living matter were already completely certain and proved by the facts which have been discovered up to now and by reasoning on those facts, and as if there were nothing in the sources of divine revelation which demands the greatest moderation and caution in this question.

To summarize, Pius generally accepts the NOMA principle of nonoverlapping magisteria in permitting Catholics to entertain the hypothesis of evolution for the human body so long as they accept the divine infusion of the soul. But he then offers some (holy) fatherly advice to scientists about the status of evolution as a scientific concept: the idea is not yet proven, and you all need to be especially cautious because evolution raises many troubling issues right on the border of my magisterium. One may read this second theme in two different ways: either as a gratuitous incursion into a different magisterium or as a helpful perspective from an intelligent and concerned outsider. As a man of good will, and in the interest of conciliation, I am happy to embrace the latter reading.

In any case, this rarely quoted second claim (that evolution remains both unproven and a bit dangerous)—and not the familiar first argument for the NOMA principle (that Catholics may accept the evolution of the body so long as they embrace the creation of the soul)—defines the novelty and the interest of John Paul's recent statement.

John Paul begins by summarizing Pius's older encyclical of 1950, and particularly by reaffirming the NOMA principle—nothing new here, and no cause for extended publicity:

> In his encyclical Humani Generis (1950), my predecessor Pius XII had already stated that there was no opposition between evolution and the doctrine of the faith about man and his vocation.

To emphasize the power of NOMA, John Paul poses a potential problem and a sound resolution: How can we reconcile science's claim for physical continuity in human evolution with Catholicism's insistence that the soul must enter at a moment of divine infusion:

> With man, then, we find ourselves in the presence of an ontological difference, an ontological leap, one could say. However, does not the posing of such ontological discontinuity run counter to that physical continuity which seems to be the main thread of research into evolution in the field of physics and chemistry? Consideration of the method used in the various branches of knowledge makes it possible to reconcile two points of view which would seem irreconcilable. The sciences of observation describe and measure the multiple manifestations of life with increasing precision and correlate them with the time line. The moment of transition to the spiritual cannot be the object of this kind of observation.

The novelty and news value of John Paul's statement lies, rather, in his profound revision of Pius's second and rarely quoted claim that evolution, while conceivable in principle and reconcilable with religion, can cite little persuasive evidence, and may well be false. John Paul states—and I can only say amen, and thanks for noticing—that the half century between Pius's surveying the ruins of World War II and his own pontificate heralding the dawn of a new millennium has witnessed such a growth of data, and such a refinement of theory, that evolution can no longer be doubted by people of good will:

> Pius XII added...that this opinion [evolution] should not be adopted as though it were a certain, proven doctrine....Today, almost half a century after the publication of the encyclical, new knowledge has led to the recognition of more than one hypothesis in the theory of evolution. It is indeed remarkable that this theory has been progressively accepted by researchers, following a series of discoveries in various fields of knowledge. The convergence, neither sought nor fabricated, of the results of work that was conducted independently is in itself a significant argument in favor of the theory.

In conclusion, Pius had grudgingly admitted evolution as a legitimate hypothesis that he regarded as only tentatively supported and potentially (as I suspect he hoped) untrue. John Paul, nearly fifty years later, reaffirms the legitimacy of evolution under the NOMA principle—no news here—but then adds that additional data and theory have placed the factuality of evolution beyond reasonable doubt. Sincere Christians must now accept evolution not merely as a plausible possibility but also as an effectively proven fact. In other words, official Catholic opinion on evolution has moved from "say it ain't so, but we can deal with it if we have to" (Pius's grudging view of 1950) to John Paul's entirely welcoming "it has been proven true; we always celebrate nature's factuality, and we look forward to interesting discussions of theological implications." I happily endorse this turn of events as gospel—literally "good news." I may represent the magisterium of science, but I welcome the support of a primary leader from the other major magisterium of our complex lives. And I recall the wisdom of King Solomon: "As cold waters to a thirsty soul, so is good news from a far country" (Prov. 25:25).

Just as religion must bear the cross of its hard-liners, I have some scientific colleagues, including a few prominent enough to wield influence by their writings, who view this rapprochement of the separate magisteria with dismay. To colleagues like me—agnostic scientists who welcome and celebrate thc rapprochement, especially the pope's latest statement—they say: "C'mon, be honest; you know that religion is addle-pated, superstitious, old-fashioned b.s.; you're only making those welcoming noises because religion is so powerful, and we need to be diplomatic in order to assure public support and funding for science." I do not think that this attitude is common among scientists, but such a position fills me with dismay—and I therefore end this essay with a personal statement about religion, as a testimony to what I regard as a virtual consensus among thoughtful scientists (who support the NOMA principle as firmly as the pope does).

I am not, personally, a believer or a religious man in any sense of institutional commitment or practice. But I have enormous respect for religion, and the subject has always fascinated me, beyond almost all others (with a few exceptions, like evolution, paleontology, and baseball). Much of this fascination lies in the historical paradox that throughout Western history organized religion has fostered both the most unspeakable horrors and the most heart-rending examples of human goodness in the face of personal danger. (The evil, I believe, lies in the occasional confluence of religion with secular power. The Catholic Church has sponsored its share of horrors, from Inquisitions to liquidations—but only because this institution held such secular power during so much of Western history. When my folks held similar power more briefly in Old Testament times, they committed just as many atrocities with many of the same rationales.)

I believe, with all my heart, in a respectful, even loving concordat between our magisteria—the NOMA solution. NOMA represents a principled position on moral and intellectual grounds, not a mere diplomatic stance. NOMA also cuts both ways. If religion can no longer dictate the nature of factual conclusions properly under the magisterium of science, then scientists cannot claim higher insight into moral truth from any superior knowledge of the world's empirical constitution. This mutual humility has important practical consequences in a world of such diverse passions.

Religion is too important to too many people for any dismissal or denigration of the comfort still sought by many folks from theology. I may, for example, privately suspect that papal insistence on divine infusion of the soul represents a sop to our fears, a device for maintaining a belief in human superiority within an evolutionary world offering no privileged position to any creature. But

I also know that souls represent a subject outside the magisterium of science. My world cannot prove or disprove such a notion, and the concept of souls cannot threaten or impact my domain. Moreover, while I cannot personally accept the Catholic view of souls, I surely honor the metaphorical value of such a concept both for grounding moral discussion and for expressing what we most value about human potentiality: our decency, care, and all the ethical and intellectual struggles that the evolution of consciousness imposed upon us.

As a moral position (and therefore not as a deduction from my knowledge of nature's factuality), I prefer the "cold bath" theory that nature can be truly "cruel" and "indifferent"—in the utterly inappropriate terms of our ethical discourse—because nature was not constructed as our eventual abode, didn't know we were coming (we are, after all, interlopers of the latest geological microsecond), and doesn't give a damn about us (speaking metaphorically). I regard such a position as liberating, not depressing, because we then become free to conduct moral discourse—and nothing could be more important—in our own terms, spared from the delusion that we might read moral truth passively from nature's factuality.

But I recognize that such a position frightens many people, and that a more spiritual view of nature retains broad appeal (acknowledging the factuality of evolution and other phenomena, but still seeking some intrinsic meaning in human terms, and from the magisterium of religion). I do appreciate, for example, the struggles of a man who wrote to the *New York Times* on November 3, 1996, to state both his pain and his endorsement of John Paul's statement:

> Pope John Paul II's acceptance of evolution touches the doubt in my heart. The problem of pain and suffering in a world created by a God who is all love and light is hard enough to bear, even if one is a creationist. But at least a creationist can say that the original creation, coming from the hand of God was good, harmonious, innocent and gentle. What can one say about evolution, even a spiritual theory of evolution? Pain and suffering, mindless cruelty and terror are its means of creation. Evolution's engine is the grinding of predatory teeth upon the screaming, living flesh and bones of prey.... If evolution be true, my faith has rougher seas to sail.

I don't agree with this man, but we could have a wonderful argument. I would push the "cold bath" theory: he would (presumably) advocate the theme of inherent spiritual meaning in nature, however opaque the signal. But we would both be enlightened and filled with better understanding of these deep and ultimately unanswerable issues. Here, I believe, lies the greatest strength and necessity of NOMA, the nonoverlapping magisteria of science and religion. NOMA permits—indeed enjoins—the prospect of respectful discourse, of constant input from both magisteria toward the common goal of wisdom. If human beings are anything special, we are the creatures that must ponder and talk. Pope John Paul II would surely point out to me that his magisterium has always recognized this distinction, for "in principio, erat verbum"—"In the beginning was the Word."

---

64. Phillip Johnson

Phillip Johnson received his bachelor of arts in English literature at Harvard in 1961, and he most certainly knew of Stephen Jay Gould even if he took no courses from him. Johnson was not at that time much interested in religion. Raised as a "nominal Christian" in a liberal Congregationalist church in Illinois, he became a secular, "nominal agnostic."

From Harvard, Johnson studied law at the University of Chicago and served as law clerk for Supreme Court Chief Justice Earl Warren. Highly recruited by the faculty of Yale and Berkeley, he chose the latter, joining the Boalt Law School faculty in 1967. He retired in 2000.

Johnson, following a divorce and remarriage, converted to Christianity at Berkeley. On sabbatical in London, he read Dawkins's antitheistic *The Blind Watchmaker*, "which seemed fairly convincing on the first reading but full of holes on the second." Shortly after, he turned his focus on the battle against naturalism—particularly evolution.

In "Darwin on Trial," he redefined "intelligent design" in its current form and founded the ID movement we encounter today. He became cofounder and advisor of the Center for Science and Culture at Seattle's Discovery Institute. The institute is the predominant antievolution organization in the United States. Johnson's participation in a symposium of evolutionists and creationists on Southern Methodist University's Dallas campus in 1992 led him to create the infamous "wedge strategy," a systematic plan for replacing conventional science with creationist philosophy. He was author of the Santorum Amendment to the 2001 Senate funding bill, which would have fostered the teaching of ID in public schools.

The Unraveling of Scientific Materialism

The following article presents Johnson's case against the scientific paradigm.

In a retrospective essay on Carl Sagan in the January 9, 1997 New York Review of Books, Harvard Genetics Professor Richard Lewontin tells how he first met Sagan at a public debate in Arkansas in 1964. The two young scientists had been coaxed by senior colleagues to go to Little Rock to debate the affirmative side of the question: "RESOLVED, that the theory of evolution is as proved as is the fact that the earth goes around the sun." Their main opponent was a biology professor from a fundamentalist college, with a Ph.D. from the University of Texas in Zoology. Lewontin reports no details from the debate, except to say that "despite our absolutely compelling arguments, the audience unaccountably voted for the opposition."

Of course, Lewontin and Sagan attributed the vote to the audience's prejudice in favor of creationism. The resolution was framed in such a way, however, that the affirmative side should have lost even if the jury had been composed of Ivy League philosophy professors. How could the theory of evolution even conceivably be "proved" to the same degree as "the fact that the earth goes around the sun"? The latter is an observable feature of present-day reality, whereas the former deals primarily with non-repeatable events of the very distant past. The appropriate comparison would be between the theory of evolution and the accepted theory of the origin of the solar system.

If "evolution" referred only to currently observable phenomena like domestic animal breeding or finch-beak variation, then winning the debate should have been no problem for Lewontin and Sagan even with a fundamentalist jury. The statement "We breed a great variety of dogs," which rests on direct observation, is much easier to prove than the statement that the earth goes around the sun, which requires sophisticated reasoning. Not even the strictest biblical literalists deny the bred varieties of dogs, the variation of finch beaks, and similar instances within types. The more controversial claims of large-scale evolution are what arouse skepticism. Scientists may think they have good reasons for believing that living organisms evolved naturally from nonliving chemicals, or that complex organs evolved by the accumulation of micromutations through natural selection, but having reasons is not the same as having proof. I have seen people, previously inclined to believe whatever "science says," become skeptical when they realize that the scientists actually do seem to think that variations in finch beaks or peppered moths, or the mere existence of fossils, proves all the vast claims of "evolution." It is as though the scientists, so confident in their answers, simply do not understand the question.

Carl Sagan described the theory of evolution in his final book as the doctrine that "human beings (and all the other species) have slowly evolved by natural processes from a succession of more ancient beings with no divine intervention needed along the way." It is the alleged absence of divine intervention throughout the history of life—the strict materialism of the orthodox theory—that explains why a great many people, only some of whom are biblical fundamentalists, think that Darwinian evolution (beyond the micro level) is basically materialistic philosophy disguised as scientific fact. Sagan himself worried about opinion polls showing that only about 10 percent of Americans believe in a strictly materialistic evolutionary process, and, as Lewontin's anecdote concedes, some of the doubters have advanced degrees in the relevant sciences. Dissent as widespread as that must rest on something less easily remedied than mere ignorance of facts.

Lewontin eventually parted company with Sagan over how to explain why the theory of evolution seems so obviously true to mainstream scientists and so doubtful to much of the public. Sagan attributed the persistence of unbelief to ignorance and hucksterism and set out to cure the problem with popular books, magazine articles, and television programs promoting the virtues of mainstream science over its fringe rivals. Lewontin, a Marxist whose philosophical sophistication exceeds that of Sagan by several orders of magnitude, came to see the issue as essentially one of basic intellectual commitment rather than factual knowledge.

The reason for opposition to scientific accounts of our origins, according to Lewontin, is not that people are ignorant of facts, but that they have not learned to think from the right starting point. In his words, "The primary problem is not to provide the public with the knowledge of how far it is to the nearest star and what genes are made of.... Rather, the problem is to get them to reject irrational and supernatural explanations of the world, the demons that exist only in their imaginations, and to accept a social and intellectual apparatus, Science, as the only begetter of truth." What the public needs to learn is that, like it or not, "We exist as material beings in a material world, all of whose phenomena are the consequences of material relations among material entities." In a word, the public needs to accept materialism, which means that they must put God (whom Lewontin calls the "Supreme Extraterrestrial") in the trash can of history where such myths belong.

Although Lewontin wants the public to accept science as the only source of truth, he freely admits that mainstream science itself is not free of the hokum that Sagan so often found in fringe science. As examples he cites three influential scientists who are particularly successful at writing for the public: E. O. Wilson, Richard Dawkins, and Lewis Thomas,

> each of whom has put unsubstantiated assertions or counterfactual claims at the very center of the stories they have retailed in the market. Wilson's Sociobiology and On Human Nature rest on the surface of a quaking marsh of unsupported claims about the genetic determination of everything from altruism to xenophobia. Dawkins' vulgarizations of Darwinism speak of nothing in evolution but an inexorable ascendancy of genes that are selectively superior, while the entire body of technical advance in experimental and theoretical evolutionary genetics of the last fifty years has moved in the direction of emphasizing nonselective forces in evolution. Thomas, in various essays, propagandized for the success of modern scientific medicine in eliminating death from disease, while the unchallenged statistical compilations on mortality show that in Europe and North America infectious diseases... had ceased to be major causes of mortality by the early decades of the twentieth century.

Lewontin laments that even scientists frequently cannot judge the reliability of scientific claims outside their fields of speciality, and have to take the word of recognized authorities on faith. "Who am I to believe about quantum physics if not Steven Weinberg, or about the solar system if not Carl Sagan? What worries me is that they may believe what Dawkins and Wilson tell them about evolution."

One major living scientific popularizer whom Lewontin does not trash is his Harvard colleague and political ally Stephen Jay Gould. Just to fill out the picture, however, it seems that admirers of Dawkins have as low an opinion of Gould as Lewontin has of Dawkins or Wilson. According to a 1994 essay in the New York Review of Books by John Maynard Smith, the dean of British neo-Darwinists, "the evolutionary biologists with whom I have discussed his [Gould's] work tend to see him as a man whose ideas are so confused as to be hardly worth bothering with, but as one who should not be publicly criticized because he is at least on our side against the creationists. All this would not matter, were it not that he is giving non-biologists a largely false picture of the state of evolutionary theory." Lewontin fears that non-biologists will fail to recognize that Dawkins is peddling pseudoscience; Maynard Smith fears exactly the same of Gould.

If eminent experts say that evolution according to Gould is too confused to be worth bothering about, and others equally eminent say that evolution according to Dawkins rests on unsubstantiated assertions and counterfactual claims, the public can hardly be blamed for suspecting that grand-scale evolution may rest on something less impressive than rock-solid, unimpeachable fact. Lewontin confirms this suspicion by explaining why "we" (i.e., the kind of people who read the New York Review) reject out of hand the view of those who think they see the hand of the Creator in the material world:

> We take the side of science in spite of the patent absurdity of some of its constructs, in spite of its failure to fulfill many of its extravagant promises of health and life, in spite of the tolerance of the scientific community for unsubstantiated just-so stories, because we have a prior commitment, a commitment to materialism. It is not that the methods and institutions of science somehow compel us to accept a material explanation of the phenomenal world, but, on the contrary, that we are forced by our a priori adherence to material causes to create an apparatus of investigation and a set of concepts that produce material explanations, no matter how counterintuitive, no matter how mystifying to the uninitiated. Moreover, that materialism is absolute, for we cannot allow a Divine Foot in the door. The eminent Kant scholar Lewis Beck used to say that anyone who could believe in God could believe in anything. To appeal to an omnipotent deity is to allow that at any moment the regularities of nature may be ruptured, that miracles may happen.

That paragraph is the most insightful statement of what is at issue in the creation/evolution controversy that I have ever read from a senior figure in the scientific establishment. It explains neatly how the theory of evolution can seem so certain to scientific insiders, and so shaky to the outsiders. For scientific materialists the materialism comes first; the science comes thereafter. We might more accurately term them "materialists employing science." And if materialism is true, then some materialistic theory of evolution has to be true simply as a matter of logical deduction, regardless of the evidence. That theory will necessarily be at least roughly like neo-Darwinism, in that it will have to involve some combination of random changes and law-like processes capable of producing complicated organisms that (in Dawkins' words) "give the appearance of having been designed for a purpose."

The prior commitment explains why evolutionary scientists are not disturbed when they learn that the fossil record does not provide examples of gradual macroevolutionary transformation, despite decades of determined effort by paleontologists to confirm neo-Darwinian presuppositions. That is also why biological chemists like Stanley Miller continue in confidence even when geochemists tell them that the early earth did not have the oxygen-free atmosphere essential for producing the chemicals required by the theory of the origin of life in a prebiotic soup. They reason that there had to be some source (comets?) capable of providing the needed molecules, because otherwise life would not have evolved. When evidence showed that the period available on the early earth for the evolution of life was extremely brief in comparison to the time previously posited for chemical evolution scenarios, Carl Sagan calmly concluded that the chemical evolution of life must be easier than we had supposed, because it happened so rapidly on the early earth.

That is also why neo-Darwinists like Richard Dawkins are not troubled by the Cambrian Explosion, where all the invertebrate animal groups appear suddenly and without identifiable ancestors. Whatever the fossil record may suggest, those Cambrian animals had to evolve by accepted neo-Darwinian means, which is to say by material processes requiring no intelligent guidance or supernatural input. Materialist philosophy demands no less. That is also why Niles Eldridge, surveying the absence of evidence for macroevolutionary transformations in the rich marine invertebrate fossil record, can observe that "evolution always seems to happen somewhere else," and then describe himself on the very next page as a "knee-jerk neo-Darwinist." Finally, that is why Darwinists do not take critics of materialist evolution seriously, but speculate instead about "hidden agendas" and resort immediately to ridicule. In their minds, to question materialism is to question reality. All these specific points are illustrations of what it means to say that "we" have an a priori commitment to materialism.

The scientific leadership cannot afford to disclose that commitment frankly to the public. Imagine what chance the affirmative side would have if the question for public debate were rephrased candidly as "RESOLVED, that everyone should adopt an a priori commitment to materialism." Everyone would see what many now sense dimly: that a methodological premise useful for limited purposes has been expanded to form a metaphysical absolute. Of course people who define science as the search for materialistic explanations will find it useful to assume that such explanations always exist. To suppose that a philosophical preference can validate a cherished scientific theory is to define "science" as a way of supporting prejudice. Yet that is exactly what the Darwinists seem to be doing, when their evidence is evaluated by critics who are willing to question materialism.

One of those critics, bearing impeccable scientific credentials, is Michael Behe, who argues that complex molecular systems (such as bacterial and protozoan flagella, immune systems, blood clotting, and cellular transport) are "irreducibly complex." This means that the systems incorporate elements that interact with each other in such complex ways that it is impossible to describe detailed, testable Darwinian mechanisms for their evolution. (My review of Behe's Darwin's Black Box appeared in FT, October 1996.) Never mind for now whether you think that Behe's argument can prevail over sustained opposition from the materialists. The primary dispute is not over who is going to win, but about whether the argument can even get started. If we know a priori that materialism is true, then contrary evidence properly belongs under the rug, where it has always duly been swept.

For Lewontin, the public's determined resistance to scientific materialism constitutes "a deep problem in democratic self-governance." Quoting Jesus' words from the Gospel of John, he thinks that "the truth that makes us free" is not an accumulation of knowledge, but a metaphysical understanding (i.e., materialism) that sets us free from belief in supernatural entities like God. How is the scientific elite to persuade or bamboozle the public to accept the crucial starting point? Lewontin turns for guidance to the most prestigious of all opponents of democracy, Plato. In his dialogue the Gorgias, Plato reports a debate between the rationalist Socrates and three sophists or teachers of rhetoric. The debaters all agree that the public is incompetent to make reasoned decisions on justice and public policy. The question in dispute is whether the effective decision should be made by experts (Socrates) or by the manipulators of words (the sophists).

In familiar contemporary terms, the question might be stated as whether a court should appoint a panel of impartial authorities to decide whether the defendant's product caused the plaintiff's cancer, or whether the jury should be swayed by rival trial lawyers each touting their own experts. Much turns on whether we believe that the authorities are truly impartial, or whether they have interests of their own. When the National Academy of Sciences appoints a committee to advise the public on evolution, it consists of persons picked in part for their scientific outlook, which is to say their a priori acceptance of materialism. Members of such a panel know a lot of facts in their specific areas of research and have a lot to lose if the "fact of evolution" is exposed as a philosophical assumption. Should skeptics accept such persons as impartial fact-finders? Lewontin himself knows too much about cognitive elites to say anything so naive, and so in the end he gives up and concludes that "we" do not know how to get the public to the right starting point.

Lewontin is brilliantly insightful, but too crankily honest to be as good a manipulator as his Harvard colleague Stephen Jay Gould. Gould displays both his talent and his unscrupulousness in an essay in the March 1997 issue of Natural History, entitled "Nonoverlapping Magisteria" and subtitled "Science and religion are not in conflict, for their teachings occupy distinctly different domains." With a subtitle like that, you can be sure that Gould is out to reassure the public that evolution leads to no alarming conclusions. True to form, Gould insists that the only dissenters from evolution are "Protestant fundamentalists who believe that every word of the Bible must be literally true." Gould also insists that evolution (he never defines the word) is "both true and entirely compatible with Christian belief." Gould is familiar with nonliteralist opposition to evolutionary naturalism, but he blandly denies that any such phenomenon exists. He even quotes a letter written to the New York Times in answer to an op-ed essay by Michael Behe, without revealing the context. You can do things like that when you know that the media won't call you to account.

The centerpiece of Gould's essay is an analysis of the complete text of Pope John Paul's statement of October 22, 1996 to the Pontifical Academy of Sciences endorsing evolution as "more than a hypothesis." He fails to quote the Pope's crucial qualification that "theories of evolution which, in accordance with the philosophies inspiring them, consider the spirit as emerging from the forces of living matter or as a mere epiphenomenon of this matter, are incompatible with the truth about man." Of course, a theory based on materialism assumes by definition that there is no "spirit" active in this world that is independent of matter. Gould knows this perfectly well, and he also knows, just as Richard Lewontin does, that the evidence doesn't support the claims for the creative power of natural selection made by writers such as Richard Dawkins. That is why the philosophy that really supports the theory has to be protected from critical scrutiny.

Gould's essay is a tissue of half-truths aimed at putting the religious people to sleep, or luring them into a "dialogue" on terms set by the materialists. Thus Gould graciously allows religion to participate in discussions of morality or the meaning of life, because science does not claim authority over such questions of value, and because "Religion is too important to too many people for any dismissal or denigration of the comfort still sought by many folks from theology." Gould insists, however, that all such discussion must cede to science the power to determine the facts, and one of the facts is an evolutionary process that is every bit as materialistic and purposeless for Gould as it is for Lewontin or Dawkins. If religion wants to accept a dialogue on those terms, that's fine with Gould—but don't let those religious people think they get to make an independent judgment about the evidence that supposedly supports the "facts." And if the religious people are gullible enough to accept materialism as one of the facts, they won't be capable of causing much trouble.

The debate about creation and evolution is not deadlocked. Propagandists like Gould try to give the impression that nothing has changed, but essays like Lewontin's and books like Behe's demonstrate that honest thinkers on both sides are near agreement on a redefinition of the conflict. Biblical literalism is not the issue. The issue is whether materialism and rationality are the same thing. Darwinism is based on an a priori commitment to materialism, not on a philosophically neutral assessment of the evidence. Separate the philosophy from the science, and the proud tower collapses. When the public understands this clearly, Lewontin's Darwinism will start to move out of the science curriculum and into the department of intellectual history, where it can gather dust on the shelf next to Lewontin's Marxism.

65. Michael Ruse

Michael Ruse is retired professor of history and philosophy at the University of Guelph, Canada, and currently philosopher of biology at Florida State University (Tallahassee). He has long been on the forefront of evolution's defense as an articulate writer and speaker. With Phillip Johnson, he was a keynote speaker at the 1992 Dallas symposium, "Darwinism: Scientific Inference or Philosophical Preference?" (Buell and Hearn 1994). A primary critique of evolution by creationists has been the claim that evolution is founded as much on myth as truth, and constitutes a secular substitute for religion—the two being equally based on faith. Ruse addresses the issue in the following article.

Is Evolution Just another Religion?

The noted American evolutionist and popular science writer Stephen Jay Gould argues strenuously that science and religion occupy different domains and speak to different issues. Properly understood therefore they cannot be in conflict: the ongoing American controversy about the literal truth of Genesis versus the claims of the evolutionists must be predicated on mistaken understandings of either religion or science or both. In his ecumenical tolerance, Gould apparently stands in sharp contrast to the no-less-noted British evolutionist and popular science writer Richard Dawkins. He sees science and religion in stark opposition, finding that a "cowardly flabbiness of the intellect afflicts otherwise rational people confronted with long-established religions." He thinks that Darwinian evolution (natural selection in particular) makes Christianity quite untenable, and he concludes that life is quite without ultimate meaning. Closer examination of the writings of the two evolutionists suggests that on the science/religion relationship they may not be as far apart as a surface reading suggests. Gould certainly thinks that achieving mutual respect and understanding will be a one-way process: the retreat of religious believers from just about every existence claim that they hold dear and sacred. Not only will such believers have to give up the literal truth of Genesis "which of course most Christians have long relinquished" but they will even have to forgo general speculations about God's method of creation. When Arthur Peacocke—both a physical chemist and an Anglican priest—made the neo-Augustinian inference that evolution suggests that God created sequentially rather than all at once, Gould sharply commented that there is here an illicit transgression of the science/religion boundary: "Is Mr. Peacocke's God just retooling himself in the spiffy language of modern science?" Even religious claims about the objects of creation themselves will have to be rethought. Gould is famous for his insistence that human evolution is a one-off occurrence which came purely by chance. Little comfort for the Jew or Christian who thinks that we are in some wise made in the image of God and that our appearance here on earth was very much not accidental.

It seems, in fact, that Gould allows little more role for religion than that of (as the poet Matthew Arnold once put it) "ethics heightened, enkindled, lit up by feeling": something perhaps acceptable to Christians on the extreme liberal wing but hardly to others. If this be science extending "a hand of fellowship" through "the right combination of education and humility," then perhaps the believer might prefer to stay friendless but with pride and belief intact. But, whether or not Gould is being entirely forthright in his intentions and commitments, people are entitled to their opinions. If evolutionists want to exclude what many would regard as meaningful religion, then that is evolutionists' business. In the West, at least, it is a free world.

Yet there is somewhat more to this whole issue, as literalistic evangelicals complain bitterly. And even though one may have little sympathy for the general position of these critics "to deny the basic fact of evolution in this day and age is scientifically silly and religiously unnecessary" one ought to be prepared to listen to objections whatever the source. In particular, argue these conservative Christians, evolutionists are not content with simply asserting the truth of their theory and the falsity of religion: they want to substitute their theory as an alternative religion. Its proponents promote evolution as a kind of secular alternative for more traditional faith. This is not merely a science versus religion issue, but a religion versus religion issue.

Of course, no one claims that evolutionists want to found an alternative religion rather as the positivists in the last century founded a whole new church, with secular saints and worship and other paraphernalia of the social side to religion. Rather, the claim is that evolution becomes more than mere science and turns into a source of meaning and optimism and renewal and so forth—not to mention moral dicta—to help us to move forward through life. And to be honest, *prima facie*, given the writings of today's evolutionists, one fears that the critics may have a point. A recent editorial by Gould in *Science*, for all that he tells us that "factual nature cannot, in principle, answer the deep questions about ethics and meaning," is redolent with language that would fit comfortably in a preacher's sermon. We learn that evolution "liberates the human spirit," that "for sheer excitement" evolution "beats any myth of human origins by light years," that evolution combines "truth value and visceral thrill," and (quoting Darwin at the end of the *Origin* on the grandeur of the evolutionary view) that we should "praise this evolutionary nexus a far more stately mansion for the human soul than any pretty or parochial comfort ever conjured by our swollen neurology to obscure the source of physical being..."

Is this simply overblown rhetoric or is it symptomatic of something more significant? As evolutionists, we have learnt that the secrets of the present lie often in the past. Turning to the history of evolutionary theory, we find that at the beginning "the speculations by eighteenth-century thinkers about organic origins" there is good reason to think that evolutionism (if anachronistically one may so call it) was rather less than sober professional science and rather more a vehicle for ideology and philosophy and those aspirations which one does associate with religious yearnings. Erasmus Darwin—British physician, friend of industrialists, widely read poet, and grandfather of Charles—speculated that all life developed upwards through time from primordial blobs, and that civilized (that is, white, Anglo-Saxon, Protestant, English-speaking) humans are the apotheosis of the process. Relying on scattered bits of real evidence, about fossils for instance, and on a great deal of non-evidence "the *phalli*, which were hung round the necks of Roman ladies, or worn in their hair, might have effect of producing a greater proportion of male children" infused by an enthusiasm for social and scientific progress, backed by a deistic belief in a god who is an unmoved mover working through unbroken law, through the medium of rhyming couplets Darwin presented his gallimaufry of fact and fiction to an eager public.

> "Imperious man, who rules the bestial crowd,
> Of language, reason, and reflection proud,
> With brow erect who scorns this earthy sod,
> And styles himself the image of his God;
> Arose from rudiments of form and sense,
> An embryon point, or microscopic ens!"

This was not merely a story about the past but an eschatological vision of how the world might be perfected if everyone adopted the tools and attitudes of the British capitalist in order to raise the general welfare of all. Moreover, the real point is not simply that we today might judge this stuff to be the excrescence of a kind of secular religion (not that secular actually), but that that was the reaction of Erasmus Darwin's contemporaries. The great comparative anatomist Georges Cuvier, himself incidentally a practicing Protestant, was scathing about the very form and context and intent of evolutionism. Rather than progress brought on by human effort, he thought that our real salvation lies in God's Providential grace bestowed in response to our prayers. He thought evolution was empirically wrong; but, with an eye to the greater danger, Cuvier could see that it was intended as a religion substitute and not just as mere science.

Surely Charles Darwin changed all of this? In a sense he did, for with the *Origin* evolution was upgraded to the status of something that one could in theory judge on its empirical merits. No longer was evolution a mere pseudoscience, a faith masquerading as a reflection of true reality. But, the *Origin* itself had hardly shaken off all traces of the past. Darwin shared with his grandfather a deistic-backed optimism about the evolutionary process: all of those comments about the Creator were meant sincerely, and the end passage highlighted by Gould about the grandeur of the evolutionary perspective was modified from a natural theological passage by the Scottish man of science, David Brewster. More significantly, never forget that it was not Darwin who was the chief Victorian proponent of evolutionary thought. That role fell to his self-appointed 'bulldog,' episcopal debater, Thomas Henry Huxley. And here we come to one of the most fascinating aspects of evolution's history. Although (having first been its sternest critic) Huxley adopted evolutionism with the enthusiasm of Saint Paul for Christianity, as with Saint Paul Huxley's version of the truth was not exactly identical to that of its founder. As Paul took Jesus' teaching and adapted it for his audience, so Huxley took Darwin's teaching and adapted it for his audience with significant consequences for the future status of evolutionary thought.

Huxley was a great professionalizer of science, biology in particular, and he realized that for success he had to find support money for the disciplines and jobs for the students. Physiology, Huxley sold to the medical profession: doctors were desperate to change from killing patients to curing them, and (more self-interestedly) they were not insensible to the social advantages of professional scientific training possessed only by those within the fold. Morphology, Huxley sold to the teaching profession as a proper substitute for the out-dated classics: it is no accident that Huxley sat on the London School Board and that he sponsored summer schools for teachers. Evolution was different. Knowledge about the dinosaurs does not mend broken legs and in any case the subject was too metaphysically loaded for simple classroom use. But it fit perfectly into another role. Huxley and his friends wanted to do more than provide jobs for scientists. They wanted to reform Victorian Britain: to clean up the civil service, to introduce universal education, to promote science-backed sanitation and urbanization generally, and much more. Evolution was the perfect platform for the ideology "secular religion if you like" that Huxley and his friends could use, as an alternative to the established Anglicanism that they saw supporting the establishment they wanted to overthrow.

And so it proved. To the amazement of his students, Huxley never talked about evolution in class, but he lectured non-stop on the subject from podia in working men's clubs and at general meetings like those of the British Association for the Advancement of Science and in the popular media. He promoted an evolutionism that is upward-looking and optimistic and a philosophy for a new age. There is an intentionally Biblical ring to his famous dictum that one should sit down before the facts, without preconception, as a little child. It was equally intentional that Huxley's famous clash with the Bishop of Oxford "where supposedly, upon being asked whether he was descended from monkeys on his grandfather's side or his grandmother's side, he replied that he would rather be descended from a monkey than a bishop of the church of England" achieved mythical status. (Like most myths, the reality of the debate was quite other than it later seemed.)

Since Darwin's writings really did not provide much by way of moral guidance, it is no chance that an evolutionist who did write at length on morality "fellow Englishman Herbert Spencer, enthusiast for laissez-faire, fanatic for progress" was taken up and cherished and promoted. (Later in life, in tune with a general *fin de siécle* sense of decline, Huxley's vision grew more sombre and the two old friends fell out.) It is no chance either that, as an alternative to the cathedrals of the Church, museums of natural history were just then being founded. Stocked with wonderful displays of fossils (including the newly uncovered American dinosaurs) and conveying unambiguous messages of progress, they were excellent places of sober entertainment and instruction for the young and impressionable. Expectedly, the British Museum (Natural History) was headed by one Huxley protégé, E. Ray Lankester, and the American Museum of Natural History was to be headed by another Huxley disciple, former student Henry Fairfield Osborn. It is little wonder that, in the popular press, Darwin's bulldog was known as 'Pope Huxley.' (Being, like all good Englishmen, deeply prejudiced against foreigners, he would probably have preferred to have been Archbishop of Canterbury.)

Although there was a weak and rather unsuccessful German-based evolutionary morphology which existed alongside all of this popularization and value promotion, essentially evolution kept its non-professional, religion-like status right up through to the third and fourth decades of this century. Then, thanks to the mathematical work of the great theoretical populational geneticists "notably R.A. Fisher and J.B.S. Haldane in Britain and Sewall Wright in America" evolution was ready to upgrade from its museum-based, low standing. And this it did, thanks to the synthesizers and students of the empirical biological world men like Julian Huxley and E.B. Ford in Britain and Theodosius Dobzhansky and his associates in America. Finally, in neo-Darwinism or the Synthetic Theory—as modern evolutionism became known—one had a science which could stand on and only on our understanding of the world of nature and which did not need or much want an infusion of ideology or philosophy to give it meaning or purpose.

But this does not mean that evolution-as-religion got up and went away. Indeed, almost to a person (Ford was perhaps one exception), the synthesizers of the 1930s and 1940s wanted both their professional value-free science and at the same time the old ideology-laden evolutionism which would give significance to life and existence. This is well born out by the fact that, along with their professional works, the neo-Darwinians turned out one popular book after another, full of philosophy and meaning. The titles speak for themselves: *The Biology of Ultimate Concern* (Dobzhansky), *The Biological Basis of Human Freedom* (Dobzhansky), *The Meaning of Evolution* (G.G. Simpson, the paleontologist), *The Basis of Progressive Evolution* (G.L. Stebbins, the botanist), *Religion without Revelation* (Julian Huxley), and more.

And so it continues down to the present. However one interprets the somewhat ambiguous position of Stephen Jay Gould, his fellow Harvard evolutionist Edward O. Wilson is categorical in his determination to make of evolution a religion: one which will challenge and replace the older conventional religions like Christianity.

> The evolutionary epic is mythology in the sense that the laws it adduces here and now are believed but can never be definitively proved to form a cause-and-effect continuum from physics to the social sciences, from this world to all other worlds in the visible universe, and backward through time to the beginning of the universe. Every part of existence is considered to be obedient to physical laws requiring no external control. The scientist's devotion to parsimony in explanation excludes the divine spirit and other extraneous agents. Most importantly, we have come to the crucial stage in the history of biology when religion itself is subject to the explanations of the natural sciences...
>
> If this interpretation is correct, the final decisive edge enjoyed by scientific naturalism will come from its capacity to explain traditional religion, its chief competition, as a wholly material phenomenon. Theology is not likely to survive as an independent intellectual discipline. [Wilson, *The Diversity of* Life, 1978, 192]

As with Herbert Spencer 'a Wilson hero' evolution is used to support a moral message. Less the laissez-faire individualism of the nineteenth century, and more a concern for such things as biodiversity and the preservation of tropical rain forests. Wilson believes that we humans have evolved in symbiotic relationship with the rest of living nature. Save we preserve such nature, we ourselves will perish and die. With something of the enthusiasm of the Baptist preachers of his childhood, in a nigh-dispensationalist fashion Wilson begs us to repent our profligate ways before it is too late.

So what do we conclude? There is surely today a dimension to evolutionary thought and work which is regular professional science, value free and ideology past. One thinks for instance of the speculations of William Hamilton about the significance of parasitism for the biology of sexuality. Hamilton may or may not be right, but his claims are simple science. Nothing more. Conservative

Christian critics who argue that the whole of evolutionary thought is merely a religion alternative or substitute are wrong, willfully so. There is however another dimension to evolutionary thought: a dimension which does go beyond straight science. Here one might indeed say that values and philosophy and ideology—perhaps even a religion or religion substitute—thrives and gives people meaning and purpose. This is not necessarily synonymous with popular writing about evolution, but it is work which does tend to fall in the popular domain, or in places (like the commentary section of *Nature*!) where scientists can be a little more expansive in their thinking.

Is it bad that there is this religious-like dimension to evolutionary thought? I see no reason to say that it is. If someone like Wilson—who feels strongly that our nature will always demand a religious perspective and commitment—wants to make a religion of his science, then this is surely his right as much as it is our right to accept or reject it as we will. But it is important to realize that evolutionary thought does have this dimension and not to confuse this with other, more ideology-free work. And, although we may and must let others know of our enthusiasm for the wonderful idea of evolution, we must make sure that others are not led into confusion either.

Ideas to Think About

1. Evaluate Gould's position from the perspective of theology. Is there any reason to believe that the scientific "magisterium" actually dictates the terms of the separation? Is there reason for religion to fear that the scientific niche—as carved out by Gould—is actually intrusive on the religious magisterium?
2. While Johnson decries science's a priori commitment to materialism, does he address religion's a priori commitment to supernaturalism? Both science and religion are founded on a priori acceptance of untestable assumptions. How might this be justified? What is a rational alternative?
3. Distinguish between "facts" as identified by a materialistic philosophy and "facts" as identified by a nonmaterialistic philosophy. How might such facts be confirmed and applied to a model of reality? Would this constitute a new paradigm, and if so, what would it look like?
4. Evaluate the complaint that evolutionists make claims that lie beyond the evidence on which they rely. Is Dawkins, in his insistence that scientific evidence not only dismisses the necessity of God but actually denies God's existence, encroaching on Gould's religious magisterium?
5. Ruse accepts the ideologically laden philosophy of evolution as a quasi-religious belief system. In what way might evolutionary belief be considered an alternative religion? In what way is it not a religion? How might it be considered a religious substitute?

For Further Reading

Evolution as a Religion, Mary Midgley. Rev. Ed. New York: Routledge (2002). *The Guardian* characterizes philosopher Midgley as "the foremost scourge of scientific pretension," and "someone whose wit is admired even by those who think she sometimes oversteps the mark." This iconoclastic little book famously indicts the more radical evolutionary ideologues.

Reason in the Balance, Phillip E. Johnson. Downers Grove, Ill.: InterVarsity Press (1995). Johnson remains the most lucid and intellectual spokesperson for the creationist camp. This book treats what he calls the "intellectual superstitions" of the Western world, branching beyond evolution to take on sex education, abortion, and education.

The God Delusion, Richard Dawkins. New York: Houghton Mifflin (2006). Evolutionist and atheist Dawkins is a proud iconoclast of all things religious. In this infamous treatise, he indicts belief in God as irrational in the face of evolutionary science, and even claims it to be a threat to rational culture.

UNIT 14

Nature-Nurture Revisited: The Rise of Sociobiology

> Biology is the key to human nature, and social sciences cannot afford to ignore its rapidly tightening principles. But the social sciences are potentially far richer in content. Eventually they will absorb the relevant ideas of biology and go on to beggar them. The proper study of man is, for reasons that now transcend anthropocentrism, man.
>
> **Edward O. Wilson**, *On Human Nature*

Sociobiology is not exactly old wine in new bottles, but its forerunning ideas can trace back at least to 1871. That year, Darwin published his *Descent of Man and Selection in Relation to Sex*. To Darwin, sexual selection, including competition among males for female sexual access, and female choice among suitors, was allied to but not identical to natural selection, as it focused on fertility rather than mortality.

Sociobiological theory encompasses but reaches beyond sexual selection and has become a hotly debated topic on both scientific and political fronts. The issue is thus somewhat like the scientific and political-religious battles fought against natural selection in the years following its 1859 introduction.

Neither is sociobiology a descendant of the nature-nurture debate, as the title might suggest. Its historical context is actually the challenge of explaining the evolution of behavior as part of the broader evolutionary paradigm. What the title references is the nature-nurture category into which the politicization of sociobiology thrust it during the contentious times of its genesis.

This application of biology to human behavior has generated the greatest heat, and the ignition came in response to Edward O. Wilson's *Sociobiology: The New Synthesis*, published in 1975. Wilson's colleagues—at Harvard and elsewhere—quickly chose sides and generated a battle that continues. Opponents recalled the earlier nature-nurture theory that ended in the eugenics movement and saw—or thought they saw—a disturbing parallel. The great tension at Harvard is recounted in Wilson's article included in this unit.

The heritage of sociobiology, however, lies in ethological patterns in nonhuman species. Its heroes are Nikolaas Tinbergen (1907–88), Konrad Lorenz (1903–89), and Karl von Frisch (1886–1982), who pioneered behavior studies in insects and birds in the mid-twentieth century. The three shared the 1975 Nobel Prize in Medicine or Physiology.

Sociobiology is thus largely European in origin, more particularly British, and most particularly at Oxford University: Tinbergen was at Oxford. Richard Dawkins was his student. W. D. Hamilton taught there from 1984, after receiving his early training from John Maynard Smith. Smith, whose work in behavioral ecology influenced the development of sociobiology, was variously at University College, London, and the University of Sussex.

The modern incarnation of sociobiology lies in renewed interest in Darwin's sexual selection and related behavior-based selective mechanisms in the 1960s: such survival-enhancing behavior as altruism, kin selection, and selfishness. Early advocates were Robert Trivers, William D. Hamilton, and John Maynard Smith, and so long as the studies applied to insects and lower vertebrates—where experimental design and hypothesis testing was more scientifically rigorous—there was effectively no debate.

It was the human application that aroused doubts, both on its scientific methodology and on the implicit genetic presuppositions such application made. For example, in non-kin-related altruism, the reciprocity involved may in fact exist independently of any conferred genetic advantage, so the burden of proof is not that such altruism is common, but there is reproductive advantage among those who practice it. This may be evidenced in nonhuman (particularly insect) social groups, but whether such demonstration can equate the observed altruistic behavior of wasps to that in humans is seriously debated.

Opposition to the application of sociobiological concepts to human societies (see, for example, Sahlins 1976) has extended beyond claims of flaws in the scientific methodology to claims of ideological bias—notably the assertion that human sociobiology reflects a philosophy of social determinism. In this regard, the issue has been raised from scientific to social concern.

In the current century, *evolutionary psychology* has emerged as a composite of sociobiological concepts and the application of natural selection to the mind. Like the former, it is adaptationist in focus, but its practitioners emphasize the psychological—rather than the behavioral—adaptations arising in human evolution. Understanding the nature of human psychology, in this respect, involves discovering the emotional and cognitive adaptations in the human past.

The selections that follow represent some of the early explanations of sociobiological concepts, as well as their initial application to human behavior. We conclude with Wilson's account of the politicization of his work in the heated debates that began at Harvard.

---

66. ROBERT L. TRIVERS

Parental Investment and Sexual Selection

[141] *What governs the operation of sexual selection is the relative parental investment of the sexes in their offspring.* Competition for mates usually characterizes males because males usually invest almost nothing in their offspring. Where male parental investment per offspring is comparable to female investment one would expect male and female reproductive success to vary in similar ways and for female choice to be no more discriminating than male choice (except as noted below). Where male parental investment strongly exceeds that of the female (regardless of which sex invests more in the sex cells) one would expect females to compete among themselves for males and for males to be selective about whom they accept as a mate.

Note that it may not be possible for an individual of one sex to invest in only part of the offspring of an individual of the opposite sex. When a male invests less per typical offspring than does a female but more than one-half what she invests (or vice-versa) then selection may not favor male competition to pair with more than one female, if the offspring of the second female cannot be parceled out to more than one male. If the net reproductive success for a male investing in the offspring of one female is larger than that gained from investing in the offspring of two females, then the male will be selected to invest in the offspring of only one female.

---

67. W. D. HAMILTON

The genetic evolution of social behavior. I

In the following excerpts, Hamilton makes the general case for a probable genetic basis for social behavior by arguing in favor of "inclusive fitness," and how it might have evolved. The data come from social insects. The first article provides a mathematical analysis, from which behaviors ranging from purely selfish to fully altruistic are examined. Without including the equations, the model may be represented as discriminating among these behaviors by the following diagram:

		Neighbors gain	Neighbors lose
Individual	gains	*Selected*	Selfish behavior [?]
	loses	Altruistic Behavior [?]	*Counter selected*

(8) [*Inclusive fitness is*] production of adult offspring... stripped of all components which can be considered as due to the individual's social environment, leaving the fitness he would express if not exposed to any of the harms or benefits of that environment...augmented by certain fractions of the quantities of harm and benefit which the individual himself causes to the fitness of his neighbors. The fractions in question are simply the coefficients of relationship *[i.e., ½ for full sibs, ¼ for half-sibs, ⅛ for first cousins, etc.]*.

[16] [F]or a hereditary tendency to perform an action of this kind *[e.g., inclusive fitness]* to evolve, the benefit to a sib must average at least twice the loss to the individual, the benefit to a half-sib must be at least four times the loss, to a cousin eight times and so on.... [W]e expect to find that no one is prepared to sacrifice his life for any single person but that everyone will sacrifice it when he can thereby save more than two brothers, or four half-brothers, or eight first cousins...

The genetic evolution of social behavior. II

[17] In brief outline, the theory points out that for a gene to receive positive selection it is not necessarily enough that it should increase the fitness of its bearer above the average if this tends to be done at the heavy expense of related individuals,

because relatives, on account of their common ancestry, tend to carry replicas of the same gene: and conversely that a gene may receive positive selection even though disadvantageous to its bearers if it causes them to confer sufficiently large advantages on relatives.... [18] In general, it has been shown that Wright's Coefficient of Relationship *r* approximates closely to the chance that a replica will be carried. Thus if an altruistic trait is in question more than $1/r$ units of reproductive potential or "fitness" must be endowed on a relative of degree *r* for every one unit lost by the altruist if the population is to gain on average more replicas than it loses. Similarly, if a selfish trait is in question, the individual must receive and use at least a fraction *r* of the quantity of "fitness" deprived from his relative if the causative gene is to be selected.... The social behavior of a species evolves in such a way that in each distinct behavior-evoking situation the individual will seem to value his neighbors' fitness against his own according to the coefficients of relationship appropriate to that situation.

Robert Trivers's major contribution to altruism theory was to explore the concept of reciprocity as an integral part of its selective advantage, even when there is no biological or social relationship among participants. In the following selection, he presents a model to support this, which he applies to human social behavior and to social organization.

68. Robert L. Trivers

The Evolution of Reciprocal Altruism

[35] One human being saving another, who is not closely related and is about to drown, is an instance of altruism. Assume that the chance of the drowning man dying is one-half if no one leaps in to save him, but that the chance that his potential rescuer will drown if he leaps in to save him is much smaller, say, one in [36] twenty. Assume that the drowning man always drowns when his rescuer does and that he is always saved when the rescuer survives the rescue attempt. Also assume that the energy costs involved in rescuing are trivial compared to the survival probabilities. Were this an isolated event, it is clear that the rescuer should not bother to save the drowning man. But if the drowning man reciprocates at some future time, and if the survival chances are then exactly reversed, it will have been to the benefit of each participant to have risked his life for the other. Each participant will have traded a one-half chance of dying for about a one-tenth chance. If we assume that the entire population is sooner or later exposed to the same risk of drowning, the two individuals who risk their lives to save each other will be selected over those who face drowning on their own. Note that the benefits of reciprocity depend on the unequal cost/benefit ratio of the altruistic act,... cost and benefit being defined here as the increase or decrease in chances of the relevant alleles propagating themselves in the population. *[Trivers here discusses the disadvantages of cheating—being saved but later refusing to save.]*

[45] Reciprocal altruism in the human species takes place in a number of contexts and in all known cultures.... Any complete list of human altruism could contain the following types of altruistic behavior:

1. helping in times of danger (e.g. accidents, predation, intraspecific aggression);
2. sharing food;
3. helping the sick, the wounded, or the very young and old;
4. sharing implements; and
5. sharing knowledge.

All of these forms of behavior often meet the criterion of small cost to the giver and great benefit to the taker.

69. Edward O. Wilson

The biographical footnote to the following article reads "Edward O. Wilson is Pellegrino University Professor at Harvard University and Curator in Entomology at the Museum of Comparative Zoology (Agassiz Museum), Harvard University, Cambridge, Massachusetts 02138. Presented here is his keynote address to the November 1994 convention of the National Association of Scholars in Cambridge, MA. A portion was taken in slightly modified form from Professor Wilson's autobiography, Naturalist *(Washington, D.C. and Covelo, Calif.: Island Press-Shearwater Books, 1994)."*

Wilson is currently research professor emeritus and honorary curator in entomology in the department of organismic and evolutionary biology at Harvard. Richard Lewontin, also at Harvard, is currently professor of biology emeritus and Alexander Agassiz Professor of Zoology in the Museum of Comparative Zoology, emeritus.

The article should be required reading for all scientists. It not only describes the radical opposition to Sociobiology *and the polarization of sides on ideological—indeed socio-political—grounds; it also pays fascinating tribute to the fact that scientists are human like everyone else, and given to strong beliefs. The passions of Lewontin and Gould for the "nurture" side of the equation are mirrored in the equal passions for "nature" felt so strongly by Wilson and Dawkins. Hull (2000) has described this with alacrity.*

When this debate translates into "free will" versus "biological determinism," we see the stirrings of ideological protest. When it then morphs into the political domain, it easily co-opts others who may be less concerned with the strictly scientific disagreements. It then becomes a battle of the new Marxism against the new liberalism. This intellectual-emotional metamorphosis belies the question of cause and effect: Do political beliefs influence one's science, or does a particular view of science induce a political position? What it does do is to reveal promise and risk in the practice of academic science.

Science and Ideology

I have composed this text with considerable humility because it is addressed to scholars and scientists many of whom speak more authoritatively on the history and philosophy of science than I. My own preferred reading list on the subject would include Gerald Holton's *Science and Anti-Science* and the wonderfully scorching book, *Higher Superstition*, by Paul Gross and Norman Levitt. For a full reading list that could compose a complete college course on the subject, I would add John Passmore's *Science and Its Critics*, published in 1978, and Steven Weinberg's adamantine image of the power and ideology-demolishing reach of modern physics in his book *Dreams of a Final Theory*. In many ways I would defer to these authors.

I hope they nonetheless might agree with me that the nobility of science as a human endeavor was well encapsulated by the physicist Subrahmanyan Chandrasekhar when he used the Icarus metaphor in praise of Sir Arthur Eddington. He said, "Let us see how high we can fly before the sun melts the wax in our wings." And on the appropriateness of the rosette of the National Academy of Sciences, the other NAS, that is splendidly symbolic in this sense: the gold of science is placed solidly in the center, surrounded by the purple of natural philosophy. Members are elected primarily or solely on the basis of objective discoveries they have made, expressible in clear declarative sentences, and not by any ideological test.

By science in common parlance is meant natural science, which gathers knowledge of the world as an organized, systematic enterprise and attempts to condense it into testable laws and principles by a wide-ranging and shifting set of methods. The diagnostic features of science that distinguish it from pseudoscience are, first, repeatability: the same phenomenon is sought again, preferably by independent investigation, and the interpretation given it confirmed or discarded by means of novel analysis and experimentation. And second, economy: scientists attempt to abstract the information into the form that is simplest, most easily recalled, and most esthetically pleasing—the combination called elegance—while yielding the largest amount of information with the least amount of effort. Third, mensuration: if something can be properly measured, using universally accepted scales, generalizations about it will be rendered less ambiguous. And fourth and finally, heuristic: the best science stimulates further discovery, often in unpredictable new directions, whose content confirms or modifies the parent formulation.

Science is thus not just a profession. Nor is it a delectation of mavens. Nor is it a philosophy. It is a combination of mental operations that has increasingly become the habit of educated peoples. It's a culture of illuminations hit upon by a fortunate turn of history, of uncountable small and large steps, of adjustments to reality during the past four centuries that yielded the most powerful way of knowing about the world ever devised.

Is this triumphalism? Not, I think, in the sense of a mental force impinging on history, art, and ritual. But, blended with technology, science as a way of knowing has already transformed human existence. Being richly self-rewarding and universally distributed, it feeds upon itself and grows exponentially. Scientific knowledge doubles every ten to fifteen years, as measured by articles, new journals, and the number of professional scientists. Understanding based on the new information now reaches into virtually every sphere of human activity and every moral dilemma. One need think only momentarily on nuclear armament, the Green Revolution, genetic engineering, cloning, artificial intelligence, visits to the planets, and human activity as an atmosphere-altering force—all changes that have originated or accelerated during the past fifty years—to see where science is taking us. The future, if we are to have one, is increasingly to be in the hands of the scientifically literate, those who at least know what it is all about. There can be no multicultural solution to the genetics of cystic fibrosis; the ozone hole cannot be deconstructed; there is nothing whatsoever relativistic or culturally contextual about the dopamine transporter molecules whose blockage by cocaine gives a rush of euphoria, the kind that leads the constructivist to doubt the objectivity of science.

Science isn't easy; that's why it took so long to get started, and then mostly in one place, western Europe. Part of the reason is that the actual process of scientific discovery is relatively rare and intellectually distinct from the body of accumulated knowledge it creates. Scientists as a rule do not discover in order to know. Instead, as Alfred North Whitehead said, they know in order to discover. They learn what they have to know, often remaining poorly informed about the rest of the world, in order to get to the frontier of knowledge where they can make discoveries. They move forward, each along a deliberately narrowed sector. Make one significant discovery and you're a scientist in the true, elitist sense, the sense most desired by scientists themselves, and you go into the textbooks. No discoveries and you are nothing, no matter how much you otherwise learn and write about science. When a scientist practicing this way begins to sort out knowledge in order to look for meaning, and especially when he carries that knowledge outside the circle of discoverers, he becomes a humanist. Thus in science there exists a fundamental distinction between process and product. That is why so many important scientists are narrow, foolish people and why so many learned, wise scholars in the field are not considered by their peers to be strong scientists, unless they find something new about how the world works.

It follows that the process of discovery, the inner fire of the scientific enterprise, cannot be communicated effectively to the citizen who doesn't already know a substantial amount of science. Only when he possesses some of the content of science can he grasp its living culture. Then he can understand how scientific knowledge is validated and how best to make judgments on his own accord. Graphs and "margins of error" make sense to him. He can explain them to others. Controls, multiple competing hypotheses, and disconfirmation become habits of thought. Accounts of science in newsmagazines are read with an engrained reserve, and scientists are viewed less as savants than as the artists and lucky conjurers they are in fact. Moral-tinged controversies are weighed with close attention to testable reality in the physical world. Of course these abilities are very limited today, and that is why anti-science ideologues and other charlatans get away with so much.

Which brings me to anti-science. I know less about postmodernism than most of you here, but let me give you my impression of how it relates to science. Postmodernist critics present a Disney World representation of science, a fantasy of what science is, and how scientists work, and why they work, a distortion embellished variously by obsolete theories of psychoanalysis and the battle cries of political ideology. Within the academy, it seems to me that postmodernism and the divisive forms of multiculturalism are substantially a revolt of the proletariat, wherein second-rate scholarship is parlayed into tenured professorships and book contracts—not by quality, not by originality, but by claims of entitlement of race, gender, and moralistic ideologies. But as I will show in a moment, some of it runs deeper, to turn the minds of even a few otherwise respected scientists.

There is also a tendency to think of these toxic developments as a recent replacement within the academy of Marxism, which, as Irving Howe nicely put it, has now retreated to departments of English in hopes of dying a comfortable death. But virulent anti-science was rampant twenty years ago. Here is how Science magazine described some of the disruption by radical leftists of the 1971 meeting of the American Association for the Advancement of Science, held in Chicago: "Glenn T. Seaborg, president-elect of the AAAS, took the advice of convention officials and fled from a meeting room to avoid being 'indicted' by young radicals; Edward Teller, the so-called 'father of the hydrogen bomb,' was repeatedly badgered despite the two bodyguards who trailed him everywhere," and "Mrs. Garrett Hardin, wife of a distinguished biologist, got so angry that she poked a young radical with her knitting needle."

The podium at one meeting was seized from Philip Handler, president of the National Academy of Sciences, who was then denounced as a "lackey of the ruling class." A group calling itself the Women's International Terrorist Conspiracy from Hell (or WITCHES for short) pronounced a hex on the AAAS meeting: "Science, Technology. We declare its use a sham. And subject all who use it ill to the witches' damn."

Now, the relevance of this frivolity to later attacks on science from the political extreme Left is closer than it might at first seem. The demonstrations were planned and executed by Science for the People, an organization devoted to anti-science ideology, and a key figure in the demonstrations was Richard Lewontin, then at the University of Chicago and soon to join the faculty at Harvard.

I introduce the organization and the man because they were to be key forces in my own life during the sociobiology controversy of the late 1970s, which I'll now relate briefly to you as a case history of ideologically motivated anti-science. There is no substitute for personal experience.

The sociobiology episode was one of the most conspicuous in the history of political correctness in academic life in the dark time before the expression, p.c., was coined and before the National Association of Scholars or any other form of organized resistance arose to blunt its excesses.

Let me say that sociobiology, the study of the biological basis of social behavior, was never in dispute when applied to animals. The discipline was accepted from the beginning by biologists. In 1989 the members of the Animal Behavior Society, the main international organization in the discipline, ranked my 1975 text *Sociobiology: The New Synthesis* the most important book on animal behavior ever published. But my speculative extension of the theories of sociobiology to human beings in the twenty-seventh and final chapter of the book received an entirely different reception, at least outside biology. The prevailing view in the social sciences in the seventies was essentially that there is no biologically based human nature, that human behavior is almost entirely sociocultural in origin, and therefore that the genes play little or no role except in bestowing intellectual and emotional capacity. I urged the contrasting view, that biology plays a larger role, in close concert with culture, and that human behavior cannot be understood without biology. I think it fair to say that this perception, as heretical as it was in the 1970s, is mainstream today. Actually, the opposing view never had the intensity displayed by Science for the People and a few other extreme critics on the Left. Among more than 200 scholarly books on human sociobiology and closely related subjects published since 1975, those generally favorable outnumber those mostly critical twenty to one. Social theorists who disagreed with sociobiology for the most part just steered around it.

The radical activists, however, went ballistic on this issue. Shortly after the publication of *Sociobiology*, Richard Lewontin organized fifteen scientists, teachers, and students in the Boston area as the Sociobiology Study Group, which then affiliated with Science for the People. The latter, larger aggregate of radical activists was begun in the 1960s to expose the misdeeds of scientists and technologists, including especially thinking considered to be politically dangerous. It was and remains nation wide, although greatly attenuated in its tone and influence.

What was correct political thinking? That has been made clear by Lewontin during the debate and afterward. "There is nothing in Marx, Lenin, or Mao," he wrote with his fellow Marxist Richard Levins, "that is or can be in contradiction with a particular set of phenomena in the objective world." True science, in other words, must be defined intrinsically to be forever separate from political thought. Ideology can then be constructed as a mental process insulated from science.

In formulating sociobiology, I wanted to move evolutionary biology into every potentially congenial subject, including human behavior and even political behavior, roughshod if need be and as quickly as possible. Lewontin obviously did not.

By adopting a narrow criterion of acceptable research deserving the title of science, Lewontin freed himself to pursue a political agenda unencumbered by science. He purveyed the postmodernist view that accepted truth, unless based upon unassailable fact, is no more than a reflection of dominant ideology and political power. After his turn to political activism, around 1970, he worked to promote his own accepted truth: the Marxian view of holism, envisioning a mental universe within which social systems ebb and flow in response to the forces of economics and class struggle. He disputed the idea of reductionism in evolutionary biology, even though it was and is the virtually unchallenged linchpin of the natural sciences as a whole. And most particularly, he rejected it for human social behavior. He said, in 1991, "By reductionism, we mean the belief that the world is broken up into tiny bits and pieces, each of which has its own properties and which combine together to make larger things. The individual makes society, for example, and society is nothing but the manifestation of the properties of individual human beings. Individual properties are the causes and the properties of the social whole are the effects of those causes."

Now this reductionism, as Lewontin expressed and rejected it, is precisely my view of how the world works. It forms the basis of human sociobiology as I construed it. But it is not science, Lewontin insisted. It cannot be made into science. And according to his own political beliefs, expressed over many years, sociobiology or any other social theory based on the biology of individuals cannot even possibly be true. Here is how he summarized his postmodernist argument: "This individualistic view of the biological world is simply a reflection of the ideologies of the bourgeois revolutions of the eighteenth century that placed the individual as the center of everything."

That much being understood, Lewontin concluded, and the shackles of bourgeois ideology cast aside, we are then freed to proceed along more progressive—that is to say, Marxist—political guidelines. These do not require scientific validation, at least not by any connection with genetics, neurobiology, or evolutionary theory. The genes, Lewontin declared, "have been replaced by an entirely new level of causation, that of social interaction with its own laws and its own nature that can be understood and explored only through that unique form of experience, social action." Hence the inviolable wisdom of Marx, Lenin, and Mao to which he alluded elsewhere.

Now I can come to the essence of the radical science movement. As loopy as it all may seem today, and especially after the collapse of world socialism, the argument has to be taken seriously, since it has been accepted to varying degrees by a few influential scientists, including Stephen Jay Gould, Richard Levins, and Ruth Hubbard, who are highly regarded in the public eye as scientists, even as they continue to promote a Marxian view.

Here then is the argument in its raw form: only an anti-reductionist, non-bourgeois science can help humanity attain the highest goal, which is a socialist world. In the 1984 book *Not in Our Genes*, Lewontin, Steven Rose, and Leon Kamin, all worthies of radical science philosophy, explained their purpose as follows:

> We share a commitment to the prospect of the creation of a more socially just—a socialist—society. And we recognize that a critical science is an integral part of the struggle to create that society, just as we also believe that the social function of much of today's science is to hinder the creation of that society by acting to preserve the interests of the dominant class, gender, and race. This belief—in the possibility of a critical and liberatory science—is why we have each in our separate ways and to varying degrees been involved in the development of what has become known over the 1970s and 1980s, in the United States and Britain, as the radical science movement.

That well respected scientists, two of whom, Lewontin and Levins, had been elected to the National Academy of Sciences (and soon removed themselves in ideological protest) could advocate an approach to science guided by a radically sociocultural version of Marxism may seem odd today given recent history. But it helps to explain the distinctive flavor of the controversy at Harvard in the 1970s, in an atmosphere of unfettered political correctness. In the standard leftward frameshift of academia prevailing at that time, Lewontin and members of Science for the People were classified as progressives, admittedly a bit extreme in their methods, while I—Roosevelt liberal turned pragmatic centrist—was cast well to the right.

Now to return to my story. Although the unofficial headquarters of the Sociobiology Study Group was Lewontin's office, located directly below my own at the Museum of Comparative Zoology, I was completely unaware of its deliberations. After meeting for three months, the group arrived at its foreordained verdict. In a letter published in the New York Review of Books (one might ask, where else?) on 13 November 1975, the members declared that human sociobiology was not only unsupported by evidence but also politically dangerous. All hypotheses attempting to establish a biological basis of social behavior, they wrote,

> tend to provide a genetic justification of the status quo and of existing privileges for certain groups according to class, race, or sex. Historically, powerful countries or ruling groups within them have drawn support for the maintenance or extension of their power from these products of the scientific community...Such theories provided an important basis for the enactment of sterilization laws and restrictive immigration laws by the United States between 1910 and 1930 and also for the

> eugenics policies which led to the establishment of gas chambers in Nazi Germany.

I learned of the letter when it reached the newsstands on 13 November. An editor at Harvard University Press called me to say that word about it was spreading fast and might prove a sensation. For a group of scientists to declare so publicly that a colleague has made a technical error is serious enough. To link him with racist eugenics and Nazi policies was, in the overheated academic atmosphere of the 1970s, far worse. And the purpose of the letter was not so much to correct alleged technical errors as to destroy credibility. Furthermore, the position of the Sociobiology Study Group was ethical in tone and therefore very difficult to challenge. The idea of human sociobiology as a field of study is both intellectually and morally wrong, the critics in Science for the People said. It would of course have been impolite for me to point to the imperfections of the Soviet Stalinists or to question the Marxist view. That in the view of many would have been a revival of McCarthyism and confirm my critics' opinion that I had a political agenda.

In the liberal dovecotes of Harvard University of the seventies and eighties, a reactionary professor was like an atheist in a Benedictine monastery. As the weeks passed and winter snows began to fall, I received little support from my colleagues on the Harvard faculty. Several friends spoke up in interviews and public radio forums to oppose Science for the People. They included Ernst Mayr, Bernard Davis, Ralph Mitchell, and my close friend and collaborator Bert Holldobler. But mostly what I got was silence, even when the internal Harvard dispute became national news. I know now after many private conversations that the majority of my fellow natural scientists on the Harvard faculty were sympathetic to my biological approach to human behavior but confused by the motives and political aims of the Science for the People study group. Both Lewontin, who was chairman of my department, and Gould, a respected member, continued to be treated with deference. The department members may also have thought that where there is smoke, there is fire. So they stuck to their work and kept a safe distance.

I had been blindsided by the attack. Having expected some frontal fire from social scientists on primarily evidential grounds, I had received instead a political enfilade from the flank. A few observers were surprised that I was surprised. John Maynard Smith, a senior British evolutionary biologist and former Marxist, said that he disliked the last chapter of Sociobiology himself and "it was also absolutely obvious to me—I cannot believe Wilson didn't know—that this was going to provoke great hostility from American Marxists, and Marxists everywhere." But it was true that I didn't know. I was unprepared perhaps because, as Maynard Smith further observed, I am an American rather than a European. In 1975 I was a political naif: I knew almost nothing about Marxism as either a political belief or a mode of analysis; I had paid little attention to the dynamism of the activist Left, and I had never heard of Science for the People. I was not an intellectual in the European or New York/Cambridge sense.

After the Sociobiology Study Group exposed me as a counterrevolutionary adventurist, and as they intensified their attacks in articles and teach-ins, other radical activists in the Boston area, including the violence-prone International Committee against Racism, conducted a campaign of leaflets and teach-ins of their own to oppose human sociobiology. As this activity spread through the winter and spring of 1975–76, I grew fearful that it might reach a level embarrassing to my family and the university. I briefly considered offers of professorships from three universities—in case, their representatives said, I wished to leave the physical center of the controversy. But the pressure was tolerable, since I was a senior professor with tenure, with a reputation based on other discoveries, and in any case could not bear to leave Harvard's ant collection, the world's largest and best. For a few days a protester in Harvard Square used a bullhorn to call for my dismissal. Two students from the University of Michigan invaded my class on evolutionary biology one day to shout slogans and deliver antisociobiology monologues. I withdrew from department meetings for a year to avoid embarrassment arising from my notoriety, especially with key members of Science for the People present at these meetings. In 1979 I was doused with water by a group of protestors at the annual meeting of the American Association for the Advancement of Science, possibly the only incident in recent history that a scientist was physically attacked, however mildly, for the expression of an idea. In 1982 I went to the Science Center at Harvard University under police escort to deliver a public lecture, because of the gathering of a crowd of protestors around the entrance, angered because of the title of my talk: "The coevolution of biology and culture."

Gerald Holton has warned of the rivulets of unreason that can come together at different times and different circumstances to form a threatening floodstream. The evidences remind us that, in Bertrand Russell's words, the mass of people would rather believe than know. Holton, Gross, Levitt, and others have shown that politicized antiscience is a flourishing trade within the academy. I will add with conviction that on occasion it can take root in the very entrails of science. And not just in a totalitarian state, exemplified by Soviet Lysenkoism and Nazi eugenics, but in a democracy and promoted by people who feel they are doing the morally right thing.

The sociobiology controversy as an example is not unique in recent history, although I wish it were. On 16 May 1986, a group of academic luminaries, including Robert Hinde, John Paul Scott, and several other prominent behavioral scientists, issued the Seville Declaration (following a conference in Spain), declaring invalid any theories or claims that aggression and war have a genetic basis. Such thinking is according to them, "scientifically incorrect." "Wars," the Declaration said, "begin in the minds of men." Warfare is a product of culture; biology contributes only in providing language and the capacity to invent wars. Case closed. The authors of the Declaration suggested, in effect, that if you have any thoughts otherwise about these matters, keep your mouth shut. The Seville Declaration was adopted that same year as the official policy of the American Anthropological Association. Eighty percent of the members who returned ballots on the motion to adopt voted in favor. Virtually all the main premises and conclusions of the Seville group are contradicted by the evidence, but no matter—the Declaration seemed to its signers and ratifiers the politically and morally correct thing to do. All the participants must have felt good about supporting it.

But as we shall see as the new IQ wars develop over the coming months [they have since proved virulent on the anti-genetic side—Author], as ideologues on both sides spring into their accustomed positions, feeling good is not what science is all about. Getting it right, and then basing social decisions on tested and carefully weighed objective knowledge, is what science is all about.

---

70. David Sloan Wilson and Edward O. Wilson

Rethinking the Theoretical Foundation of Sociobiology

The following article presents a summary of the theoretical basis for sociobiology and its historical transformations, including some of its misunderstood assertions. In addition, the article provides a good discussion of the several specific frameworks in which sociobiological theory has been cast.

[327] Darwin perceived a fundamental problem of social life and its potential solution in the following famous passage from *Descent of Man* (1871:166): It must not be forgotten that although a high standard of morality gives but a slight or no advantage to each individual man and his children over the other men of the same tribe... an increase in the number of well-endowed men and an advancement in the standard of morality will certainly give an immense advantage to one tribe over another.

The problem is that for a social group to function as an adaptive unit, its members must do things for each other. Yet, these group-advantageous behaviors seldom maximize relative fitness within the social group. The solution, according to Darwin, is that natural selection takes place at more than one level of the biological hierarchy. Selfish individuals might out-compete altruists within groups, but internally altruistic groups out-compete selfish groups. This is the essential logic of what has become known as multilevel selection theory.

Darwin's insight would seem to provide an elegant theoretical foundation for sociobiology, but that is not what happened, as anyone familiar with the subject knows. Instead, group selection was widely rejected in the 1960s and other theoretical frameworks were developed to explain the evolution of altruism and cooperation in more individualistic terms....

[329] During evolution by natural selection, a heritable trait that increases the fitness of others in a group (or the group as a whole) at the expense of the individual possessing the trait will decline in frequency within the group. This is the fundamental problem that Darwin identified for traits associated with human morality, and it applies with equal force to group-advantageous traits in other species....

Something more than natural selection [330] within single groups is required to explain how altruism and other group-advantageous traits evolve by natural selection. For Darwin, in the passage quoted above, that "something" was between-group selection. Group-advantageous traits do increase the fitness of groups, relative to other groups, even if they are selectively neutral or disadvantageous within groups. Total evolutionary change in a population can be regarded as a final vector made up of two component vectors, within and between-group selection, that often point in different directions....

These issues began to occupy center stage among evolutionary biologists in the 1960s, especially under the influence of George C Williams's (1966) *Adaptation and Natural Selection.* Williams began by *affirming* the importance of multilevel selection as a theoretical framework, agreeing with Darwin and the population geneticists that group-level adaptations require a process of group-level selection. He then made an additional claim that between-group selection is almost invariably weak compared to within-group selection (both posi- [331] tions are represented in the above-quoted passage). It was this additional claim that turned multilevel selection theory into what became known as "the theory of individual selection." Ever since, students have been taught that group selection is possible in principle, but can be ignored in practice. Seemingly other-oriented behaviors must be explained as forms of self-interest that do not invoke group selection, such as by helping one's own genes in the bodies of others (kin selection), or by helping others in expectation of return benefits (reciprocity). The concept of average effects in population genetics theory, which averages the fitness of alleles across all genotypic, social, and environmental contexts, was elaborated by both Williams and Richard Dawkins (1976) into the "gene's eye view" of evolution, in which everything that evolves is interpreted as a form of "genetic selfishness."

All of the early models assumed that altruistic and selfish behaviors are caused directly by corresponding genes, which means that the only way for groups to vary *behaviorally* is for them to vary *genetically.* Hardly anyone regards such strict genetic determinism as biologically realistic, and this was assumed in the models primarily to simplify the mathematics. Yet, when more complex genotype-phenotype relationships are built into the models, the balance between levels of selection can be easily and dramatically altered. In other words, it is possible for modest amounts of genetic variation among groups to result in substantial amounts of heritable phenotypic variation among groups (D S Wilson 2004)....

[334] Inclusive fitness theory (also called kin selection theory), evolutionary game theory (including the concept of reciprocal altruism), and selfish gene theory were all developed explicitly as alternatives to group selection.

[336] The developments outlined above have led to a situation that participants of the controversy in the 1960s would have difficulty recognizing. The theories that were originally regarded as alternatives, such that one might be right and another wrong, are now seen as equivalent in the sense that they all correctly predict what evolves in the total population. They differ, however, in how they partition selection into component vectors along the way. The frameworks are largely intertranslatable and broadly overlap in the kinds of traits and population structures that they consider....

The central issue addressed by Williams in *Adaptation and Natural Selection* was whether adaptations can evolve at the level of social groups and other higher level units. The problem, as recognized by Darwin and affirmed by Williams, was that traits that are "for the good of the group" are usually not favored by selection within groups—what we have called the fundamental problem of social life. When Williams and others rejected group selection, they were rejecting the possibility that adaptations evolve above the level of individual organisms. This is not a matter of perspective, but a fundamental biological claim. If true, it is every bit as momentous as it appeared to be in the 1960s. If false, then its retraction is equally momentous.

A sample of issues debated by contemporary theorists and philosophers of biology will show that, whatever the merits of pluralism, they do not deny the fundamental problem of social life or provide a solution other than between-group selection....

[338] There is a need for all perspectives to converge upon a core set of empirical claims, including the following:

1. There *is* a fundamental problem that requires a solution in order to explain the evolution of altruism and other group-level adaptations. Traits that are "for the good of the group" are seldom selectively advantageous within groups....
2. If a trait is locally disadvantageous wherever it occurs, then the only way for it to evolve in the total population is for it to be advantageous at a larger scale. Groups whose members act "for the good of the group" must contribute more to the total gene pool than groups whose members act otherwise....
3. Higher-level selection cannot be categorically ignored as a significant evolutionary force. Instead, it must be evaluated separately and on a case-by-case basis. Furthermore, all of the generalities about the likelihood of group selection that became accepted in the 1960s need to be reexamined....
4. The fact that a given trait evolves in the total population is not an argument against group selection. Evaluating levels of selection requires a nested series of relative fitness comparisons; between genes *within* individuals, between individuals *within* groups, between groups *within* a population of groups, and so on, each presenting traits that are separate targets for selection....

...We think that items 1–4 above can become the basis for a new consensus about when adap- [339] tations evolve at any given level of the biological hierarchy, restoring clarity and unity to sociobiological theory.

Ideas to Think About

1. In human societies, where mating has been socially transformed into marriage, kinship often restricts and proscribes reproductive bonds. How might the concept of parental investment and sexual selection apply in the human case?
2. Compare Hamilton's argument for altruism based on degree of relationship to Trivers's reciprocal altruism argument, which requires no relationship. Do both of these appear equally applicable to nonhuman social animals? How might each apply to humans?
3. What major experimental challenges exist in attempting to confirm the genetic basis for altruism in humans? Are the challenges similar for confirming a nongenetic basis?
4. Evaluate Wilson's discussion of the political agenda of his opponents. Are instances in recent scientific behavior that reflect similar predispositions or biases (e.g., global warming, the Food and Drug Administration) similar?
5. Overall, how does the rise of sociobiological thought reflect the evolving view of the place of humankind in the natural order, as well as the relative roles of culture and biology in distinguishing the human and nonhuman?

For Further Reading

On Human Nature, Edward O. Wilson. Cambridge, MA: Harvard University Press (1978). This is a later and more fully argued case for sociobiology than his initial work. In it, he argues for a transformational philosophy of humanness, in which ethical choices and our basic moral concepts are rooted in our biology.

Defenders of the Truth: The Sociobiology Debate, Ullica Segerstrale. New York: Oxford University Press (2001). The author presents a fascinating historical account of the scientific, political, and ideological debate surrounding Wilson's 1975 *Sociobiology*. It is a treatment sympathetic to Wilson, but the author gives a dispassionate account of the history and intercontinental contributions to the debate.

The Selfish Gene, Richard Dawkins. New York: Oxford University Press (1989). Based on the notion that we humans are mainly our genes' way of reproducing themselves, Dawkins argues the case that evolution is principally about the selfishness of genes in attempting to promote their survival by building—through natural selection—better transport systems. Enter the *meme:* the unit of cultural transmission that is critical to the survival of that system. A heady and fascinating read.

UNIT 15

Current Challenges to the Synthetic Theory

> The proponents of the synthetic theory maintain that all evolution is due to the accumulation of small genetic changes, guided by natural selection, and that transspecific evolution is nothing but an extrapolation and magnification of the events that take place within populations and species.
>
> —Ernst Mayr, *1963*

It is a frequent misconception that the evolutionary synthesis of the mid-twentieth century resulted in a monolithic model of evolution that finally united the genetic and organismic levels of analysis. Although the synthesis resolved misunderstandings between the experimentalists and naturalists, there were still unresolved problems, and the synthesis revealed new dilemmas, as well, in attempting to understand the details of evolutionary processes.

The older problems—such as the nature of speciation and the relationship between microevolution and macroevolution—as well as the newer ones regarding organism-environment coevolution and nongenic evolution—began to be fleshed out during the same period as sociobiology emerged, and for the same reasons.

The multilevel selection questions raised in the previous unit, regarding focal points in behavioral selection at the individual, intragroup, and intergroup levels, are analogous to the hierarchies discussed here: evolutionary changes at the local population level, within species, in the process of speciation, and at higher macroevolutionary levels. Behavior is involved alike in both, but the focus in this unit is the influence of these levels of analysis on biological species.

The fundamental philosophical question is whether the Darwinian paradigm of natural selection may be sufficiently broadened to include these newer concepts or altered with minimal damage to the original model to envelope an adjusted model, or, alternately, whether a new paradigm is demanded.

Perhaps the first challenge raising this philosophical question was the 1971 publication of the "punctuated equilibrium" explanation of speciation (Eldridge and Gould 1972). The explanation—that most species transformation occurs in small geographically constrained areas, where a species population undergoes relatively brief adaptive change after a long period of stasis—was quickly misunderstood through exaggerating its premises: It became seen as a saltational theory (a sudden leap resulting in new species) that countered Darwinian gradualness. It posited speciation from cladogenesis (a branching evolutionary tree with a limited number of new twigs), opposing Darwin's anagenesis (the succession of new species within major lineages). In fact, in geological time, the punctuation is not "sudden," and Darwin never denied branching evolution from populations within species.

But the punctuational model did alter our complacent narrowness. It recognized that probably little speciation is ploddingly and incrementally slow. It recognized that most speciation involves adaptive radiation into new ecological niches. It recognized that macroevolution is not simply microevolution writ large.

Above all, it recognized that speciation is not a single process, does not always demand new genes, and involves a dynamic role of both phenotype and environment. This unit reflects two of these new ideas: niche construction as a parallel process to natural selection, and evolutionary developmental biology (evo-devo) as a non-Darwinian source for macroevolution.

These two conceptual realms in biology reflect Darwin's own salient contribution to biological thought. He separated *variation* as an internally derived feature of organisms from *adaptation* as the external feature of the organism's survival. Thus, *natural selection* became the discovered process that knits the two features into a whole.

In a delightful little book, *The Triple Helix* (Harvard University Press, 2000), Richard Lewontin provides immense insight into how these alternative views of the organism lead us today into the two conceptual realms.

"Darwin's alienation of the outside from the inside was an absolutely essential step in the development of modern Biology," Lewontin writes. But "the time has come when further progress in our understanding of nature requires that we reconsider the relationship between the outside and the inside, between organism and environment" (47). This leads us to niche construction concepts.

"In contrast," Lewontin notes, "for developmental biologists the variation between individual organisms, and even between species, is not of interest. On the contrary, such variation is an annoyance and is ignored wherever possible. What is at the center of interest is the set of mechanisms that are common to all individuals" (9).

"The concentration on developmental processes that appear to be common to all organisms results in a concentration on those causal elements which are also common. But such common elements must be internal to the organism, part of its fixed essence, rather than coming from accidental and variable forces of the external milieu" (10). This leads us to evo-devo concepts.

We have seen—and will see further—that Darwin was concerned with both of these approaches: external relationships and internal development. We begin with niche construction.

I: Niche Construction

We clearly now know that the environment is not an independent set of external features to which the organism responds. Each responds to the other in an ongoing and highly linked series of connections. "The metaphor of adaptation," writes Lewontin, "while once an important heuristic for building evolutionary theory, is now an impediment to a real understanding of the evolutionary process and needs to be replaced by another. Although metaphors are dangerous, the actual process of evolution seems best captured by the process of *construction*" (2000, 48).

---

71. Rachel Day, Kevin N. Laland, and John Odling-Smee

Rethinking Adaptation: The Niche-Construction Perspective

"Niche construction," then, is a relatively new approach to the organism-environment relationship that focuses on the capacity of organisms "to construct, modify, and select important components of their local environments," write the authors in the selection that follows. It has been long recognized that organisms modify—often even destroying—their environments. This has been conventionally interpreted, however, as a product of natural selection, although the authors note that many scientists in this field now consider niche construction as a parallel process to natural selection.

[80] Through niche construction organisms not only influence the nature of their world, but also in part determine the selection pressures to which they and their descendants are exposed, and they do so in a non-random manner.

[81] In many instances, the changes brought about in the local environment through the niche construction of an organism will also be experienced by the organism's offspring or other descendants. Thus generations of organisms not only acquire genes from their ancestors but also an *ecological inheritance*, that is, a legacy of a sub-set of natural selection pressures that have been modified by the niche construction of their genetic or ecological ancestors (Odling-Smee 1988). Thus the niche-construction perspective incorporates two kinds of descent (genetic and ecological inheritance), and two kinds of modifying process (natural selection and niche construction) each of which is potentially capable of generating a complementary match between organism and environment....

[82] Niche-constructing traits are more than just adaptations, because they play the additional role of modifying natural selection pressures, frequently in a directed manner, and in doing so they change the evolutionary dynamic. According to the niche-construction perspective, the changes to the evolutionary process brought about by niche construction and ecological inheritance are sufficiently

important and occur sufficiently frequently to warrant an overhaul in evolutionary thinking.

[93] There is now considerable evidence that niche construction is widespread across all taxonomic groups of organisms and that it regularly modifies selection pressures. Theoretical models have demonstrated that the feedback from niche construction will almost certainly bring about important changes in the evolutionary process. However, the niche-construction perspective is not merely a more accurate portrayal of evolution. A major benefit provided by the explicit introduction of niche construction into evolutionary theory is that it is likely to generate new hypotheses and stimulate new empirical work. It is fruitful to regard the dynamic complementary match between organisms and environments as a product of reciprocal interacting processes of natural selection and niche construction.

Lewontin (2000) notes that, whereas environments certainly exist independently of organisms, niches do not. They emerge together, because all niches and organisms reflect mutual alterative forces.

[51] To arrive at a concept of the environment that will be correct and useful for our understanding of past evolution,... we need to clarify several facets of the relation between organism and environment. First, organisms determine which elements of the external world are put together to make their environments and what the relations are among the elements that are relevant to them....

[54] A second facet of the relation between organism and environment that needs to be clarified is this: organisms not only determine what aspects of the outside world are relevant to them by peculiarities of their shape and metabolism, but they actively construct, in the literal sense of the word, a world around themselves....

Third, organisms not only determine what is relevant and [55] create a set of physical relations among the relevant aspects of the outer world, but they are in a constant process of altering their environment. Every species, not only *Homo sapiens,* is in the process of destroying its own environment by using resources that are in short supply and transforming them into a form that cannot be used again by the individuals of the species. Food is turned into poisonous waste products by every metabolizing cell. Plants suck up water from the soil and transpire it into the air. Although water is returned to the soil, its local rate of replenishment is essentially independent of its local rate of extraction, so that plants in a particular place are creating their own drought.

For human evolution, the niche-construction model is allied to the gene-culture co-evolution, an application of sociobiology. Across biology, however, it has contributed new and fruitful ways of examining the evolutionary process.

II: Evo-Devo

A nagging, unrelenting question has long encouraged a developmental approach to understanding both species diversity and their common descent: How can we explain the broad diversity of phyla—and particularly the diversity among Metazoans having a bilateral symmetry and, most characteristically, an internal bony skeleton—by simply referencing a combination of random mutation and nonrandom natural selection? It has been a stretch. The wide differences in specific structure were awash in a surprisingly common body plan. Can the differences reflect different genetic organization, while the commonalities reflect identical gene systems? If so, how can one explain the apparent paradox? Of course, the related issue of reconciling continuities within species with discontinuities among them was as old as Darwinism itself. Evolutionists uncomfortable with conventional answers fidgeted uneasily.

The anticipatory "body plan" ideas of Charles Bonnet in the eighteenth century and Geoffroy Saint-Hilaire in the nineteenth, which we encountered in Units 4 and 6, respectively, could not yet be integrated into a gene-based paradigm, nor could the application of embryology recommended by Huxley and by Darwin himself.

Despite intervening theories ranging from Goldschmidt's "hopeful monsters" (Gould 1977), resulting from macromutations, to various non-Darwinian mechanisms (for example, meiotic drive—see Sandler and Novitsky [1957]), an empirically developed basis for morphogenesis had been elusive in the twentieth century.

One approaching answer, evolutionary developmental biology (evo-devo), comes in revolutionary experiments in the genetics of embryology. This is the new razor's edge of discovery in biology, and it is indeed building into a third scientific revolution: It sends a century of conventional genetic wisdom and a host of genetic models of speciation into a tailspin. It replaces them with a counterintuitive alternative.

Beginning with Mendel's rediscovery, in 1900, genetic experiments—principally on the fruit fly *Drosophila melanogaster*—suggested that radical phenotypic change involved radical gene change through mutation. It was logically assumed that any basic change to animal form, such as the transition from one species to another and particularly the more profound transition from one body plan to another, involves a transition from old genes to new genes, whether through replacement or addition. Logical it was, but truth often defies logic.

With the vastly improved understanding of the gene at the molecular level, and of its functions at the physiological level, studies of embryonic development radically challenged that assumption. Instead of new genes, the emergence of new forms has often required only alteration in the control or sequence of expression of existing gene complexes. In the closing decades of the twentieth century, the family of genes that control morphological pattern in both vertebrates and arthropods—the homeotic, or *Hox*, genes—was identified. Depending on how the expression of these genes is regulated, they can produce profound alterations in the morphology of body segments, their number, and their pattern.

The fact that many of these genes are common to vertebrates, as well as to those invertebrates that show bilateral symmetry, suggests an ancient source and thus a common origin. Saint-Hilaire's earlier unification of insects, living within their skeletons, with chordates, whose skeletons lie within them, seems prophetic in retrospect.

The science of evo-devo is arcane, and almost any discussion of its experimental and comparative results requires a reasonable understanding of the three fields of paleontology, embryology, and molecular biology. I have selected articles that provide a reasonable overview of the history and important evolutionary conclusions of this new field, with clarity and a minimum of specialized terminology. I have necessarily omitted the more cryptic terms and esoteric concepts, while attempting to clarify others. My comments are italicized and in brackets.

To introduce this material, I return to Darwin's *Origin* and to his treatment of embryology. The selections continue with William Bateson's early recommendation to focus on sources of morphological variation in attempting to study the origin of discontinuity among species. Sean Carroll's classic review article on the commonality and function of *Hox* genes in both insects and vertebrates comes next, as well as the 1997 review by Shubin et al. of the ancestral circumstances uniting both chordate and arthropod limb evolution. Finally, we include portions of the report by Grenier at al. on the ingenious efforts to link the *Hox* genes in present arthropods and vertebrates to those of pre-Cambrian life forms.

---

72. CHARLES DARWIN

On the origin of species by means of natural selection, Ch. 13

[439] It has already been casually remarked that certain organs in the individual, which when mature become widely different and serve for different purposes, are in the embryo exactly alike. The embryos, also, of distinct animals within the same class are often strikingly similar: a better proof of this cannot be given, than a circumstance mentioned by Agassiz, namely, that having forgotten to ticket the embryo of some vertebrate animal, he cannot now tell whether it be that of a mammal, bird, or reptile....

The points of structure, in which the embryos of widely different animals of the same class resemble each other, often have no direct relation to their condi- [440] tions of existence. We cannot, for instance, suppose that in the embryos of the vertebrata the peculiar loop-like course of the arteries near the branchial slits are related to similar conditions, in the young mammal which is nourished in the womb of its mother, in the egg of the bird which is hatched in a nest, and in the spawn of a frog under water. We have no more reason to believe in such a relation, than we have to believe that the same bones in the hand of a man, wing of a bat, and fin of a porpoise, are related to similar conditions of life....

[442] How, then, can we explain these several facts in embryology, namely the very general, but not universal difference in structure between the embryo and the adult; of parts in the same individual embryo, which ultimately become very unlike and serve for diverse purposes, being at this early period of growth alike; of embryos of different species within the same class, generally, but not universally, resembling each other; of the structure of the embryo not being closely related to its conditions of existence, except when the [443] embryo becomes at any period of life active and has to provide for itself; of the embryo apparently having sometimes a higher organisation than the mature animal, into which it is developed. I believe that all these facts can be explained, as follows, on the view of descent with modification....

It is commonly assumed, perhaps from monstrosities often affecting the embryo at a very early period, that slight variations necessarily appear at an equally early period. But we have little evidence on this head—indeed the evidence rather points the other way; for it is notorious that breeders of cattle, horses, and various fancy animals, cannot positively tell, until some time after the animal has been born, what its merits or form will ultimately turn out. We see this plainly in our own children; we cannot always tell whether the child will be tall or short, or what its precise features will be. The question is not, at what period of life any variation has been caused, but at what period it is fully displayed. The cause may have acted, and I believe generally has acted, even before the embryo is formed; and the variation may be due to the male and female sexual elements having been affected by the conditions to which either parent, or their ancestors, have been exposed. Nevertheless an effect thus caused at a very early period, even before the formation of the embryo, may appear late in life; as when an hereditary disease, which appears in old age alone, has been communicated to the offspring from the reproductive element of one parent....

[445] As the evidence appears to me conclusive, that the several domestic breeds of pigeon have descended from one wild species, I compared young pigeons of various breeds, within twelve hours after being hatched; I carefully measured the proportions (but will not here give details) of the beak, width of mouth, length of nostril and of eyelid, size of feet and length of leg, in the wild stock, in pouters, fantails, runts, barbs, dragons, carriers, and tumblers. Now some of these birds, when mature, differ so extraordinarily in length and form of beak, that they would, I cannot doubt, be ranked in distinct genera, had they been natural productions. But when the nestling birds of these several breeds were placed in a row, though most of them could be distinguished from each other, yet their proportional differences in the above specified several points were incomparably less than in the full-grown birds. Some characteristic points of difference for instance, that of the width of mouth—could hardly be detected in the [446] young. But there was one remarkable exception to this rule, for the young of the short-faced tumbler differed from the young of the wild rock-pigeon and of the other breeds, in all its proportions, almost exactly as much as in the adult state....

Now let us apply these facts and the above two principles which latter, though not proved true, can be shown to be in some degree probable to species in a state of nature. Let us take a genus of birds, descended on my theory from some one parent-species, and of which the several new species have become modified through natural selection in accordance with their diverse habits. Then, from the many slight successive steps of variation having supervened at a rather late age, and having been inherited at a corresponding [447] age, the young of the new species of our supposed genus will manifestly tend to resemble each other much more closely than do the adults, just as we have seen in the case of pigeons. We may extend this view to whole families or even classes. The fore-limbs, for instance, which served as legs in the parent-species, may become, by a long course of modification, adapted in one descendant to act as hands, in another as paddles, in another as wings; and on the above two principles namely of each successive modification supervening at a rather late age, and being inherited at a corresponding late age the fore-limbs in the embryos of the several descendants of the parent-species will still resemble each other closely, for they will not have been modified. But in each individual new species, the embryonic fore-limbs will differ greatly from the fore-limbs in the mature animal; the limbs in the latter having undergone much modification at a rather late period of life, and having thus been converted into hands, or paddles, or wings. Whatever influence long-continued exercise or use on the one hand, and disuse on the other, may have in modifying an organ, such influence will mainly affect the mature animal, which has come to its full powers of activity and has to gain its own living; and the effects thus produced will be inherited at a corresponding mature age. Whereas the young will remain unmodified, or be modified in a lesser degree, by the effects of use and disuse....

[448] As all the organic beings, extinct and recent, which [449] have ever lived on this earth have to be classed together, and as all have been connected by the finest gradations, the best, or indeed, if our collections were nearly perfect, the only possible arrangement, would be genealogical. Descent being on my view the hidden bond of connexion which naturalists have been seeking under the term of the natural system. On this view we can understand how it is that, in the eyes of most naturalists, the structure of the embryo is even more important for classification than that of the adult. For the embryo is the animal in its less modified state; and in so far it reveals the structure of its progenitor. In two groups of animal, however much they may at present differ from each other in structure and habits, if they pass through the same or similar embryonic stages, we may feel assured that they have both descended from the same or nearly similar parents, and are therefore in that degree closely related. Thus, community in embryonic structure reveals community of descent.... Agassiz believes this to be a law of nature; but I am bound to confess that I only hope to see the law hereafter proved true....

---

73. William Bateson

Materials for the Study of Variation

The subtitle of William Bateson's famous treatise tells us where he stands, philosophically, on the issue of evolution: new species emerge by leaps, not steps. Mendelian saltation, rather than Darwinian gradualism, guides his thoughts. And well it is for us, after all, because Bateson gave us the recognition that in almost all animals, like segments are repeated (in the earthworm, in the vertebral column)—a meristic series. But discontinuous variation may occur when some elements of a meristic series become transformed into an element of another series (a leg becomes an antenna). For this kind of saltational change—on which he supposed species evolution depends—Bateson coined the term homeosis, *from which we now get the category of* homeotic *genes. He also coined* genetics *to describe the study of variation in inheritance.*

Bateson's father was master of St. John's College, Cambridge, which William attended. William's son, Gregory, was an anthropologist, a linguist, and a creator of the science of cybernetics.

[17] The Study of Variation is essentially a study of differences between organisms, so for each observation of Variation at least two substantive organisms are required for comparison. It is proposed to confine the present treatment of the subject to a consideration of the integral steps by which Variation may proceed; hence it is desirable that the two organisms compared should be parent and offspring, and if, as is often the case, the actual parent is unknown, it is at least necessary that the normal form of the species should be known and that there must be reasonable evidence that the varying offspring is actually descended from such a normal. For this reason, evidence from a comparison of Local Races, and other established Varieties, though a very valuable part of the Study, will for the most part not be here introduced. For the belief that such races are descended from the putative normal scarcely ever rests on proof, and still more rarely is there evidence of the number of generations in which the change has been effected.

[18] Species are discontinuous; may not the Variation by which Species are produced be discontinuous too? It may be stated at once that evidence of such Discontinuous Variation does exist, and in this first consideration of the subject attention will be confined to it. The fact that Continuous Variation exists is also none the less a fact, but it is most important that the two classes of phenomena should be recognized as distinct, for there is reason to think that they are distinct essentially, and that though both may occur simultaneously and in conjunction, yet they are manifestations of distinct processes. The attempt to distinguish these two kinds of Variation from each other constitutes one of the chief parts of the study. It will not perhaps be possible to find any general expression which shall accurately differentiate between Variations which are Discontinuous and those which are Continuous, but it is possible to recognize attributes proper to each and to distinguish changes which are or may be effected in the one way from other changes which are or may be effected in the other.

[19] Not only are the bodies of all organisms heterogeneous, but in the great majority the Heterogeneity occurs in a particular way and according to geometrical rule. This character is not peculiar to a few organisms, but is common to nearly all.

Order of form will first be found to appear in the fact that in any living body the Heterogeneity is in some degree symmetrically distributed around one or more centres. In the great majority of instances these centres of symmetry are themselves distributed about other centres, so that in one or more planes the whole body is symmetrical.

[20] Symmetry then depends essentially on the fact that structures found in one part of an organism are repeated and occur again in another part of the same organism.... This phenomenon of Repetition of Parts, generally occurring in such a way as to form a Symmetry or Pattern, comes near to being a universal character of the bodies of living things. It will in cases which follow be often convenient to employ a single term to denote this phenomenon wherever and however occurring. For this purpose the term **Merism** will be used....

[84] The evidence of Meristic Variation relates essentially to the manner in which changes occur in the number of members in Meristic series. Such numerical changes may come about in two ways, which are in some respects distinct from each other. For instance, the number of legs and body-segments in *Peripatus edwardsii [this is one of the "velvet worms," of the phylum* Onychophora, *the subject of the final paper in this section.—RKW]* varies from 29 to 341: here the variation in number must be a manifestation of an original difference in the manner of division or segmentation in the progress of development. The change is strictly Meristic or divisional. On the other hand, change in number may arise by the Substantive Variation of members of a Meristic series already constituted. For example, the evidence will shew that the number of oviducal openings in *Astacus* may be increased from one pair to two or even three pairs. Here the numerical variation has come about through the assumption by the penultimate and last thoracic appendages, of a character typically proper to the appendages of the antepenultimate segment of the thorax alone. Now there is here no change in the number of segments composing the Meristic series, but by Substantive Variation the number of openings has been increased.

The case of the modification of the antenna of an insect into a foot, of the eye of a Crustacean into an antenna, of a petal into a stamen, and the like, are examples of the same kind.

It is desirable and indeed necessary that such Variations, which consist in the assumption by one member of a Meristic series, of the form or characters proper to other members of the series, should be recognized as constituting a distinct group of phenomena. In the case of plants such Variation is very common and is one of the most familiar forms of abnormality.... For this reason it is desirable that the term which denotes it should not lead to misunderstanding, and I think that a new term is demanded. [85] For the word 'Metamorphy' I therefore propose to substitute the term **Homœosis**, which is also more correct; for the essential phenomenon is not that there has merely been a change, but that something has been changed into the likeness of something else.

---

74. Sean B. Carroll

Homeotic Genes and the Evolution of Arthropods and Chordates

Through developments in molecular biology in recent decades—particularly in laboratory protocols for tracking gene expression in embryonic development—distinctions between alternative functions of existing genes and unique functions of unique genes have revolutionized biology. The days of "one gene—one protein" now seem medieval, despite having only seventy years of history.

As one of the world's preeminent investigators in evo-devo, Sean Carroll has himself participated in numerous revisions of its central ideas and has been responsible for many of its important discoveries. In this review article, he examines the evidence tying a single Hox *gene cluster to the evolutionary history of arthropods (the large phylum that includes crustaceans, insects, spiders, and others), and a duplicate cluster to the history of the chordates (predominantly the vertebrate animals). These two groups, sharing some common* Hox *genes, underwent their subsequent evolutionary history largely through modification in the regulation of these genes—not through gene mutations.*

Conserved gene sets are those that have not changed significantly through evolution. Presumably, then, they are critical to survival and unlikely to survive mutation.

[479] One of the most important biological discoveries of the past decade is that arthropods and chordates, and indeed most or all other animals, share a special family of genes, the homeotic (or *Hox)* genes, which are important for determining body pattern. The diversity of *Hox*-regulated features in arthropods (segment morphology, appendage number and pattern) and vertebrates (vertebral morphology, limb and central nervous system pattern) suggests that the *Hox* genes are implicated at some level in the morphological evolution of these animals.

...The evidence indicates that the duplication of *Hox* clusters and other developmental genes between primitive chordates and early vertebrates enabled the evolution of the anatomical complexity of vertebrates. However, the diversity of arthropods, insects and vertebrates has arisen primarily through regulatory evolution.

...Three remarkable conserved features unite the *Hox* genes of higher animals: (1) their organization in gene complexes; (2) their expression in discrete regions in the same relative order along the main (A-P) body axis[1]; and (3) their possession of a sequence of 180 base pairs (the homeobox) encoding a DNA-binding motif (the homeodomain).

[484] The available evidence suggests that primitive arthropods and chordates each possessed a single *Hox* complex containing the diverse array of *Hox* genes found in their modern descendants. Although this cluster was duplicated in the chordates, presumably in one or two early phases in their evolution, it appears that the subsequent course of vertebrate evolution from primitive bony fishes to mammals, and the entire course of arthropod evolution, was founded upon conserved sets of *Hox* genes. The phylogeny of *Hox* genes and the many examples cited above of large-scale morphological changes associated with diversity in *Hox* gene regulation and target regulation suggest that the primary genetic mechanism enabling morphological diversity among arthropods and vertebrates is regulatory evolution.

[485] Finally, it must be appreciated that the comparisons of differences between higher taxa only reveal what has changed, not how it changed. We do not know the rate at which these changes arose or the extent of variation in *Hox*-regulated characters in populations. The idea of macroevolution in a single step, the 'hopeful monster' so often insinuated in the discussion of homeotic genes[2], is widely discredited[3]. The new perspective emerging from the study of *Hox* genes in phylogeny and their regulatory roles in development needs to be integrated within the evolutionary frameworks of palaeontology and population biology.

---

75. Neil Shubin, Cliff Tabin, and Sean Carroll

Fossils, Genes and the Evolution of Animal Limbs

Although early genetic studies of insects demonstrated that wings, antennae, mouthparts, and legs are all modifications of a common embryonic form (although only recently understood at the genetic-developmental level), the question of the origin of appendages in general has long been an unanswered one. In this article, the authors explore the nature of serial homology (homology among similar segments, as in vertebrae or insect leg segments) and provide evidence of the deep evolutionary history for common ancestry in vertebrate, insect, and other arthropod limbs.

[639] The adaptive evolution of vertebrates and arthropods to aquatic, terrestrial, and aerial environments was accomplished by the invention of many novel features, especially new types of appendages. Enormous progress has been made in the past few years in understanding appendage development in both phyla. These genetic discoveries can be integrated with palaeontological data to address some of the principal events in the history of animal designs....

Vertebrate limb diversity was produced by changes in the number, position and shape of structures that can

[1]Slack, J., Holland. P. & Graham, C. *Nature* **361**, 490–492 (1993).

[2]Goldschmidt, R. The *Material Basis of Evolution* (Yale Univ. Press, New Haven. CT, 1940).

[3]Wallace, B. Q. *Rev. Biol.* **60**, 31–42 (1985).

be traced to Ordovician[2,3] (463–439 Myr) through Late Devonian[2,3,4] (409–362 Myr) fossils. The demands of feeding and locomotion in Ordovician and Silurian seas led to a surprising variability of the earliest known appendages: … [641] We propose that the temporal and spatial shift in the expression of *Hox* genes during limb development correlates with transformations inferred from the fossil record. Devonian fossils provide morphological links between structures in fins and limbs. Sarcopterygian fins *[these are the lobe-finned fish, such as the coelacanth, emerging in Upper Silurian times]* are dominated by an axis of segmented endoskeletal elements that extends from proximal to distal[2,3,4,27]. This axis is most similar to tetrapod limbs proximally, where the humerus, radius and ulna (femur, tibia and fibula) can readily be compared between taxa[2,3,4,5,15,22].

[644] Studies of *Drosophila melanogaster* have revealed that each type of appendage is typically specified by a single or a pair of *Hox* genes acting in the individual body segment that gives rise to a particular appendage[48]. For example, the *Antennapedia* gene acts in all three pairs of walking legs but the distinct morphology of the first, second and third pair of walking legs is determined by the *Sex combs reduced, Antennapedia,* and *Ultrabithorax Hox* genes[49], respectively. In the antenna, no *Hox* gene is active. If *Hox* gene function is lost or ectopically activated *[activated in an abnormal place or segment]* in individual segments, the identity of the corresponding appendage is transformed. Thus, loss of *Antennapedia* transforms [*the*] second leg to antennal structures[50] and expression of *Antennapedia* in the antenna transforms it to a leg[51]. … This demonstrates that the potential to form a limb exists in all segments, but the type of limb formed is determined by individual *Hox* genes.

[645] It is clear from the fossil record that chordates and arthropods diverged at least by the Cambrian. The appendages of these two groups are not homologous because phylogenetically intermediate taxa (particularly basal chordates) do not possess comparable structures. *[For example, the Cephalochordate, amphioxus, has no appendages.]* The most surprising discovery of recent molecular studies, however, is that much of the genetic machinery that patterns the appendages of arthropods, vertebrates and other phyla is similar.

[2]Coates, M. I. The origin of vertebrate limbs. *Development (suppl.)*169–180 (1994).

[3]Coates, M. I. Fish fins or tetrapod limbs—a simple twist of fate? *Curr. Biol.* 5, 844–848 (1995).

[4]Shubin, N. The evolution of paired fins and the origin of tetrapod limbs. *Evol. Biol.* 28, 39–85 (1995).

[5]Coates, M. I. The Devonian tetrapod *Acanthostega gunnari* Jarvik: postcranial anatomy, basal tetrapod interrelationships and patterns of skeletal evolution. *Trans. R. Soc. Edinb.* 87, 363–421 (1996).

[15]Vorobyeva, E. & Hinchliffe, J. R. From fins to limbs. *Evol. Biol.* 29, 263–311 (1996).

[22]Ahlberg, P. E. & Milner, A. R. The origin and early diversification of tetrapods. *Nature* 368, 507–512 (1994).

[27]Shubin, N. & Alberch, P. Amorphogenetic approach to the origin and basic organization of the tetrapod limb. *Evol. Biol.* 20, 318–390 (1986).

[48]Carroll, S. B. Homeotic genes and the evolution of arthropods and chordates *Nature* 376, 479–485 (1995).

[49]Struhl, G. Genes controlling segmental specification in the *Drosophila* thorax. *Proc. Natl Acad. Sci. USA* 79, 7380–7384 (1982).

[50]Struhl, G. Ahomoeotic mutation transforming leg to antenna in *Drosophila*. *Nature* 292, 635–638 (1981).

[51]Gibson, G. & Gehring, W. J. Head and thoracic transformations caused by ectopic expression of *Antennapedia* during *Drosophila* development. *Development* 102, 657–675 (1988).

76. Jennifer K. Greinier et al.

"Evolution of the Entire Arthropod *Hox* Gene Set Predated the Origin and Radiation of the Onychophoran/Anthropod Clade"

In the ingeniously conjured experiment reported in the final selection, the Hox *genes of a living* Onychophoran *(a velvet worm from Australia) are compared with those of the fruit fly. This should provide presumptive evidence, so the reasoning goes, of the amount of* Hox *gene evolution since the Cambrian. The rationale was straightforward: The fossil and living Onychopodians are not only similar in morphology, but bear the same unjointed lobopods to walk on, an obviously primitive condition.*

Further, comparing Hox *genes of the living lobopod with those of the fruit fly (and other arthropods) would say much about the Cambrian forms. As Carroll puts it in his* Endless Forms Most Beautiful *(2005, 151–55) "any genes shared between living* Onychophora *and living arthropods had to exist in their last common ancestor." In fact, the same ten* Hox *genes were found in both. "Furthermore," Carroll writes, "the bodies of all of the later arthropod designs—spiders, centipedes, insects, and all sorts of crustaceans—were sculpted by the same set of* Hox *genes."*

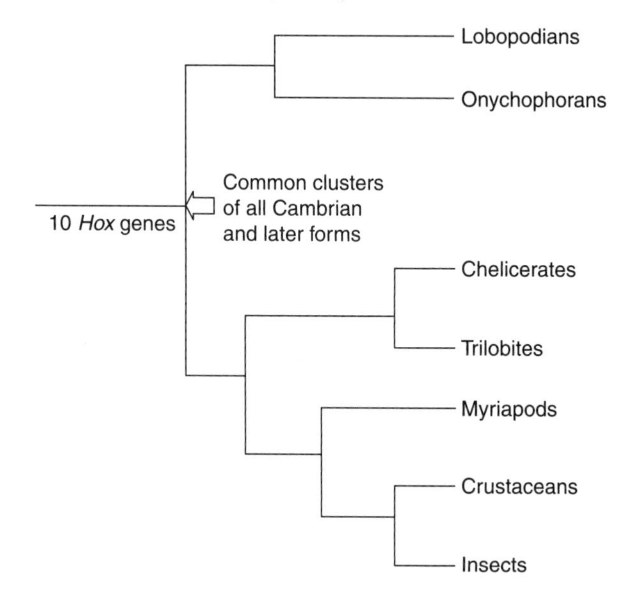

The first three authors were, at the time, students of Carroll. Whittington was then at the University of New England in New South Wales, Australia.

[547] The sudden appearance and remarkable diversity of complex animals in the Cambrian has prompted extensive paleontological (1,2), comparative (3,4) and molecular systematic (5,6) studies of animal relationships and the origin and evolution of metazoan body plans. Current debate is focused on three central issues: the origin of taxa (cladogenesis), the variety of basic designs (morphological disparity) and the evolution of the developmental control systems that regulate these designs (7–9). Comparisons of evidence from fossils and molecular studies may help to reveal whether the Cambrian marks the true origin of many higher taxa or whether major lineages diverged well before the Cambrian (7–10). In either of these scenarios, it appears that dramatic changes in body size, pattern and diversity occurred during the radiation of many taxa in the Cambrian, raising a fundamental question: which new developmental mechanisms (11) and genetic information were responsible for the evolution and diversification of larger and more complex animals?

The analysis of arthropods and their relatives may present the best opportunity for an integrated approach to the problem of the evolution of body plans: the fossil record of arthropods is relatively abundant; arthropods are the most speciose and morphologically diverse taxa; and many recent advances in developmental genetics have emerged from the study of one arthropod, *Drosophila melanogaster.* Paleontological, comparative and molecular systematic evidence suggests that arthropods are monophyletic (3, 12, 13) and descended from a lobopodian ancestor (14). The most striking trend in the evolution of the lobopodian/arthropod clade has been the diversification of segment types, from the simplicity of some Cambrian lobopodians, with four types of unjointed appendages and a uniformly patterned (homonomous) trunk (14–16), to the complexity of some extant crustaceans and insects, with as many as ten distinguishable appendage types and *a* highly diversified trunk.

As segmental diversity in highly derived insects such as *Drosophila* is regulated by eight *Hox* genes, it has been postulated that primitive arthropods possessed a more limited set of *Hox* genes which expanded during the course of arthropod and insect evolution (17, 18). Comparative studies of *Hox* genes in various metazoans indicate that many *Hox* genes predate the origin of the insects (19, 20). [548] The common ancestor of arthropods and vertebrates possessed five or six *Hox* genes (21–23), and most of the insect *Hox* genes were present in the annelid/arthropod ancestor. Importantly, arthropods have two unique *Hox* genes, *Ultrabithorax (Ubx)* and *abdominal-A (abd-A),* which are not found in vertebrates (21) or annelids (24, 26, 27). The *Ubx* and *abd-A* genes are present in both crustaceans (28) and insects (19, 20), the two arthropod classes that have the greatest segmental diversity. It is possible that the evolution of these two *Hox* genes facilitated the diversification of trunk segments during the evolution of diverse arthropods from an ancestor with homonomous trunk segments, a process that included the subdivision of the insect body plan into thorax and abdomen (tagmosis). In this study, we ask the following questions: when did all of the arthropod *Hox* genes arise, and was this before or during the arthropod radiation? And, if *Ubx* and *abd-A* are present in animals with homonomous trunks, how are these genes deployed?

...In order to trace the origin of the *Ubx* and *abd-A* genes further back through evolution, we cloned the *Hox*-genes from the Australian onychophoran *Acantfiokara kiiputensix [commonly, the "velvet worm"].* As the Onychophora are considered to be a sister group to the arthropods (2, 3) with a fossil record extending back to the late Early Cambrian, the complement of *Hox* genes shared between Onychophora and arthropods would reflect the condition prior to the origin and radiation of the arthropods and perhaps the entire lobopodian/arthropod clade. A PCR survey and the analysis of larger genomic clones revealed that this

[1]S Conway Morris, The fossil record and the early evolution of the Metazoa, *Nature* **361** (1993), pp. 219–225.

[2]J Valentine, D Erwin and D Jablonski, Developmental evolution of metazoan bodyplans: The fossil evidence. Review, *Dev Biol* **173** (1996), pp. 373–381.

[3]W Wheeler, P Cartwright and C Hayashi, Arthropod phylogeny: a combined approach, *Cladistics* **9** (1993), pp. 1–39.

[4]D Eernisse, J Albert and F Anderson, Annelida and arthropoda are not sister taxa: a phlyogenetic analysis of spiralian metazoan morphology, *Syst Biol* **41** (1992), pp. 305–330.

[5]R Raff, C Marshall and J Turbeville, Using DNA sequences to unravel the Cambrian radiation of the animal phyla, *Annu Rev Ecol Syst* **25** (1994), pp. 351–375.

[6]J Lake, Origin of the Metazoa, *Proc Natl Acad Sci USA* **87** (1990), pp. 763–766.

[7]R Fortey, The Cambrian evolutionary explosion decoupling cladogenesis from morphological disparity, *Biol J Linn Soc* **57** (1996), pp. 13–33.

[8]D Erwin, J Valentine and K Jablonski, The origin of animal body plans, *Am Sci* **85** (1997), pp. 126–137.

[9]S Conway Morris, Molecular clocks: defusing the Cambrian 'explosion'?, *Curr Biol* **7** (1997), pp. R71–R74.

[10]G Wray, J Levinton and L Shapiro, Molecular evidence for deep precambrian divergences among metazoan phyla, *Science* **274** (1996), pp. 568–573.

[11]E Davidson, K Peterson and R Cameron, Origin of bilateran body plans: evolution of developmental regulatory mechanisms, *Science* **270** (1995), pp. 1319–1325.

[12]J Boore, T Collins, D Stanton, L Daehler and W Brown, Deducing the pattern of arthropod phylogeny from mitochondrial DNA rearrangements, *Nature* **376** (1995), pp. 163–165.

[13]M Friedrich and D Tautz, Ribosomal DNA phylogeny of the major extant arthropod classes and the evolution of myriapods, *Nature* **376** (1995), pp. 165–167.

[14]G Budd, The morphology of *Opabinia regalis* and the reconstruction of the arthropod stem-group, *Lethaia* **29** (1996), pp. 1–14.

onychophoran possesses candidate orthologs of all of the *Hox* genes found in *Drosophila*. [*Orthologous genes are the same genes in different species due to common ancestry.*]

[551] The *Hox* genes control segmental identity by regulating the impression of downstream target genes. In insects, for example, the *Ubx* and *abd-A* gene products suppress limb formation in the abdomen via repression of the *Distal-less (Dll)* gene (29); however, *Dll* is not repressed in the *Ubx/abd-A* domains in centipede or onycohophoran trunks at the stages we surveyed, nor in crustacean trunks (37, 39), providing evidence that *Hox*-mediated repression of *Dll* evolved in the insect lineage to sculpt the distinctive limbless insect abdomen, Thus, the evolution of *Hox* regulation of target genes is a second developmental mechanism underlying the diversification of arthropod body plans.

...Our results demonstrate that the entire set of arthropod *Hox* genes was present in the onychophoran/arthropod ancestor, and that no new *Hox* genes were required to catalyze the radiation of the arthropods. Instead, changes in the regulation of trunk *Hox* genes along the anteroposterior axis and in *Hox* regulation of major downstream target genes appear to have enabled the morphological diversification of arthropod body patterns. Indeed, if certain phylogenies that depict Onychophora as a primitive lobopodian are correct (14), the diversification of arthropod *Hox* genes would predate the lobopodian radiation as well. The fossil record of Onychophora, other lobopodians and arthropods extends to the Late Early Cambrian (~530 million years ago) (14, 15) and the appearance of these taxa is contemporaneous with the 'explosion' of other taxa. *Hox* gene diversification must then have predated this period and could extend to the base of the Cambrian (~543 million years ago) or earlier.

III: A Final Word

77. STEPHEN JAY GOULD

Is a New and General Theory of Evolution Emerging?

The Mayr quote used as the epigram to this unit comes from Stephen J. Gould's article extracted here. Gould writes, "if Mayr's characterization of the synthesis theory is accurate, then that theory, as a general proposition, is effectively dead, despite its persistence as textbook orthodoxy." Gould here claims that a new evolutionary theory—focused on different processes than Darwinian ones—is necessary to explain higher levels of evolutionary emergence.

[121] The modern synthesis embodies a strong faith in reductionism. It advocates a smooth extrapolation across all levels and scales—from the base substitution to the origin of higher taxa....

[15] X Hou and J Bergstrüom, Cambrian lobopodians—ancestors of extant onychophorans?, *Zoo J Linn Soc* **114** (1995), pp. 3–19.

[16] L Ramsküold and H Xianguang, New early Cambrian animal and onychophoran affinities of enigmatic metazoans, *Nature* **351** (1991), pp. 225–228.

[17] EB Lewis, A gene complex controlling segmentation in *Drosophila*, *Nature* **276** (1978), pp. 565–570.

[18] M Akam, The molecular basis for metameric pattern in the *Drosophila* embryo, *Development* **101** (1987), pp. 1–22.

[19] M Akam, M Averof, J Castelli-Gair, R Dawes, F Falciani and D Ferrier, The evolving role of Hox genes in arthropods, *Development* **Supplement** (1994), pp. 209–215.

[20] S Carroll, Homeotic genes and the evolution of arthropods and chordates (Review), *Nature* **376** (1995), pp. 479–485 .

[21] R Krumlauf, *Hox* genes in vertebrate development, *Cell* **78** (1994), pp. 191–201.

[22] F Schubert, K Nieselt-Struwe and P Gruss, The *Antennapedia*-type homeobox genes have evolved from three precursors separated early in metazoan evolution, *Proc Natl Acad Sci USA* **90** (1993), pp. 143–147.

[23] J Garcia-Fernández and P Holland, Archetypal organization of the amphioxus *Hox* gene cluster, *Nature* **370** (1994), pp. 563–566.

[24] S Irvine and M Martindale, Cellular and molecular mechanisms of segmentation in annelids, *Cell Dev Biol* **7** (1996), pp. 593–604.

[25] M Dick and L Buss, A PCR-based survey of homeobox genes in *Ctenodrilus serratus* (Annelida: Polychaeta), *Mol Phylogen Evol* **3** (1994), pp. 146–158.

[26] V Wong, G Aisemberg, G Wen-Biao and E Macagno, The leech homeobox gene *Lox4* may determine segmental differentiation of identified neurons, *J Neurosci* **15** (1995), pp. 5551–5559.

[27] M Kourakis, VA Master, DK Lokhorst, D Nardelli-Haefliger, CJ Wedeen, MQ Martindale and M Shankland, Conserved anterior boundaries of *Hox* gene expression in the central nervous system of the leach Helobdella, *Dev Biol* (1997) in press.

[28] M Averof and M Akam, *HOM/Hox* genes of *Artemia*: implications for the origin of insect and crustacean body plans, *Curr Biol* **3** (1993), pp. 73–78.

[29] G Vachon, B Cohen, C Pfeifle, M McGuffin, J Botas and S Cohen, Homeotic genes of the *Bithorax* complex repress limb development in the abdomen of the *Drosophila* embryo through the target gene, *Cell* **71** (1992), pp. 437–450.

The general alternative to such reductionism is a concept of hierarchy—a world constructed not as a smooth and seamless continuum, permitting simple extrapolation from the lowest level to the highest, but as a series of ascending levels, each bound to the one below it in some ways and independent in others.... The basic processes—mutation, selection, etc.—may enter into explanations at all scales (and in that sense we may still hope for a general theory of evolution), but they work in different ways on the characteristic materials of divers levels....

The modern synthesis drew most of its direct conclusions from studies of local populations and their immediate adaptations. It then extrapolated the postulated mechanism of these adaptations—gradual, allelic substitution—to encompass all larger scale events. The synthesis is now breaking down on both sides of this argument. Many geneticists now doubt exclusive control by selection upon genetic change within local populations. Moreover,... we now doubt that the same style of change controls events at the two major higher levels: speciation and macroevolution.

[122]... [S]peciation occurs at too high a level to be observed directly in nature or produced by experiment in most cases. Therefore, theories of speciation have been based on analogy, extrapolation and inference. Darwin himself focused on artificial selection and geographic variation. He regarded subspecies as incipient species and viewed their gradual, accumulating divergence as the primary mode of origin for new taxa....

I have no doubt that many species originate in this way; but it now appears that many, perhaps most, do not. The new models stand at variance with the synthetic proposition that speciation is an extension of microevolution within local populations. Some of the new models call upon genetic variation of a different kind, and they regard reproductive isolation as potentially primary and non-adaptive rather than secondary and adaptive....

[126]... Just as mutation is random with respect to the direction of change within a population, so too might speciation be random with respect to the direction of a macroevolutionary trend. A higher form of selection, acting directly upon species through differential rates of extinction, may then be the analog of natural selection working within the population through the differential mortality of individuals.

[128]... I think I can see what is breaking down in evolutionary theory—the strict construction of the modern synthesis with its belief in pervasive adaptation, gradualism, and extrapolation.... I do not know what will take its place as a unified theory, but I would venture to predict some themes and outlines.

[129] The new theory will be rooted in a hierarchical view of nature. It will not embody the depressing notion that levels are fundamentally distinct and necessarily opposed to each other in their identification of causes.... It will possess a common body of causes and constraints, but will recognize that they work in characteristically different ways upon the material of different levels....

As a second major departure from current orthodoxy, the new theory will restore to biology a concept of organism .... Organisms are not billiard balls, struck in deterministic fashion by the cue of natural selection, and rolling to optimal positions on life's table. They influence their own destiny in interesting, complex, and comprehensible ways. We must put this concept of organism back into evolutionary biology.

Ideas to Think About

1. Within the more recent frameworks of evolutionary study is the recognition that the gene, the organism, and the environment are not simply different levels of abstraction; they are different and separable realities in evolution. What are the implications of this for theories of biology and culture?
2. In the context of niche construction theory, discuss the explanatory usefulness of human language, religion, art, and economics. Does this perspective cast Adam Smith and Thomas Malthus in a new light?
3. Darwin writes, "community in embryonic structure reveals community of descent." In what way does this have meaning in the context of the findings of evo-devo?
4. Darwin reasoned that differences in adult forms following near identities in their embryos had to do with inherent differences in developmental schemes that only revealed themselves later. In view of recent embryological research, is this an accurate statement?
5. Conventional Darwinian gradualism suggests that higher taxonomic levels (e.g., classes, orders) develop out of the increasing isolation and adaptive radiation of lower levels (e.g., genera and species). Does the evidence from analysis of *Hox*-gene regulation change this?
6. The influence of the common genetic tool kit on diverse classes of animals suggests that an alteration in our phylogenetic common ancestral "tree" is in order. Do you see in this change the possible need to alter the taxonomy, as well?
7. Does Gould make a compelling case for a new theory of evolution? To what extent is his outline of the elements of such a theory essentially non-Darwinian?

For Further Reading

The Triple Helix, Richard Lewontin. Cambridge, MA: Harvard University Press (2000). Quoted extensively in this unit, this well-written book packs enormous insightful information about the new horizons of evolutionary theory into its 130-plus pages.

Evolution in Four Dimensions, Eva Jablonka and Marion J. Lamb. Cambridge, MA: MIT Press (2006). This clearly written book explores genetic, epigenetic, behavioral, and symbolic systems of "inheritance" and describes how each can provide a basis for action by natural selection. Colorfully illustrated and thoughtfully summarized each step of the way.

Endless Forms Most Beautiful: The New Science of Evo Devo, Sean Carroll. New York: Norton (2005). This is a classic text by one of the founders of this new science. It is a provocative discussion on the application to primate and human evolution.

Your Inner Fish: A Journey into the 3.5-Billion Year History of the Human Body, Neil Shubin. New York: Pantheon (2008). This is a concise yet thorough treatment of the role of evolutionary embryology in the evolutionary road from fish to man. Shubin is one of the leaders in this cutting-edge research into the genes we share through common descent.

WORKS CITED

Barrett, Paul, ed., *The Collected Papers of Charles Darwin.* Vol. 1. Chicago: University of Chicago Press, 1977.

Behe, Michael, *Darwin's Black Box.* New York: The Free Press, 1996.

Bendyshe, Thomas, trans. and ed., *The Anthropological Treatises of Johann Friedrich Blumenbach.* London: Longman, Green, Longman, Egberts, and Green, 1865.

Bowler, Peter J., *Evolution: The History of an Idea.* 3rd ed. Berkeley: University of California Press, 2003.

Browne, Janet, *Charles Darwin: The Power of Place.* Princeton, NJ: Princeton University Press, 2002.

Buell, John, and Virginia Hearn, eds., *Darwinism: Science or Philosophy?* Richardson, TX: Foundation for Thought and Ethics, 1994.

Carroll, Sean, *Endless Forms Most Beautiful: The New Science of Evo Devo.* New York: Norton, 2005.

Dembski, William, *Intelligent Design.* Downers Grove, IL: InterVarsity Press, 1999.

———, *The Design Inference.* New York: Cambridge University Press, 1998.

Dobzhansky, Theodosius, "The Birth of the Genetic Theory of Evolution in the Soviet Union in the 1920s." In *The Evolutionary Synthesis,* edited by Ernst Mayr and William B. Provine, 229–42. Cambridge, MA: Harvard University Press, 1980.

Eiseley, Loren, *The Firmament of Time.* New York: Atheneum, 1966.

Gardiner, Martin, ed., *Great Essays in Science.* New York: Prometheus Books, 1994.

Gould, Stephen Jay, "The Return of Hopeful Monsters," *Natural History Magazine* 86, no. 6 (1977): 22–24.

Grant, Edward, "Science and the Medieval University." In *Rebirth, Reform and Resilience: Universities in Transition, 1300–1700,* edited by James M. Kittelson and Pamela J. Transue, 68–102. Columbus: Ohio State University Press, 1984.

Hull, David L., "Activism, Scientists and Sociobiology," *Nature* 407 (2000):673–74.

Johnson, Steven, *The Ghost Map.* New York: Riverhead Books, 2006.

Lewontin, Richard C., "Billions and Billions of Demons," *New York Review of Books,* January 9, 1997.

———, "Theoretical Population Genetics in the Evolutionary Synthesis." In *The Evolutionary Synthesis,* edited by Ernst Mayr and William B. Provine, 58–68. Cambridge, MA: Harvard University Press, 1980.

———, *The Triple Helix.* Cambridge, MA: Harvard University Press, 2000.

Lindberg, David C., *The Beginning of Western Science.* 2nd ed. Chicago: The University of Chicago Press, 2007.

Manchester, William, *A World Lit Only by Fire: The Medieval Mind and the Renaissance.* Boston: Little Brown, 1992.

Marx, K. F. H., "Memoir of J. F. Blumenbach, 1840." In *Anthropological Treatises,* Bendyshe, 1865.

Mayr, Ernst, "Prologue: Some Thoughts on the History of the Evolutionary Synthesis." In *The Evolutionary Synthesis,* edited by Ernst Mayr and William B. Provine, 1–48. Cambridge, MA: Harvard University Press, 1980.

———, *The Growth of Biological Thought,* Cambridge, MA: Belknap Harvard, 1982.

Neiman, Susan, *Moral Clarity: A Guide for Grown-Up Idealists.* Rev. ed. Princeton, NJ: Princeton University Press, 2009.

Odling-Smee, F. J., "Niche constructing phenotypes." In *The Role of Behavior in Evolution,* edited by H. C. Plotkin. Cambridge, MA: MIT Press, 1988.

Reynolds, Terry S., *Stronger than a Hundred Men: A History of the Vertical Water Wheel.* Baltimore: The John Hopkins University Press, 1983.

Rubenstein, Richard E., *Aristotle's Children.* New York: Harcourt, Inc., 2003.

Sahlins, Marshall, *The Use and Abuse of Biology: An Anthropological Critique of Sociobiology.* Ann Arbor: University of Michigan Press, 1976.

Sandler, L., and E. Novitsky, "Meiotic Drive as an Evolutionary Force," *The American Naturalist* 91 (1957): 105–10.

Smith, Norman Erik, "William Graham Sumner as an Anti-Social Darwinist," *Pacific Sociological Review* 22, no. 3 (July 1979): 332–47.

Sober, Elliott, *Conceptual Issues in Evolutionary Biology.* Cambridge, MA: MIT Press,1984. Completely revised edition, 1993.

Spitz, Lewis W., "The Importance of the Reformation for the Universities: Culture and Confession in the Critical Years." In *Rebirth, Reform and Resilience: Universities in Transition, 1300–1700*, edited by James M. Kittelson and Pamela J. Transue, 42–67. Columbus: Ohio State University Press, 1984.

Thomson, Keith, *The Watch on the Heath.* London: Harper Collins, 2005.

Uglo, Jenny, *The Lunar Men.* London: Faber and Faber, 2002.

CREDITS

Text Credits

Unit 1

4. Plato, *Dialogues*, 3rd ed., 5 vols., trans. Benjamin Jowett. Oxford: Oxford University Press (1892).
5. Hippocrates, *The Genuine Writings of Hippocrates*, Francis Adams, ed. and trans., New York: W. Wood and Company (1849).
6. Ross, W.D., and J.A. Smith, eds., *Aristotle*, Oxford: Clarendon Press (1908).
13. Bacon, Roger, *On Experimental Science*, [etc.]
17. Copernicus, Nicholas, *On the Revolutions* [etc.]
17. Copernicus, Nicholas, *The Revolutions* [etc.]
19. Of the Deluge and of Marine Shells. In *The Notebooks of Leonardo Da Vinci, Volume 2: Physical Geography*, translated by Jean Paul Richter. New York: Dover Publications, 1970 (first published in 1888).

Unit 2

22. Bacon, Francis, *Novum Organum* (1620), Second Part, Preface. In *The Works of Francis Bacon*, translated by Basil Montagu, Vol. III, 343–44 Philadelphia: A. Hart, 1851.
24. Bacon, Francis, *Of the Proficience and Advancement of Learning, Human and Divine*. In *The Works of Francis Bacon*, translated by Basil Montagu, Vol. I, Bk. II, 194–95, 208. Philadelphia: A. Hart, 1851.
24. Bacon, Francis, *Novum Organum*, Aphorisms. In *The Works of Francis Bacon*, translated by Basil Montagu, Vol. III, Bk. I, 345–47, 354, Philadelphia: A. Hart, 1851.
24. Bacon, Francis, *Novum Organum*, Vol. III, Bk. 1, Aphorisms, pp. 345–47, 354
27. Descartes, René, Rules for the Direction of the Mind. In *The Philosophical Works of Descartes*, rendered into English by Elizabeth S. Haldane and G. R. T. Ross, vol. 1, 3–8, 20–21. Cambridge: The University Press, 1911.
29. Newton, Isaac, *Opticks*, Bk. III, Part 1, 3rd ed. London: William and John Innys, 1721 [originally published in 1704].
30. Voltaire, Letter XIV: On Descartes and Sir Isaac Newton. In Voltaire, *Letters Concerning the English Nation*, London: for T. Pridden, 1776 [first published in 1734].

Unit 3

36. Steno, Nicholas, *The Prodromus to a Dissertation concerning Solids naturally Contained within Solids*. In *Steno: Geological Papers*, edited by Gustav Scherz, 141, 143, 145. Odense, Denmark: Odense University Press, 1969 [first published in 1669]. Reprinted with permission of the University Press of Southern Denmark. (Note: even numbered pages are in Latin; translation by Alex J. Pollock).
39. Ray, John, *The Correspondence of John Ray: consisting of selections from the philosophical letters published by Dr. Derham, and original letters of John Ray in the British Museum*. London: The Ray Society Publications, 1848.
39. Ray, John, "Experiments Concerning the Motion of the Sap in Trees, Made this Spring by Mr. Wiliugby, and Mr. Wray, Fellowes of the R. Society: and communicated to the Publisher of the Inquiries touching that subject in Numb 40," *Philosophical Transactions, The Royal Society of London* 4 (1669): 963–65.
40. Gunther, Robert W. T., *Further Correspondence of John Ray*. London: The Ray Society, 1928. Reprinted with permission of The Ray Society.
40. Burnet, Thomas, *The Sacred Theory of the Earth*. 4th ed., London: printed for John Hooke, 1719.
41. Harvey, William, *Anatomical Exercises on the Generation of Animals* (1651). In *The Works of William Harvey*, translated by Robert Willis. London: Printed for the Sydenham Society, 1847.
44. Hooke, Robert, *Micrographia, or some Physiological Descriptions of Minute Bodies, made by Magnifying Glasses, with Observations and Inquiries thereupon*. London: the Royal Society, 1664.
44. Hooke, Robert, *Micrographia*, Observ. XVIII. Of the Schematisme or Texture of Cork, and of the Cells and Pores of some other such frothy Bodies . London: the Royal Society, 1664.
45. Hooke, Robert, *Lectures and Discourses of Earthquakes and Subterraneous Eruptions*. New York: Arno Press, 1978; reprinted from *The Posthumous Works of Robert Hooke*. London: 1705. [This selection was originally published in 1668.]

Unit 4

48. Bonnet, Charles, *The Contemplation of Nature, Translated from the French of C. Bonnet*. London: printed for T. Longman in Paternoster Row; and T. Becket and P. A. de Hondt in the Strand, 1766.
49. Linnaeus, Ćarolus, *Reflections on the Study of Nature; and a Dissertation on the Sexes of Plants*. Dublin: printed by Luke White, 1786.
50. Linnaeus, Ćarolus, *Lachesis Lapponica; or, a tour of Lapland, now first published from the original manuscript journal of... Linnaeus*, edited by James Edward Smith, vol. 1. London: White and Cochrane, 1811.

51. Buffon, Comte de, *Natural History, General and Particular, by the Count de Buffon*, translated by William Smellie, 2nd ed., 3 vols. London: 1785, printed for W. Strahan and T. Cadell.
53. Blumenbach, Johann, *On the Natural Varieties of Mankind*, 1st ed., Göttingen: 1775. In *The Anthropological Treatises of Johann Friedrich Blumenbach*, translated and edited by Thomas Bendyshe, 82, 84. London: Longman, Green, Longman, Egberts, and Green, 1865.
53. Blumenbach, Johann, *Contributions to Natural History*, Part 1. Göttingen: 1806. in Bendyshe, *Anthropological Treatises.*
55. *Abstract of a Dissertation read in the Royal Society of Edinburgh, upon the Seventh of March, and Fourth of April, M,DCC,LXXV, Concerning the System of the Earth, its Duration, and Stability.* Edinburgh: 1785.
55. *Theory of the Earth*, Transactions of the Royal Society of Edinburgh, Vol. I, Part II. In *James Hutton's System of the Earth, 1785; Theory of the Earth, 1788; Observations on Granite, 1794, together with Playfair's Biography of Hutton.* Darien, CT: Hafner Publishing Co., 1970.

Unit 5

58. *Zoonomia; or the Laws of Organic Life*, Vol. 1. 2nd ed., Corrected. London: Printed for J. Johnson, in St. Paul's Church-Yard, 1796 [first published in 1794].
61. *Zoological Philosophy*, translated by Hugh Elliot. London: Macmillan and Co., 1914 [original publication 1809].
62. *Natural Theology: or, Evidences of the Existence and Attributes of the Deity.* 12th ed. London: Printed for J. Faulder, 1809 [first published in 1802].
63. *An Essay on the Principle of Population, as it affects the future improvement of society, with remarks on the speculations of Mr. Godwin, M. Condorcet, and other writers.* London: St. Paul's Church-Yard, 1798.
64. *An Inquiry Into the Nature and Causes of the Wealth of Nations (1776)*, edited by Edwin Cannan, Bk. 1, Ch. 8. London: Methuen & Co. (1904).

Unit 6

69. *Essay on the Theory of the Earth*, 5th ed. Edinburgh: W. Blackwood, 1827 [first published in 1821].
70. *Memoirs of Baron Cuvier*, edited by R. Lee. London: Longman, 1833. This selection is a quote in the Memoirs from his *Histoire Naturel des Poissons* (11 vols., 1828–48).
71. *Anatomical Philosophy, on the Respiratory Organs with Respect to the Determination and Identity of their Body Parts.* Paris: Crevor, 1818. In Hervé Le Guyader, *Étienne Geoffroy Saint-Hilaire, 1772–1844*, translated by Marjorie Grene. Chicago: University of Chicago Press, 2004. Copyright © 2004 by the University of Chicago. Reprinted with permission of the University of Chicago.
71. *Memoirs sur l'organisation des insectes.* Paris: Crevot, 1824. In Le Guyader, *Étienne Geoffroy Saint-Hilaire.* Copyright © 2004 by the University of Chicago. Reprinted with permission.
72. Richard Owen, *On the Archetype and Homologies of the Vertebrate Skeleton*, printed from the Hunterian Lectures, 1844–46, and given at the British Association for the Advancement of Science in 1846. London: John van Voorst, Paternoster Row, 1848.
74. *Principles of Geology*, vol. I. London: John Murray, 1830.
74. *Principles of Geology*, vol. II. London: John Murray, 1832.
75. *Principles of Geology*, vol. III. London: John Murray, 1833.
75. Robert Chambers, *Vestiges of the Natural History of Creation*, ch. 12. London: John Churchill, 1844.
77. *Vestiges of the Natural History of Creation*, ch. 14. London: John Churchill, 1844.

Unit 7

81. Darwin, Charles, *The Correspondence of Charles Darwin*, Vol. 1: 1821-1836, edited by Frederick Burkhardt, Sydney Smith, David Kohn, and William Montgomery. Copyright 1985, Cambridge University Press. Reprinted with permission of Cambridge University Press.
83. Francis Darwin, Ed., *The Life and Letters of Charles Darwin, including an autobiographical chapter.* Vol. 1. London: John Murray, 1887.
83. Darwin, Charles, *The Correspondence of Charles Darwin*, Vol. 1: 1821-1836, edited by Frederick Burkhardt, Sydney Smith, David Kohn, and William Montgomery. Copyright 1985, Cambridge University Press. Reprinted with permission of Cambridge University Press.
85. *Darwin, C. R., Narrative of the surveying voyages of His Majesty's Ships Adventure and Beagle between the years 1826 and 1836, describing their examination of the southern shores of South America, and the Beagle's circumnavigation of the globe. Journal and remarks. 1832–1836.* London: Henry Colburn, 1839.
85. Darwin, C. R., *Journal of researches into the natural history and geology of the countries visited during the voyage of H.M.S. Beagle round the world, under the Command of Capt. FitzRoy, R.N.* 2nd ed. London: John Murray, 1845.
85. Darwin, Charles, *The Zoology of the Voyage of H.M.S. Beagle, under the command of Captain FitzRoy, R.N., during the years 1832–1836.* Edited and superintended by Charles Darwin, Part III: Birds. London: Smith, Elder and Co., 1841.
86. *Darwin's Ornithological Notes (1836).* Edited with an introduction, notes and appendix by Nora Barlow. *Bulletin, British Museum (Natural History), Historical Series 2*, no. 7 (1963).
86. Darwin, Charles, *Journal of researches into the natural history and geology of the countries visited during the voyage of H.M.S. Beagle round the world, under the Command of Capt. Fitz Roy, R.N.* 2nd ed. London: John Murray, 1845.
88. Darwin, Charles, *Narrative of the surveying voyages of His Majesty's Ships Adventure and Beagle between the years 1826 and 1836. Journal and Remark, 1832–1836.* London: Henry Colburn, 1839.

Unit 8

90. Barlow, Nora, ed. *The autobiography of Charles Darwin 1809–1882. With the original omissions restored. Edited and with appendix and notes by his grand-daughter Nora Barlow.* London: Collins, 1958.
91. Herbert, Sandra, ed., The red notebook of Charles Darwin. *Bulletin of the British Museum (Natural History) Historical Series* 7 (24 April 1980): 1–164. [Bracketed comments taken from Herbert's footnotes.]
91. de Beer, Gavin, ed., Darwin's notebooks on transmutation of species. Part I. First notebook [B] (July 1837–February 1838). From *Bulletin of the British Museum (Natural History). Historical Series* 2, No. 2, Copyright 1960, 23–73. Reprinted with permission of Cambridge University Press.
91. Barlow, *The autobiography of Charles Darwin 1809–1882.*
92. Darwin, Charles, *The Correspondence of Charles Darwin*, Vol. 6: 1856-1857, edited by Frederick Burkhardt and Sydney Smith. Copyright 1990, Cambridge University Press. Reprinted with permission of Cambridge University Press.

93. Darwin, Charles, *The Correspondence of Charles Darwin*, Vol. 7: 1858-1859, edited by Frederick Burkhardt and Sydney Smith. Copyright 1990, Cambridge University Press. Reprinted with permission of Cambridge University Press.
95. Darwin, Charles, and Alfred Wallace, "On the Tendency of Species to form Varieties," and "On the Perpetuation of Varieties and Species by Natural Means of Selection," *Journal of the Proceedings of the Linnean Society, Zoology 3* (20 August 1858):45-62 (Read July 1, 1858).
96. I. Extract from an unpublished Work on Species, by C. Darwin, Esq., consisting of a portion of a Chapter entitled, "On the Variation of Organic Beings in a state of Nature; on the Natural Means of Selection; on the Comparison of Domestic Races and true Species."
97. II. Abstract of a Letter from C. Darwin, Esq., to Prof. Asa Gray, Boston, U.S., dated Down, September 5th, 1857.
98. III. On the Tendency of Varieties to depart indefinitely from the Original Type. By Alfred Russel Wallace.
102. Darwin, Charles, *The Correspondence of Charles Darwin*, Vol. 7: 1858-1859, edited by Frederick Burkhardt and Sydney Smith. Copyright 1990, Cambridge University Press. Reprinted with permission of Cambridge University Press.

Unit 9

106. Darwin, Charles, *On the Origin of Species by Means of Natural Selection, or the preservation of favoured races in the struggle for life.* 1st ed. London: John Murray, 1859.

Unit 10

123. Darwin, Charles, *The Correspondence of Charles Darwin*, Vol. 7: 1858-1859, edited by Frederick Burkhardt and Sydney Smith. Copyright 1990, Cambridge University Press. Reprinted with permission of Cambridge University Press.
124. Owen, Richard [Anonymous], "Darwin on the Origin of Species," *Edinburgh Review* 3 (April, 1860): 487–532.
125. Darwin, Charles, *The Correspondence of Charles Darwin*, Vol. 8: 1860, edited by Frederick Burkhardt, Janet Browne, Duncan M. Porter, and Marsha Richmond. Copyright 1993, Cambridge University Press. Reprinted with permission of Cambridge University Press.
126. Wilberforce, Samuel [Anonymous]. "On the Origin of Species, by means of Natural Selection; or the Preservation of Favoured Races in the Struggle for Life. By Charles Darwin, M. A., F.R.S. London, 1860," *Quarterly Review*, 1860, 102:225–64.
128. Huxley, Thomas [Anonymous], "Darwin on the Origin of Species," *Westminster Review* 17 (1860): 541–70.
131. Jenkin, Fleeming, "The Origin of Species," *The North British Review* 46 (June 1867): 277–318. (Review).
133. St. George Mivart, *On the Genesis of Species*. London: MacMillan and Co., 1871.
133. Darwin, C. R., *The Descent of Man, and Selection in Relation to Sex*. vol. 1, 1st ed. London: John Murray, 1871.

Unit 11

136. Mendel, Gregor, Versuche über Pflanzen-hybriden, *Verhand-lungen des naturforschenden Vereines in Brünn, Bd. IV für das Jahr 1865, Abhand-lungen*, 3–47, 1866. ["Experiments in Plant Hybridization," *Proceedings of the Natural History Society of Brünn*.] Courtesy Electronic Scholarly Publishing Project, © 1996.
138. Hardy, G. H., "Mendelian Proportions in a Mixed Population," *Science*, N.S., 28 (1908):49–50.
139. Morgan, Thomas Hunt, *The Scientific Basis of Evolution*, 2nd ed. New York: W. W. Norton, 1935. Copyright renewed 1962 by Howard K. Morgan, Lilian Scherp, Isabel M. Mountain, and Edith Morgan Whitaker. Used by permission of W. W. Norton & Co., Inc.
141. Simpson, George Gaylord, *The Meaning of Evolution*. New Haven, CT: Yale University Press, 1964. This selection is taken from pp. 271–79 of the paperback edition of 1964. Copyright © 1964 by Yale University Press. Used with permission of Yale University Press.
144. Mayr, Ernst, "Happy Birthday: 80 Years of Watching the Evolutionary Scenery," *Science* 305, no. 5680 (July 2, 2004): 46–47. Copyright 2004 by The American Association for the Advancement of Science. Used with permission.

Unit 12

150. Spencer, Herbert, "Progress: Its Law and Cause," *Westminster Review* 67, no. 132 (April 1857): 445–85.
152. Sumner, William Graham, *The Challenge of Facts and Other Essays*, edited by Albert Galloway Keller. New Haven: Yale University Press, 1914.
155. Galton, Francis, "Eugenics: Its Definition, Scope, and Aims," *The American Journal of Sociology* 10, no. 1 (July 1904): 1–25.
157. 274 U.S. 200, *Buck v. Bell*, May 2, 1927. Mr. Justice Holmes, Opinion of the Court.

Unit 13

160. Gould, Stephen Jay, "Nonoverlapping Magisteria," *Natural History* 106 (March 1997): 16–22. Used with permission of Rhonda Roland Shearer, Director Art Science Research Laboratory.
165. Johnson, Phillip, "The Unraveling of Scientific Materialism," *First Things: A Monthly Journal of Religion and Public Life* 77 (Nov. 1997): 22–25. From *First Things*, Copyright 1997. Reprinted with permission.
169. Ruse, Michael, "Is Evolution Just another Religion?" *The Global Spiral*, Metanexus Institute, e-publication, 1997–2008 (www.metanexus.net). Reprinted with permission of Michael Ruse.

Unit 14

174. Trivers, Robert L., "Parental Investment and Sexual Selection." In *Sexual Selection and the Descent of Man, 1871–1971*, edited by Bernard C. Campbell, 136–79. Chicago: Aldine Publishing Co., 1979.
174. Hamilton, W. D. "The genetic evolution of social behavior. I," *Journal of Theoretical Biology* 7 (1964): 1–16. Copyright 1964 by Elsevier Publishing Co. Reprinted with permission.
174. Hamilton, W. D., "*The genetic evolution of social behavior. II*," Journal of Theoretical Biology 7 (1964): 17–52. Copyright 1964 by Elsevier Publishing Co. Reprinted with permission.
175. Trivers, Robert L., "The Evolution of Reciprocal Altruism," *The Quarterly Review of Biology* 46 (1971): 35–57. Copyright 1971 by The University of Chicago Press. Reprinted with permission.
176. Wilson, Edward O., "Science and Ideology," *Academic Questions* 8 (Summer 1995): 73–81. With kind permission from Springer Science and Business Media: Academic

Questions, "Science and Ideology," Vol. 8, Summer, 1995, pp. 73–81, Edward O. Wilson.

180. David Sloan Wilson and Edward O. Wilson, "Rethinking the Theoretical Foundation of Sociobiology," *The Quarterly Review of Biology*, 84:327–348 (2007). Copyright © 2007, The University of Chicago Press. All rights reserved. Reprinted with permission.

Unit 15

184. Day, Rachel L., Kevin N. Laland and John Odling-Smee. "Rethinking Adaptation: The Niche-Construction Perspective." *Perspectives in Biology & Medicine* 46:1 92003), 80–95. Copyright © 2003 The Johns Hopkins University Press. Reprinted with permission of The Johns Hopkins University Press.
186. Darwin, C. R., *On the origin of species by means of natural selection, or the preservation of favoured races in the struggle for life*. 1st ed. London: John Murray, 1859. [Ch. 13 excerpts].
188. Bateson, William, *Materials for the Study of Variation, treated with especial regard to discontinuity in the origin of species*. New York: Macmillan and Co., 1894.
189. Carroll, Sean B., "Homeotic genes and the evolution of arthropods and chordates," *Nature* 376 (1995):479–85. Reprinted by permission from Macmillan Publishers, Ltd. [Endnotes renumbered; illustrations omitted].
189. Shubin, Neil, Cliff Tabin, and Sean Carroll. "Fossils, genes and the evolution of animal limbs," *Nature* 388 (1997): 639–48. Reprinted by permission from Macmillan Publishers, Ltd: *Nature* 388: 639–648, copyright 1997. [Illustrations omitted; comments italicized in brackets are mine—RKW].
191. Grenier, Jennifer K., Theodore L. Garber, Robert Warren, Paul M. Whitington, and Sean Carroll, "Evolution of the entire arthropod *Hox* gene set predated the origin and radiation of the onychophoran/anthropod clade," *Current Biology* 7 (1997): 547–53. [Figures omitted; numbers in parentheses are endnotes; comments italicized in brackets are mine—RKW].
192. Gould, Stephen J., "Is a new and general theory of evolution emerging?" *Paleobiology* 6, no. 1 (1980): 119–30. Copyright 1980, The Paleontological Society, Inc. Reprinted by permission from The Paleontological Society, Inc.

Figure Credits

86–87. Courtesy DeGolyer Library, Southern Methodist University, Dallas, Texas, QH11.D2 1946 v.2.

INDEX

Numbers in **bold** refer to pages with primary focus on the subject.